Parallel Robots: Theory and Applications

Series Editors

Since a few decades, parallel robots have attracted the attention of industrials and researchers. For industrials, they are interesting due to their inherent advantages which are their large payload-to-mass ratio, intrinsic stiffness properties and great acceleration capacities. For researchers, their inherent complexity leading to issues in terms of design, modelling and control allowed the definition of many interesting scientific problems to be solved. We reached now a point where the research on parallel robots has led to the definition of a quantity of scientific results which, for the moment, miss to be compiled to be easily usable by researchers, teachers, professionals and students for their formation and future works. This collection aims at filling this gap, by proposing a series of book devoted to all the scientific and technological fields required for having a better understanding of the behavior of parallel robots and/or for helping the future parallel robot designer in his work. Therefore, we intend to cover all topics on parallel robots, from their (geometric, kinematic, dynamic, elastic, etc.) modelling, to their advanced control, by also considering the singularity analysis, the calibration and model parameter identification problems, the definition of advanced algebraic tools necessary for their study, their optimal design, and many other scientific and technological aspects.

Springer and the series editors welcome book proposals. Potential authors are invited to contact Thomas Ditzinger, Editorial Director Interdisciplinary Applied Science, at thomas.ditzinger@springer.com.

Indexed by zbMATH.

Alexandr Klimchik · Anatol Pashkevich ·
Damien Chablat

Stiffness Modeling of Parallel Robots

Springer

Alexandr Klimchik
School of Engineering and Physical Sciences, Lincoln Centre for Autonomous Systems (L-CAS)
University of Lincoln
Lincoln, UK

Anatol Pashkevich
Department of Automation, Production and Computer Science
IMT Atlantique
Nantes, France

Damien Chablat
Centre national de la recherche scientifique (CNRS), Laboratoire des Sciences du Numérique de Nantes (LS2N)
Nantes, France

ISSN 2524-6232 ISSN 2524-6240 (electronic)
Parallel Robots: Theory and Applications
ISBN 978-3-031-76319-9 ISBN 978-3-031-76320-5 (eBook)
https://doi.org/10.1007/978-3-031-76320-5

This Springer imprint is published by the registered company Springer Nature Switzerland AG
The registered company address is: Gewerbestrasse 11, 6330 Cham, Switzerland

Introduction

Stiffness modelling of robotic manipulators is one of the important problems of industrial robotics [1]. It combines geometric modelling with computing the elastic responses of manipulator components to external forces and torques, which enhances robot accuracy and obtains desirable performance in various technological operations. Also, proper stiffness modelling can lead to energy-efficient robot designs and better force control [2]. Moreover, stiffness modelling allows us to optimise the design of performance-based robotic systems tuned to specific operational needs. It should be mentioned that this book mainly focuses on the stiffness modelling of conventional industrial robots, which are usually treated as quasi-rigid ones and possess relatively small elastic deflections described by Hooke's law. Such type of robots fundamentally differs from innovative soft ones where components are assumed to be quite flexible and expose considerable plastic deflections [3, 4]. However, although this book is mostly devoted to the stiffness modelling of quasi-rigid robots, some methods can also be applied to soft robots.

The problem of stiffness modelling is extremely important for industrial robots employed in contact operations like machining, friction stir welding, incremental sheet forming and some others [5]. Besides, it is essential for medical robotics, especially in surgical robots and rehabilitation devices, where safety and delicate interaction with human tissues are crucial [6]. Current challenges in this area are related to stiffness modelling of parallel, over-constrained and low mobility manipulating mechanisms with non-homogeneous and nonlinear behaviour with respect to applied force/torque magnitude and direction as well as the end-effector workspace location. For such types of mechanisms, it is always challenging to identify elastic parameters of robot components from physical and virtual experiments, estimate the end-effector deflections caused by interaction forces arising in technological processes, and implement related real-time compliance error compensation or reduction.

The main methods currently used for stiffness modelling of quasi-rigid robotic manipulators are Finite Element Analysis (FEA), Matrix Structural Analysis (MSA), and Virtual Joint Modeling (VJM). These methods have been successfully applied in various domains, including aerospace, automotive and medical robotics. The first of these methods, the FEA, ensures very precise modelling that can capture complex geometric features and physical properties of manipulator components while this technique is computationally expensive and can be hardly used in real-time applications. The current challenge of the FEA in robotic stiffness modelling is to find a reasonable trade-off between the degree of model details and computational expenses. The second method, the MSA, is less computationally demanding compared to the FEA; it is well-suited for systems with lumped (non-distributed) loads and constraints. However, in the frame of the MSA, it is rather difficult to describe the nonlinear elastic behaviours of some robotic manipulators, which is the main challenge here. The third method, the VJM, simplifies the modelling process by extending the traditional rigid geometric model with equivalent flexible virtual joints describing the elasticity of the manipulator components. This technique is efficient for both linear and nonlinear stiffness modelling of serial manipulators, but currently, it can hardly handle complex kinematic structures with internal cross-linkages.

Generally, the main assumptions behind all the above-mentioned methods involve simplifications regarding material properties and loading conditions to make the problems manageable for practical engineers [7]. The ongoing research in this area aims to relax some of these simplifications in order to create more accurate stiffness models for robotic systems. However, at present, the following typical assumptions are used:

- *Material Homogeneity and Isotropy*: it is usually assumed that the materials used in the robot's manufacturing are homogeneous and isotropic, which means that their mechanical properties are the same in all points and directions.
- *Linear Elasticity*: the majority of stiffness models assume that the materials behave according to Hooke's law, i.e. the degree of deformations is directly proportional to the applied loads, within the material elastic limits.
- *Rigidity of Some Components*: for certain components, it may be assumed that compliance is negligible, for example, the robot base and some of its joints may be treated as perfectly rigid to simplify the analysis.
- *Neglecting Joint Clearances and Friction*: in most cases, the effects of joint clearances and friction are ignored, which, in fact, can affect the accuracy of the manipulator stiffness modelling.
- *Simplified Load Properties*: most of the stiffness models consider a simplified description of the loading, assuming that only lumped (non-distributed) forces are applied and only axial tension is substantial, ignoring shear, bending, or torsional effects.
- *Ignoring Dynamic Effects*: The majority of stiffness models are derived assuming that robots are in static mode, not taking into account the dynamic loads that occur during the robot's movement.

The above assumptions are vital to make the stiffness modelling problem solvable for practical engineers, but they can cause discrepancies between the modelling results and the actual behaviour of the robotic system. For this reason, ongoing research in the field aims to relax these assumptions and develop more accurate models that can predict the real-world performance of robotic manipulators in more complex scenarios.

In this book, we explain how to use main stiffness modelling methods developed by the authors and other researchers for various kinematic structures, including complex over-constrained mechanisms with multiple cross-linkages; and how to model non-linear elastic behaviour of serial and parallel robotic manipulators.

References

1. Yang C, Li Q, Chen Q (2020) Analytical elastostatic stiffness modeling of parallel manipulators considering the compliance of the link and joint. Appl Math Model 78:322–349
2. Giraud FH, Mete M, Paik J (2022) Flexure variable stiffness actuators. Adv Intell Syst 4:2270030
3. Xu L, Wagner RJ, Liu S, He Q, Li T, Pan W, Feng Y, Feng H, Meng Q, Zou X, Fu Y, Shi X, Zhao D, Ding J, Vernerey FJ (2022) Locomotion of an untethered, worm-inspired soft robot driven by a shape-memory alloy skeleton. Sci Rep 12:12392
4. Kwok KW, Wurdemann H, Arezzo A, Menciassi A, Althoefer K (2022) Soft robot-assisted minimally invasive surgery and interventions: advances and outlook. Proc IEEE 110:871–892
5. Wu K, Li J, Zhao H, Zhong Y (2022) Review of industrial robot stiffness identification and modelling. Appl Sci 12:8719
6. Yao C, Li C, Shan Y, Zhang X, Zhang L, Zhao J (2024) Research on stiffness adaptation control of medical assistive robots based on stiffness prediction. Int J Intell Rob Appl
7. Yang C, Huang F, Ye W, Chen Q (2023) Gravity-based kinetostatic modeling of parallel manipulators using screw theory. Chin J Mech Eng 36:152

Contents

Chapter 1
Introduction to Stiffness Modelling of Robotic Manipulators

1.1 Manipulator Stiffness Modelling Problem

The mechanical stiffness of a manipulator is one of the most important indicators in the performance evaluation of robotic systems [1–3]. In particular, for industrial robots whose primary target is the precise manipulation of a technological tool, the manipulator stiffness defines the positioning errors due to the *external loading* arising during the workpiece processing. Examples of application robots in milling, drilling and friction stir welding (FSW) [4–6] are given in Fig. 1.1. Similarly, in industrial pick-and-place applications which are intended for simple but fast manipulations, the stiffness defines admissible velocity/acceleration while approaching the target point, in order to avoid undesirable displacements due to inertia forces [7]. Other examples include medical manipulators, where elastic deformations of mechanical components under the *link weight* and the task load are the primary sources of positioning errors Fig. 1.2, [8]. It is obvious that in all of these cases, the desired stiffness should be high enough to meet the requirements of the relevant application.

In contrast, for service robots, which interact directly with humans, a rather *low stiffness* is required to eliminate collisions causing an operator injury. This leads to special "soft" (biologically inspired) manipulator architectures that are based on the utilization of numerous elastic links, joints and actuators [17, 18]. Some more recent examples incorporating elastic elements are wire-driven (cable-based) robots that are potentially interesting for medical rehabilitation and rescue operations [19–21]. Another active research area that needs sophisticated stiffness analysis is micro-manipulation where the motions are generated due to piezoelectric actuators but conventional passive joints are replaced with elastic hinges [22–24].

For *parallel robots*, stiffness becomes a very important issue for several reasons. First of all, there are many practical applications where conventional parallel manipulators (Stewart platform, Machine Verne, etc., Fig. 1.3, [11–16]) are subject to essential external forces caused by extremely heavy end-platform or significant interaction forces. On the other hand, to achieve high dynamic capabilities, there is a tendency to

A. Klimchik et al., *Stiffness Modeling of Parallel Robots*, Parallel Robots: Theory and Applications, https://doi.org/10.1007/978-3-031-76320-5_1

(a) Milling with robot (b) Drilling with robot (c) FSW with robot

Fig. 1.1 Examples of robot applications where the compliance errors are caused by tool-workpiece interaction

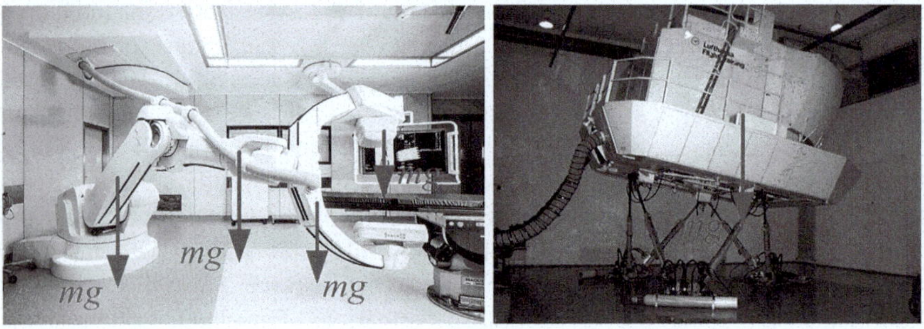

(a) Medical robot for diagnostics (b) Stewart platform used in fight simulation

Fig. 1.2 Examples of robot applications where the compliance errors are caused by gravity forces [9, 10]

reduce the link masses (and cross-sections), which obviously affects the manipulator stiffness properties. For this reason, stiffness analysis becomes a key phase in the design and application of parallel robots.

Similar to the general structural mechanics [25, 26], the *robot stiffness* analysis evaluates the manipulator resistance to the deformations caused by an external force/ torque applied to the end-effector [27]. It is clear that the manipulator stiffness, as well as other manipulator properties (kinematical, for instance), depends on the manipulator configuration [28]. Moreover, in general, the force–deflection relation is nonlinear, while for small deformations or so-called unloaded mode, they can be approximated by the linear function. Numerically this property is defined by the stiffness matrix **K** (or the compliance matrix **k**), which gives the linear relation between end-effector displacement and external loading which causes this transition (assuming that all of them are small enough) [29]. As it follows from related works,

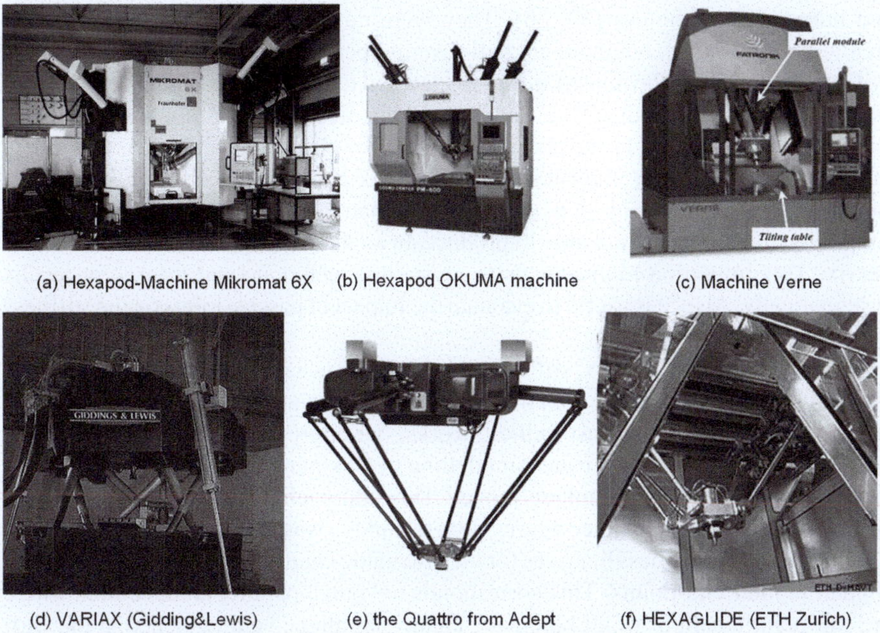

Fig. 1.3 Parallel robot with essential elastostatic deflections [11–16]

for the conservative systems, that are considered here, the stiffness matrix **K** is a 6 × 6 semi-definite non-negative symmetrical one. In general case, to represent the coupling between the translation and rotation, its structure may be non-diagonal.

In robotics, the term stiffness matrix is usually referred to as the "*Cartesian Stiffness Matrix*" $\mathbf{K}_C$ and it should be distinguished from the "*Joint-Space Stiffness Matrix*" $\mathbf{K}_\theta$ that describes forces-deflection relation in the joint coordinates [30]. These stiffness matrices can be mapped to each other using the Conservative Congruency Transformation [31]. The mathematical background for this computation originates from the work of Salisbury [32] who derived a closed-form expression for the Cartesian stiffness matrix of a serial manipulator assuming that the mechanical elasticity is concentrated in the actuated joints. Retaining this assumption, Gosselin [33] extended this result for the case of parallel manipulators (where the links were assumed to be rigid, and the passive joints to be perfect). Further development of this approach allowed us to consider link stiffness by supplementing the rigid beam with linear and torsion springs [34]. The recent modification proposed by Pashkevich et al. [35, 36] deals with 6 × 6 stiffness matrices which are computed from FEA experiments and can take into account couplings between translational and rotational deflections, the real shape of complex link and joints particularities. The latter essentially increased the accuracy of lumped modelling approaches (VJM and MSA) while preserving their high computational efficiency. At present, there are different variations of stiffness modelling approaches, which differ in modelling assumptions and numerical techniques [37, 38]. They were developed for different kinematic architectures [39, 40], but there are still several open questions to be considered.

Currently, most of the related works are devoted to the stiffness modelling of *serial chains*. They include both stiffness analysis of classical serial manipulators (without passive joints) [30, 32, 41] and also some results for serial chains with passive joints. In particular, in [36] an algorithmic solution has been proposed. This solution is capable of producing rank-deficient stiffness matrices. Despite its computational simplicity, the method is not able to produce an analytical expression for the desired matrix. This limitation motivates us to revise this technique and to present results systematically. Also, it is useful to evaluate the impact of passive joints on the stiffness matrix elements.

For *parallel manipulators*, stiffness modelling is usually performed for all kinematic chains simultaneously [33, 42], using the aggregated elastostatic equilibrium equations [43, 44]. In contrast to these works, our approach is based on a two-step procedure, which includes stiffness modelling of all kinematic chains *separately* and then *aggregates* them in a unique model. This approach has been used by several authors [45–47], but the related aggregation technique was reduced to simple summations of Cartesian stiffness matrices for the kinematic chains and the external loadings applied to their end-points. This corresponds to "pure" parallel architectures where the end-point locations of all kinematic chains are aligned and matched at the end-platform reference point. Besides, in previous works it has been implicitly assumed that the kinematic chains are perfect, i.e. their geometrical parameters are strictly nominal, and the parallel manipulator can be assembled without additional internal stresses. However, in practice, the parallel manipulator architecture is usually quite complex. In particular, the kinematic chains may be attached to different points of the end-platform. Besides, the kinematic chain geometry usually differs from the nominal one, which causes some internal forces/torques while assembling. These issues motivate us to systematize the stiffness modelling techniques for parallel manipulators.

It should be stressed that conventional stiffness analysis of robotic manipulators focuses on the so-called *unloaded mode,* which ignores the influence of the external or internal forces applied to the robot. Consequently, relevant techniques are targeted at the linearization of the "force–deflection" relation in the neighbourhood of the non-loaded equilibrium, which is topologically trivial and is perfectly described by the stiffness matrix. However, since many practical applications implicitly assume external and/or internal loading, the existing techniques must be extended to the case of the *loaded mode* where the manipulator may demonstrate essentially non-linear behaviour, which is not exposed in the unloaded case. In particular, as it follows from the general theory of elasticity, the loading may potentially lead to multiple equilibriums, bifurcations of the equilibriums and static instability of certain manipulator configurations. Some aspects of the multiple equilibrium problem for robotic manipulators were examined in the works of Duffy et al. [48, 49] who applied the Catastrophe theory for the stability analysis of the planar parallel manipulators with several flexural elements under external loading. In structural mechanics, similar phenomena are well-studied for the case of an "axially compressed column" that exhibits *buckling* (sudden loss of static stability and rapid change of the shape) if the loading exceeds the critical value. Nevertheless, in robotics, relevant problems did not attract much attention, mainly due to the high rigidity of commercially available robots. However current trends in the mechanical design of manipulators that are targeted at essential reduction of moving masses motivate relaxing this assumption and enhancement of existing stiffness analysis techniques.

1.2 Compensation for Manipulator Compliance Errors

In many robotic applications, compliance errors are undesirable because of their essential influence on the processing accuracy and quality of the final product. To overcome this difficulty, it is possible to modify either the control algorithm or the prescribed trajectory, which is used as the reference input for a control system [50]. In practice, the problem of compliance error compensation can be solved in two ways that differ in the degree of modification of the robot control software:

(a) by *modification of the manipulator model* in the robot controller (Fig. 1.4a), in order to describe both robot geometry and its elasto-static behaviour (in simple cases, it can be limited by tuning of the controller embedded model, but may also involve essential model enhancement by introducing additional parameters, if it is allowed by a robot manufacturer);
(b) by *modification of the robot control program* (Fig. 1.4b), that defines the prescribed trajectory in the Cartesian space (here, using the relevant error model, the input trajectory is generated in such a way that under the loading the output trajectory coincides with the desired one).

It is clear that the first approach can be implemented in on-line mode, while the second one requires preliminary off-line computations. It is worth mentioning that the stiffness models being used further are suitable for both of these approaches. But in practice, it is rather unrealistic to include the stiffness model in a commercial industrial controller where all transformations between the joint and Cartesian coordinates are based on the manipulator geometrical model. In contrast, the off-line error compensation, based on the second approach, is attractive for industrial applications.

For the *geometric errors*, relevant compensation techniques are already well developed. A comprehensive review of related works is given in [52, 53]. Here, if the main error sources are concentrated in the link length or the joint offsets, the compensation is reduced to straightforward modification of the manipulator parameters in the robot controller. Otherwise, if any geometric error sources are not presented in the nominal inverse/direct kinematics, relevant modification of the controller input is required. In this case, it is possible to use (in off-line mode) either an extended geometric model with additional parameters or simply a non-linear function that describes the error distribution throughout the workspace. Examples of such functions are given in [54, 55] where the neural network technique is employed.

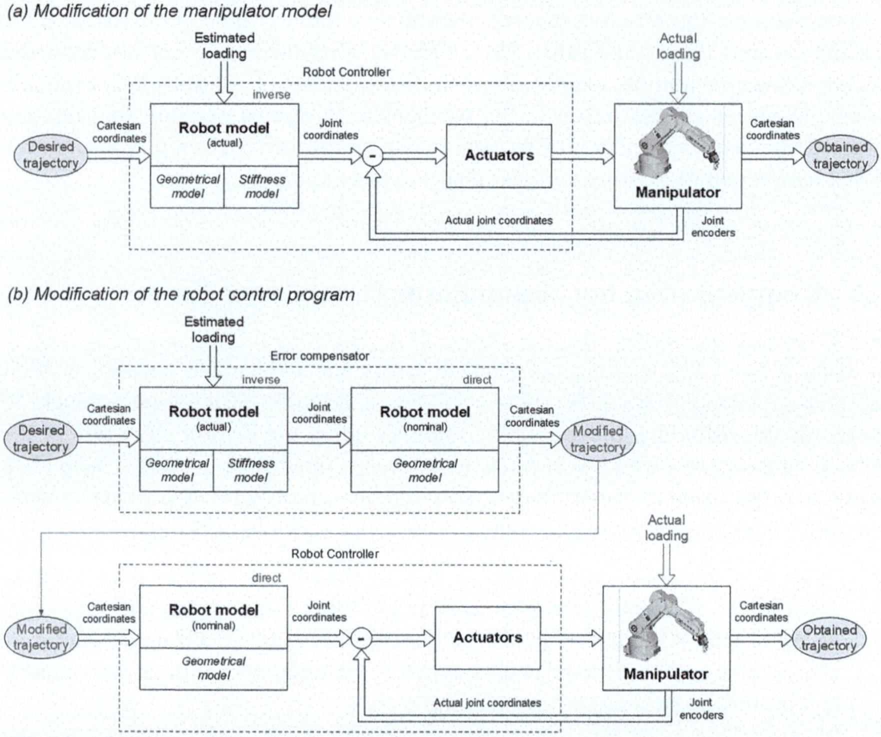

Fig. 1.4 Methods of compliance error compensation in robotic manipulators [52]

For *compliance errors*, the compensation technique must rely on two constituents. The first of them describes the distribution of the stiffness properties throughout the workspace and is defined by the stiffness matrix as a function of the joint coordinates or the Cartesian location [34]. The second constituent describes the forces/torques acting on the end-effector while the manipulator is performing its manufacturing task (manipulator *loading*). In the frame of this book, it is assumed that forces/torques are given and can be obtained either from the model of the dedicated technological process or by direct measurements using relevant sensors integrated into the end-effector. However, it is necessary to take into account that the force/torque sensors introduce additional undesirable compliance which has a direct effect on the position accuracy [56].

Because of the nonlinearity of the manipulator geometric model, the *stiffness matrix* required for the compliance error compensation highly depends on the robot configuration and essentially varies throughout the workspace. Hypothetically, it can be also approximated by a neural network, similar to the geometrical error compensation mentioned above. However, this approach is not attractive practically, so it is more convenient to compute the stiffness matrix using specially developed expressions and algorithms.

From a general point of view, full-scale compensation for compliance errors requires essential revision of the manipulator model embedded in the robot controller. In fact, instead of a conventional geometrical model that provides inverse/direct coordinate transformations from the joint to Cartesian spaces and vice versa, here it is necessary to employ the so-called *elasto-static model* [57]. The latter defines the mapping between the joint and Cartesian spaces taking into account deflections caused by external forces/torques applied to the manipulator end-effector. It is essentially more complicated than the geometric model and requires rather intensive computations.

If the compliance errors are relatively small, the composition of the conventional geometric model and the stiffness matrix gives a rather accurate approximation of the modified mapping from the joint to Cartesian space. In this case, for the first compensation scheme (see Fig. 1.4a), the elasto-static model can be easily implemented online if there is access to the control software modification. Otherwise, the second scheme (see Fig. 1.4b) can be applied. Moreover, concerning robot-based machining, there is a solution that does not require force/torque measurements or computations [50, 58]. The basic idea of this method is presented in Fig. 1.5, where at the first stage, the machining experiment gives a trajectory corrupted by compliance errors. Then, the difference between the desired and the obtained trajectories is evaluated via appropriate measurements, which give the compliance errors along the path. Using this data and assuming that the stiffness model is linear, the target trajectory for the robot controller is modified by applying the "mirror" technique (where corresponding points of the corrupted and target trajectories are symmetrical with respect to the relevant points of the desired trajectory). In order to improve accuracy in [59] it was proposed to perform the machining experiments iteratively. An evident advantage of this technique is its applicability to the compensation of all types of robot errors, including geometrical and compliance ones. However, this approach is only suitable for large-scale production where the manufacturing task and the workpiece location remain the same. In this area, it is also worth mentioning the work of Eastwood and Webb [60], who proposed a polar compensation methodology for gravitational deflection compensation for hybrid parallel kinematic machines. In some other works [61] the problems of geometric and compliance error compensation have been considered simultaneously but practical implementation of these techniques is rather difficult. At present, there are also several variations of the above-mentioned techniques that are presented below.

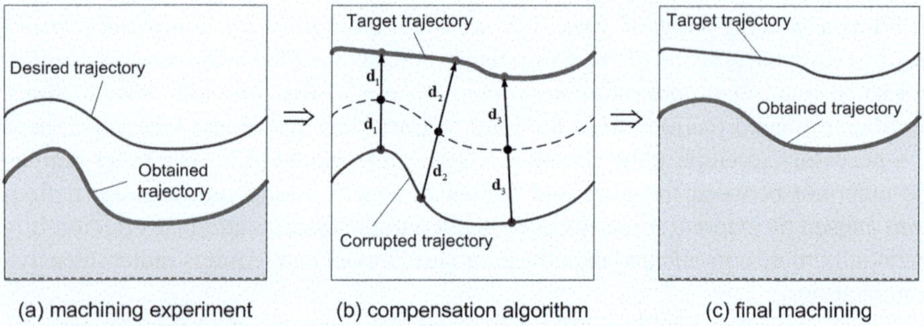

Fig. 1.5 Compliance error compensation using symmetrical trajectory approach [51]

References

1. Park J-J, Kim B-S, Song J-B, Kim H-S (2008) Safe link mechanism based on nonlinear stiffness for collision safety. Mech Mach Theory 43:1332–1348
2. De Luca A, Book W (2008) Robots with flexible elements. In: Siciliano B, Khatib O (eds) Springer handbook of robotics. Springer, Berlin Heidelberg, pp 287–319
3. Angeles J, Park FC (2008) Performance evaluation and design criteria. Springer Handbook of Robotics. Springer, pp 229–244
4. Guillo M, Dubourg L (2016) Impact and improvement of tool deviation in friction stir welding: weld quality & real-time compensation on an industrial robot. Robot Comput Integrat Manufact 39:22–31
5. Garnier S, Subrin K, Waiyagan K (2017) Modelling of robotic drilling. Procedia CIRP 58:416–421
6. Zhu Z, Tang X, Chen C, Peng F, Yan R, Zhou L, Li Z, Wu J (2022) High precision and efficiency robotic milling of complex parts: challenges, approaches and trends. Chin J Aeronaut 35:22–46
7. Nof SY (1999) Handbook of industrial robotics. John Wiley & Sons
8. Meggiolaro MA, Dubowsky S, Mavroidis C (2005) Geometric and elastic error calibration of a high accuracy patient positioning system. Mech Mach Theory 40:415–427
9. Neugebauer R, Wieland F, Schwaar M, Karczewski Z, Hochmuth C (1998) Erfahrungen mit dem Mikromat Hexapod 6X. Internationales Parallelkinematik-Kolloquium, TH Zürich -ETH-, Institut für Werkzeugmaschinen und Fertigung, Zürich
10. http://www.mtdcnc.com/new-machine-tools/Okuma-PM-600V-Multi-Function-Milling-Machine-2010227-2049.asp
11. Kanaan D, Wenger P, Chablat D (2009) Kinematic analysis of a serial–parallel machine tool: the VERNE machine. Mech Mach Theory 44:487–498
12. Geldart M, Webb P, Larsson H, Backstrom M, Gindy N, Rask K (2003) A direct comparison of the machining performance of a variax 5 axis parallel kinetic machining centre with conventional 3 and 5 axis machine tools. Int J Mach Tools Manuf 43:1107–1116
13. Pierrot F, Shibukawa T (1999) From Hexa to HexaM. In: Boër CR, Molinari-Tosatti L, Smith KS (eds) Parallel kinematic machines. Springer, London, pp 357–364
14. Hebsacker M, Treib T, Zirn O, Honegger M (1999) Hexaglide 6 DOF and Triaglide 3 DOF parallel manipulators. In: Boër CR, Molinari-Tosatti L, Smith KS (eds) Parallel kinematic machines. Springer, London, pp 345–355
15. Filippini R, Sen S, Bicchi A (2008) Toward soft robots you can depend on. Robot Automat Magaz IEEE 15:31–41
16. http://sparc-robotics.eu/robots-in-healthcare/
17. http://en.wikipedia.org/wiki/Stewart_platform

18. Scarfogliero U, Stefanini C, Dario P (2009) The use of compliant joints and elastic energy storage in bio-inspired legged robots. Mech Mach Theory 44:580–590
19. Merlet JP (2009) Analysis of wire elasticity for wire-driven parallel robots. In: Ceccarelli M (ed) Proceedings of EUCOMES 08. Springer Netherlands, pp 471–478
20. Kozak K, Qian Z, Jinsong W (2006) Static analysis of cable-driven manipulators with non-negligible cable mass. Robotics IEEE Trans 22:425–433
21. Yuan H, Courteille E, Deblaise D (2015) Static and dynamic stiffness analyses of cable-driven parallel robots with non-negligible cable mass and elasticity. Mech Mach Theory 85:64–81
22. Lobontiu N, Garcia E (2003) Analytical model of displacement amplification and stiffness optimization for a class of flexure-based compliant mechanisms. Comput Struct 81:2797–2810
23. Byoung Hun K, Ting-Yung Wen J, Dagalakis NG, Gorman JJ (2005) Analysis and design of parallel mechanisms with flexure joints. Robot IEEE Trans 21:1179–1185
24. Yong YK, Lu T-F (2009) Kinetostatic modeling of 3-RRR compliant micro-motion stages with flexure hinges. Mech Mach Theory 44:1156–1175
25. Timoshenko S, Goodier JN (1970) Theory of elasticity. McGraw-Hill, Singapore
26. Hjelmstad KD (2007) Fundamentals of structural mechanics. Springer Science & Business Media
27. Duffy J (2007) Statics and kinematics with applications to robotics. Cambridge University Press
28. Alici G, Shirinzadeh B (2005) Enhanced stiffness modeling, identification and characterization for robot manipulators. Robot IEEE Trans 21:554–564
29. Koseki Y, Tanikawa T, Koyachi N, Arai T (2002) Kinematic analysis of a translational 3-dof micro-parallel mechanism using the matrix method. Adv Robot 16:251–264
30. Ciblak N, Lipkin H (1999) Synthesis of Cartesian stiffness for robotic applications. In: Robotics and automation, 1999. Proceedings IEEE international conference on, IEEE, pp 2147–2152
31. Chen S-F, Kao I (2000) Conservative congruence transformation for joint and Cartesian stiffness matrices of robotic hands and fingers. Int J Robot Res 19:835–847
32. Salisbury JK (1980) Active stiffness control of a manipulator in Cartesian coordinates. In: Decision and control including the symposium on adaptive processes, 1980 19th IEEE Conference on. IEEE, pp 95–100
33. Gosselin C (1990) Stiffness mapping for parallel manipulators. Robot Automat IEEE Trans 6:377–382
34. Gosselin C, Zhang D (2002) Stiffness analysis of parallel mechanisms using a lumped model. Int J Robot Autom 17:17–27
35. Pashkevich A, Klimchik A, Chablat D (2011) Enhanced stiffness modeling of manipulators with passive joints. Mech Mach Theory 46:662–679
36. Pashkevich A, Chablat D, Wenger P (2009) Stiffness analysis of overconstrained parallel manipulators. Mech Mach Theory 44:966–982
37. Majou F, Gosselin C, Wenger P, Chablat D (2007) Parametric stiffness analysis of the Orthoglide. Mech Mach Theory 42:296–311
38. Corradini C, Fauroux JC, Krut S Evaluation of a 4-degree of freedom parallel manipulator stiffness. In: Proceedings of the 11th world congress in mechanisms and machine science, Tianjin (China)
39. Arumugam HK, Voyles RM, Bapat S (2004) Stiffness analysis of a class of parallel mechanisms for micro-positioning applications. In: Intelligent robots and systems, (IROS 2004). Proceedings. 2004 IEEE/RSJ international conference on. IEEE, pp 1826–1831
40. Company O, Krut S, Pierrot F (2002) Modelling and preliminary design issues of a four-axis parallel machine for heavy parts handling. Proceed Inst Mech Eng Part K: J Multi-body Dynam 216:1–12
41. Pigoski T, Griffis M, Duffy J (1998) Stiffness mappings employing different frames of reference. Mech Mach Theory 33:825–838
42. Wei W, Simaan N (2010) Design of planar parallel robots with preloaded flexures for guaranteed backlash prevention. ASME J Mech Robot 2

43. Quennouelle C, Gosselin Cá (2008) Stiffness matrix of compliant parallel mechanisms. In: Advances in robot kinematics: analysis and design. Springer, pp 331–341
44. Yi B-J, Freeman RA (1993) Geometric analysis of antagonistic stiffness in redundantly actuated parallel mechanisms. J Robot Syst 10:581–603
45. Pashkevich A, Klimchik A, Chablat D, Wenger P (2009) Stiffness analysis of multi-chain parallel robotic systems. J Automat Mob Robot Intell Syst 3:75–82
46. Zhang D, Xi F, Mechefske CM, Lang SYT (2004) Analysis of parallel kinematic machine with kinetostatic modelling method. Robot Comput Integrat Manufact 20:151–165
47. Xi F, Zhang D, Mechefske CM, Lang SY (2004) Global kinetostatic modelling of tripod-based parallel kinematic machine. Mech Mach Theory 39:357–377
48. Carricato M, Duffy J, Parenti-Castelli V (2002) Catastrophe analysis of a planar system with flexural pivots. Mech Mach Theory 37:693–716
49. Hines R, Marsh D, Duffy J (1998) Catastrophe analysis of the planar two-spring mechanism. Int J Robot Res 17:89–101
50. Dépincé P, Hascoët J-Y (2006) Active integration of tool deflection effects in end milling. Part 1. Prediction of milled surfaces. Int J Mach Tools Manuf 46:937–944
51. Klimchik A, Pashkevich A, Chablat D, Hovland G (2013) Compliance error compensation technique for parallel robots composed of non-perfect serial chains. Robot Comput Integrat Manufact 29:385–393
52. Hollerbach J, Khalil W, Gautier M (2008) Model identification. In: Siciliano B, Khatib O (eds) Springer handbook of robotics. Springer, Berlin Heidelberg, pp 321–344
53. Elatta A, Gen LP, Zhi FL, Daoyuan Y, Fei L (2004) An overview of robot calibration. Inf Technol J 3:74–78
54. Zhong X, Lewis J, N-Nagy FL (1996) Inverse robot calibration using artificial neural networks. Eng Appl Artific Intell 9:83–93
55. Lu T-F, Lin GCI, He JR (1997) Neural-network-based 3D force/torque sensor calibration for robot applications. Eng Appl Artif Intell 10:87–97
56. Budak E (2006) Analytical models for high performance milling. Part I: cutting forces, structural deformations and tolerance integrity. Int J Mach Tools Manuf 46:1478–1488
57. Su H-J, McCarthy JM (2005) A polynomial homotopy formulation of the inverse static analysis of planar compliant mechanisms. J Mech Des 128:776–786
58. Gunnarsson S, Norrlof M, Hovland G, Carlsson U, Brogardh T, Svensson T, Moberg S (2000) Pathcorrection for an industrial robot. US Patern 7130718
59. Dumas C (2011) Développement de méthodes robotisées pour le parachèvement de pièces métalliques et composites. Université de Nantes
60. Eastwood SJ, Webb P (2010) A gravitational deflection compensation strategy for HPKMs. Robot Comput Integrat Manufact 26:694–702
61. Drouet P, Dubowsky S, Mavroidis C (1998) Compensation of geometric and elastic deflection errors in large manipulators based on experimental measurements: application to a high accuracy medical manipulator. In: Lenarčič J, Husty M (eds) Advances in robot kinematics: analysis and control. Springer, Netherlands, pp 513–522

Chapter 2
State of the Art of Manipulator Stiffness Modelling

2.1 Existing Approaches in Stiffness Modelling of Robotic Manipulators

In general, existing approaches to manipulator stiffness modelling may be roughly divided into three main groups: (i) the *Finite Elements Analysis* (FEA), (ii) the *Matrix Structural Analysis* (MSA), and (iii) the *Virtual Joint Modeling method* (VJM). Their advantages and disadvantages are briefly presented below, with special emphasis on computational complexity and accuracy.

Finite Element Analysis method (FEA). Its basic idea is to decompose the physical model of the mechanism on a number of rather small (finite) elements and to introduce compliant relations between adjacent nods described by relevant stiffness matrices. These finite elements have a standard shape (pyramids, cubes, etc.) for which the stiffness matrix can be computed analytically. Using this discretization, the static equilibrium equations for each node are derived, and they are aggregated in a global matrix expression defining relations between the applied force/torque and node deflections. Then, the obtained matrix of rather large size is inverted and is used to obtain the desired stiffness matrix by simple extraction of proper elements. In the modern CAD environment, the above process is highly automated and is integrated with the 3D modelling of mechanical structures and mechanisms. In particular, the decomposition into a set of finite elements (so-called meshing) usually needs a definition of the discretization step and the mesh type only. The latter can be either linear or parabolic, which correspond to pyramids with 4 and 10 nodes respectively (with either 6 or 12 compliance relations). Then, for this finite element model, the CAD-based tools provide both numerical data and convenient visualization defining the deflection vectors for each node and potentially dangerous areas with high stresses.

An evident advantage of the FEA modelling is its high accuracy which is limited by the discretization step only. For robotic applications, it is very attractive, since the links/joints are modelled with their true dimension and shape [1–3]. However,

A. Klimchik et al., *Stiffness Modeling of Parallel Robots*, Parallel Robots: Theory and Applications, https://doi.org/10.1007/978-3-031-76320-5_2

while increasing the number of finite elements, the problem of limited computer memory and the difficulty of the high-dimension matrix inversion becomes more and more critical. Besides the high computational efforts, this matrix inversion generates numerous accumulative round-off errors, which reduce accuracy. In robotics, this causes rather high computational expenses for repeated re-meshing and re-computing. However, in this area, the FEA method can be applied for the links stiffness matrix identification [4] which is further used in the frame of the VJM and MSA techniques. Such a combination allows us to use the advantages of the FEA while avoiding intensive computations for different manipulator configurations.

Matrix Structural Analysis method (MSA). This method incorporates the main ideas of the FEA but operates with rather large compliant elements such as beams, arcs, cables, etc. [5]. This obviously leads to the reduction of the computational expenses and, in some cases, allows us to obtain an analytical stiffness matrix for the specific tasks. Similar to the FEA modelling, the MSA method gives forces/torques and displacements for each node, but here the node has clear physical interpretation (manipulator active or passive joint), which can be useful for some tasks [6, 7]. For parallel robots, this method has been developed in works [8, 9], where a general technique for stiffness modelling of the manipulator with rigid/flexible links and passive joints was proposed. The method has been illustrated by stiffness analysis of parallel manipulator of Delta architecture where the links were approximated by regular beams. The latter causes some doubts about the model accuracy compared to the combination of the FEA and VJM techniques. Besides, this result was obtained under the assumption that the external forces/torques are relatively small (i.e. for the unloaded mode), and it is unlikely that such an approach can be enhanced to take into account particularities of manipulator behaviour in loaded mode. In addition, here there exists a problem of the stiffness matrix computation for the manipulator singular configurations.

From a computational point of view, the MSA method is less complicated than the FEA-based one. Even though the MSA still involves matrix operations of rather high dimension, it gives a reasonable trade-off between the accuracy and computational time, provided that link approximations by the beam elements are realistic. It should be also noted that, in their general formulations, the FEA and MSA methods are very close to each other: both of them present a physical system as a set of nodes with flexible connections. The main difference is that the MSA operates with true physical objects (like beams, arcs and others), while the FEA decomposes them into small finite elements. So, the MSA can be treated as a special case of the FEA that has already found its application in robotics.

On the other side, if for each link the FEA-based stiffness matrix identification technique is applied [4, 10], an advanced combination of the MSA and FEA that is suitable for the stiffness modelling of arbitrary parallel manipulators (with numerous internal loops) can be obtained.

Virtual Joint Modeling method (VJM). The core of this method is an extension of the conventional rigid-body model of the robotic manipulator, where the links are treated as rigid, but the joints are assumed to be compliant (in order to accumulate all types of existing flexibilities in the joints only). Geometrically, such approximation is equivalent to adding to the mechanism rigid model some auxiliary virtual joints (with embedded virtual springs). It is obvious that such a lumped presentation of the manipulator stiffness (that in reality is distributed) essentially simplifies modelling. So, at present, the VJM technique is the most popular stiffness analysis method in robotics. This method was first introduced by Salisbury [11] and Gosselin [12], who assumed that the main flexibility sources were concentrated in the actuated joints. The derived expression defining the relation between the joint and Cartesian stiffness matrices (Conservative Congruency Transformation) became a basis for the manipulator stiffness analysis in many research works. Later, these results have been further developed in order to take into account some specific geometrical constraints [13–15] and external loading [16, 17]. Nevertheless, external loading was assumed small enough to detect any non-linear effects discovered letter [18, 19]. Due to its computational simplicity, the VJM method has also been successfully applied to the analytical stiffness analysis of several parallel manipulators [20, 21].

A key issue of this method is how to define and evaluate the virtual spring parameters. In earlier works, it was assumed that each actuated joint is presented by a single one-dimensional virtual spring [22, 23]. Further, to take into account the link flexibilities, the number of virtual joints was increased and in each actuated or passive joint several translational and rotational virtual springs were included [20]. The latest developments in this area operate with 6-dimensional virtual springs identified using the FEA-based method [4, 24, 25]. This leads to an essential increase of the VJM method accuracy that becomes comparable with the accuracy of the FEA-based techniques but with much lower computational expenses.

The only essential disadvantage of the VJM method is related to some difficulty in the stiffness modelling of parallel manipulators. It has been shown that for strictly parallel architectures, where there are only serial chains between the base and moving platforms, the method is rather simple. In contrast for architectures with internal loops (or with parallelogram-based links), the VJM-based stiffness analysis is rather complicated. Nevertheless, considering all the advantages and disadvantages of the FEA, the MSA and the VJM techniques, the VJM method looks the most attractive in robotics.

Computational complexity. The main benchmark that allows us to evaluate the computational complexity of the above-described methods is related to the computational efforts required for the matrix inversions inside of the corresponding algorithms. It should be stressed that matrix inversion is the most time-consuming operation for both linear and non-linear stiffness modelling. Generally, for the matrix inversion of size $n \times n$, it can be expressed as $O(n^3)$ [26].

The FEA method n depends on the discretisation step and the type of finite elements used. As it follows from a relevant study, in the case of the parabolic mesh (10 nodes and 12 connections per a finite element) the value of n can be approximately computed as $n = 30\,\nu\,n_L$, where n_L is the number of manipulator links, ν is the number of finite elements per link (it should be about 10^3 to ensure desired

Table 2.1 Computational complexity of existing stiffness modelling methods [19]

Stiffness modelling method	Manipulator architecture		
	Stewart platform (6 serial chains, each with 2 links and 5 passive joints)	Delta (3 serial chains, each with 6 links and 5 passive joints)	Orthoglide (3 serial chains, each with 5 links and 4 passive joints)
FEA	$\sim 10^{16}$ ($n_L = 13$)	$\sim 10^{17}$ ($n_L = 19$)	$\sim 10^{17}$ ($n_L = 16$)
MSA	$\sim 10^{7}$ ($n_d = 5 \times 6$)	$\sim 10^{8}$ ($n_d = 11 \times 3$)	$\sim 10^{8}$ ($n_d = 13 \times 3$)
VJM	$\sim 10^{3}$ ($n_q = 5$)	$\sim 10^{3}$ ($n_q = 5$)	$\sim 10^{3}$ ($n_q = 4$)

n_L, n_d, n_q are the number of links, node points and passive joints (in a single chain) respectively

precision). For the MSA method, the upper bound of the above matrix size $n = 12n_d$ can be computed via the node number n_d. Finally, for the VJM method, the size of the matrix to be inverted is $n = 6 + n_q$, where n_q is the total number of degrees of freedom in the passive joints.

Using these formulas, the stiffness model's complexity was estimated for three types of manipulators (Stewart platform, Delta, Orthoglide). Relevant results are presented in Table 2.1. As it follows from them, the VJM method essentially overcomes FEA and MSA so it will be used below as a preferable one. Nevertheless, all the methods are presented in detail in the following chapters.

2.2 Advances in Stiffness Modelling of Serial and Parallel Manipulators

Since the VJM-based method has proved to be more computationally efficient while providing acceptable accuracy, let us consider it in more detail. Taking into account some specificities of the considered application area, the main attention will be paid to the VJM applicability to the stiffness modelling of parallel manipulators (both under-constrained and over-constrained), considering the impact of the passive joint, evaluating the influence of the external and internal forces/torques as well as accuracy improvement of the stiffness model.

VJM method background. In the frame of this method, all types of compliance existing in a real manipulator (both distributed and lumped) are replaced by localized virtual springs located in the joints. Then, for this virtual mechanism consisting of rigid links and compliance joints, the static equilibrium equations are derived and linearized in order to obtain the Cartesian stiffness matrix, which in a general case depends on the manipulator configuration. Usually, it is assumed that the elastic deflections in virtual springs are relatively small and the linearization is performed in the neighbourhood of the equilibrium configuration corresponding to zero forces and torques (unloaded mode).

This technique originates from the work of Salisbury [11] who derived a closed-form expression for the Cartesian stiffness matrix of a serial manipulator assuming that the mechanical elasticity is concentrated in the actuated joints. Using notations adopted in this book, the basic equations can be written as follows

$$\delta\mathbf{t} = \mathbf{J}_\theta \cdot \delta\boldsymbol{\theta}; \quad \boldsymbol{\tau} = \mathbf{J}_\theta^{\mathrm{T}} \cdot \mathbf{F}; \quad \boldsymbol{\tau} = \mathbf{K}_\theta \cdot \delta\boldsymbol{\theta}; \tag{2.1}$$

where $\delta\mathbf{t}$ denotes the end-effector displacement in the Cartesian space (both position and orientation), $\mathbf{F}$ is the vector of the external loading applied to the end-effector (that includes both the force and the torque components), $\delta\boldsymbol{\theta}$ is the deflection in the virtual joint coordinates $\boldsymbol{\theta}$ caused by this loading, $\boldsymbol{\tau}$ is the vector of reactions in the elastic joints, $\mathbf{K}_\theta$ is the corresponding joint stiffness matrix, $\mathbf{J}_\theta$ is the Jacobian matrix computed with respect to the elastic joints. Here, the first equation is derived from the manipulator geometrical model, the second one describes the static equilibrium condition (assuming that the load is not essential), and the third equation presents the linear elasticity relation (Hooke's law).

After relevant transformations of (2.1), the desired linear relation between the external loading $\mathbf{F}$ and the end-effector displacement $\delta\mathbf{t}$ is presented as

$$\delta\mathbf{t} = \left(\mathbf{J}_\theta \cdot \mathbf{K}_\theta^{-1} \cdot \mathbf{J}_\theta^{\mathrm{T}}\right) \cdot \mathbf{F}, \tag{2.2}$$

which gives the Cartesian stiffness matrix

$$\mathbf{K}_\mathrm{C} = \mathbf{J}_\theta^{-\mathrm{T}} \cdot \mathbf{K}_\theta \cdot \mathbf{J}_\theta^{-1} \tag{2.3}$$

In literature [27–29], the latter may be also presented in a slightly different form

$$\mathbf{K}_\theta = \mathbf{J}_\theta \cdot \mathbf{K}_\mathrm{C} \cdot \mathbf{J}_\theta^{\mathrm{T}} \tag{2.4}$$

which sometimes is referred to as the Conservative Congruency Transformation (CCT), to emphasize that it describes mapping from the joint stiffness to the Cartesian one, and vice versa.

In further works, similar equations were obtained for parallel manipulators assuming that they are not over-constrained and the elasticity is concentrated in the actuator joints while the passive joints are perfect [12]. Other contributions to this area include [23, 30–37], where the VJM method was partially extended. It is also worth mentioning the works where the VJM technique was applied to particular manipulators, such as the CaPAMan, Orthoglide and H4 robots, Stewart–Gough platforms, etc. [20, 38–43].

VJM model parameters. In the earlier works devoted to the VJM modelling, it was explicitly assumed that the main sources of elasticity are concentrated in actuated joints. Correspondingly, the links were considered rigid, and the VJM model included only one-dimensional springs. In more recent works, compliance of the links has been taken into account by introducing additional virtual joints describing their longitudinal elasticity [22] or stiffness properties in several directions [20]. Recent developments in this area use 6-dimensional virtual joints to describe the elasticity of each link [44].

At the beginning, the *stiffness parameters* of the virtual joints describing the link elasticity (and incorporated in the matrix $\mathbf{K}_\theta$) were evaluated rather roughly, using a simplified representation of the link shape by regular beams. Besides, it was assumed that all linear and angular deflections (compression/tension, bending, torsion) are decoupled and are presented by independent one-dimensional springs that produce a diagonal stiffness matrix of size 6×6 for each link. Afterwards, this elasticity model was enhanced by using a complete 6×6 non-diagonal stiffness matrix of the cantilever beam [44]. This allowed taking into account all types of translational/rotational compliance and relevant coupling between different deflections. Other enhancements include the link approximation by several beams, but it gives a rather modest improvement in accuracy.

Further advances in this direction (applicable to the links of complicated shapes) led to the *FEA-based identification technique* that involves virtual loading experiments in a CAD environment and stiffness matrix estimation using dedicated numerical routines [4] (Fig. 2.1). The latter essentially increased the accuracy of the VJM modelling while preserving its high computational efficiency. It is worth mentioning that the usual high computational expenses of the FEA are not a critical issue here, because it is applied only once for each link (in contrast to the straightforward FEA modelling for the entire manipulator, which requires complete re-computing for each manipulator configuration). As a result, this approach allowed users to achieve FEA-modelling accuracy using the VJM-modeling technique that provides high computational efficiency. The general methodology of this hybrid approach is presented in detail further.

However, despite good results obtained for some case studies, there are still a number of important issues in the FEA-based identification technique. They include

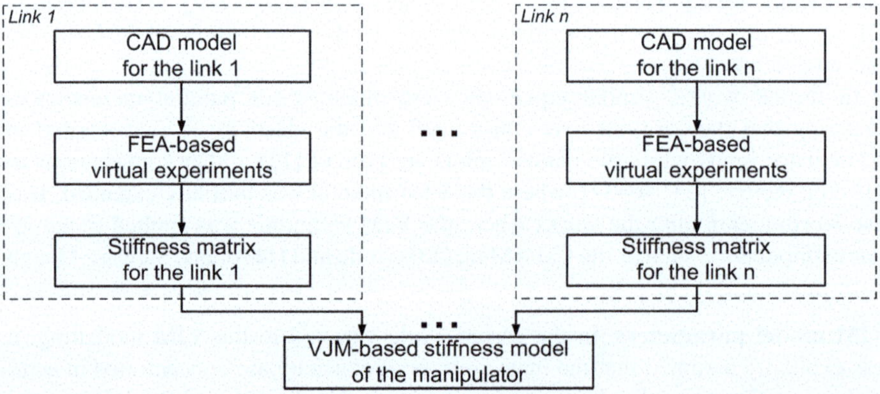

Fig. 2.1 Integration of the VJM modeling approach with the FEA-based identification technique [4]

the optimal setting of the virtual experiment (i.e. the definition of the mesh parameters, the joint contact surfaces to apply forces, etc.) and enhancement of numerical algorithms used for computation of the stiffness matrix elements (increasing robustness for FEA-modeling errors, distinguishing zero elements from small ones, etc.). These issues will be addressed in Chap. 7.

VJM model for manipulators with passive joints. Another important issue is related to taking into account the influence of the *passive joints,* which are widely used in parallel manipulators. In the simplest case, when the geometric constraints imposed by the manipulator architecture are not redundant, the passive joint coordinates may be just eliminated from the kinematic model, allowing direct application of the Salisbury formula. However, in the case of over-constrained or under-constrained manipulators, the elimination technique cannot be used directly.

For *serial kinematic chains* with passive joints, the problem has been solved for the general case [45]. In particular, it was proposed an algorithmic solution that extends the Salisbury formula and is able to produce the rank-deficient stiffness matrices describing "zero-resistance" of the end-effector to certain types of displacements, which do not require deflections in the virtual springs (due to the presence of passive joints and/or kinematic singularity of the examined posture). The relevant technique involves inversion of the dedicated square matrix of size $(n_\theta + 6) \times (n_\theta + 6)$, which is composed of the links stiffness parameters, and kinematic Jacobians of both virtual springs and passive joints (here n_θ is the number of passive joints). Then, the desired Cartesian stiffness matrix is obtained by simple extraction of an appropriate 6×6 sub-matrix from the computed inverse. The corresponding expression for stiffness matrix computation can be presented as

$$\begin{bmatrix} \mathbf{K}_C & * \\ * & * \end{bmatrix} = \begin{bmatrix} \mathbf{J}_\theta \mathbf{K}_\theta^{-1} \mathbf{J}_\theta^T & \mathbf{J}_q \\ \mathbf{J}_q^T & \mathbf{0} \end{bmatrix}^{-1} \tag{2.5}$$

where $\mathbf{J}_\theta$ is Jacobian related to the virtual springs $(6 \times n_\theta)$, $\mathbf{J}_q$ is Jacobian related to the passive joints $(6 \times n_q)$. The main advantage of this method is its computational simplicity since the number of virtual springs does not influence the size of the matrix to be inverted. Besides, the method does not require manual elimination of the redundant springs corresponding to the passive joints, since this operation is inherently included in the numerical algorithm.

However, for *parallel manipulators* with passive joints, solutions were obtained for less general cases. They include "pure" parallel architectures where the base and the end platform are connected by strictly serial kinematic chains. Here, the total stiffness matrix can be presented as the sum of partial matrices corresponding to separate chains (computed using the above-described technique), so the passive joints are taken into account easily. Besides, in this case, the manipulator over-constraining does not create additional difficulties. For instance, for the *over-constrained* manipulator of the Orthoglide family [46], each of the parallel chains yields the stiffness matrix of rank 4 while their aggregation gives the matrix of full rank 6. But if there exists a cross-linking between the parallel chains, this method cannot be applied directly. For this case, some interesting results are presented in [13, 47] where the geometrical constraints were treated in a general way but detailed computational techniques were not developed.

VJM modelling of parallel manipulators. Even though the VJM technique was originally developed for serial manipulators, it can be efficiently applied to parallel robots. The basic idea here is to obtain first the stiffness models of each kinematic chain separately, and after, to integrate them in a united model corresponding to the parallel manipulator. This idea was partially implemented in [32, 48, 49], where the manipulator structure was assumed to be strictly parallel (i.e. without internal loops) and the kinematic chains were assembled in the same end-point.

Under such assumptions, the stiffness matrix of the parallel manipulator can be computed via a straightforward summation of the chain stiffness matrices

$$\mathbf{K}_C = \sum_i \mathbf{K}_\mathrm{C}^{(i)}; \quad \begin{bmatrix} \mathbf{K}_\mathrm{C}^{(i)} & * \\ * & * \end{bmatrix} = \begin{bmatrix} \mathbf{J}_{\theta i}\mathbf{K}_{\theta i}^{-1}\mathbf{J}_{\theta i}^T & \mathbf{J}_{\mathrm{q}i} \\ \mathbf{J}_{\mathrm{q}i}^\mathrm{T} & \mathbf{0} \end{bmatrix}^{-1} \tag{2.6}$$

where the index i defines the kinematic chains and n is the total number of chains. However, in more general (and practically important) cases where the kinematic chains are connected to different points of the end-platform, this formula cannot be applied directly. Besides, for parallel manipulators with parallelogram-based links, some essential modifications are required.

Another limitation of the existing results devoted to the parallel manipulators is related to the assumption that the assembling does not produce any internal stresses. But in practice, numerous errors are accumulated in serial chains [50] and they cause non-negligible internal forces in manipulator joints (even if the external force applied to the end-effector is equal to zero). The internal forces may essentially change the manipulator behaviour (modify the stiffness matrix, change the end-effector location, etc.) and should be obviously taken into account in the stiffness model. Another important issue is associated with the stiffness modelling of parallel manipulators under loading, which will be also addressed in this book.

2.3 Stiffness Behaviour of Robotic Manipulator Under the Loading

Manipulator stiffness modelling in the loaded mode is a relatively new research area, which is worth to be considering separately. This Section presents an analysis of related works and defines some important research problems that will be the focus of this book. Some of them are based on the analogy that can be established between robotics and structural mechanics (it concerns buckling phenomena).

Types of loadings. Manipulator loading may be of a different nature. For the stiffness modeling, it is reasonable to distinguish two main types of loading, external and internal ones. The *external loading* is caused mainly by an interaction between the robot end-effector and the workpiece, which is processed or transported in the considered technological process [16, 27]. Another type of external loading exists due to the gravity influence on the manipulator links, for many heavy manipulators the link weight is not negligible [14, 15]. Besides, to compensate for a certain degree of gravity influence, some manipulators include special mechanisms (counterbalances) generating external forces/torques in the opposite direction. It is worth mentioning that the external loading generated by a technological process is always applied to the manipulator end-effector while others may be applied at intermediate points [51] (at joints, for instance). Besides, the external loadings caused by gravity have a distributed nature.

In addition to the above-mentioned forces/torques, *internal loading* in some joints may exist. For instance, to eliminate backlash, the joints may include preloaded springs, which generate the force or torque even in the "mechanical zero" configuration [52, 53]. Another example of a mechanism producing the internal loading is the so-called spring-based or pneumatic gravity compensator widely used in heavy manipulators [54, 55]. Though the internal forces/torques do not influence the global equilibrium equations, they may change the equilibrium configuration and influence the manipulator stiffness properties. For this reason, internal preloading is used sometimes to improve the manipulator stiffness, especially in the neighbourhood of kinematic singularities. Other cases where the internal loadings exist by default are related to over-constrained manipulators that are subject to the so-called antagonistic actuating [56–59]. For these manipulators, the redundant actuators generate internal forces and torques that are equilibrated in the frame of closed loops.

Some examples of the internal/external loadings of robotic manipulators are presented in Fig. 2.2. It is obvious that they essentially influence the manipulator equilibrium configuration and, consequently, may modify the stiffness properties. So, all of the above-mentioned loadings must be taken into account while developing the stiffness model of the robot manipulator.

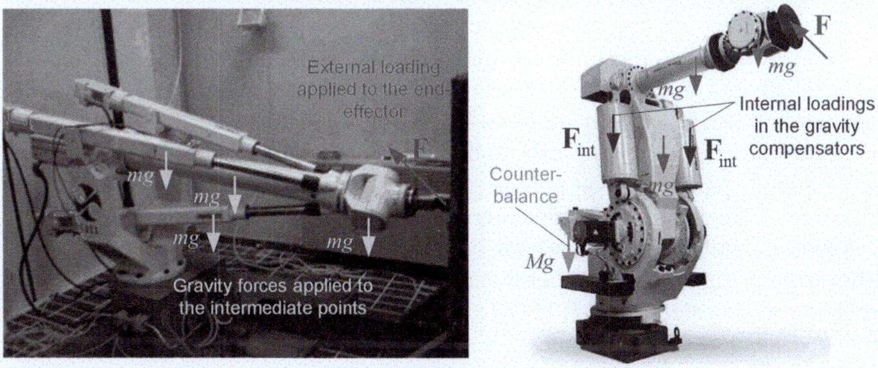

Fig. 2.2 Examples of internal and external loadings of robotic manipulators

VJM model for the loaded mode. At present, most related works evaluate the manipulator stiffness in an unloaded configuration, without the influence of external and internal loadings. There are a very limited number of publications that directly address the loaded mode case (or so-called case of "large deflections"), where in addition to the conventional "*elastic stiffness*" in the joints it is necessary to take into account the "*geometrical stiffness*" arising due to the change in the manipulator configuration under the load. Although the existence of this additional stiffness component for elastic structures has been known for a long time, the importance of this problem for robotic manipulators has been highlighted rather recently. The most essential results in this area were obtained in [16, 17, 60] where some theoretical issues and several case studies were presented for serial and parallel manipulators. Several authors [52, 53, 56] addressed the problem of stiffness analysis for the manipulators with internal preloading or antagonistic actuating, but in relevant equations some of the second-order kinematic derivates were neglected. Using the notation adopted in this work and summarizing existing results [16, 17, 61], the manipulator stiffness matrix for the loaded mode can be expressed as follows

$$\mathbf{K}_{\mathrm{C}} = \mathbf{J}_{\theta}^{-\mathrm{T}} \left(\mathbf{K}_{\theta} - \mathbf{K}_{\mathrm{F}}\right) \mathbf{J}_{\theta}^{-1} \tag{2.7}$$

where a new term $\mathbf{K}_{\mathrm{F}}$ is a complimentary stiffness matrix induced by external loading, which is not presented in the previous Eq. (2.3). This matrix depends on the Jacobian $\mathbf{J}_{\theta}$ derivatives and the vector of the loading $\mathbf{F}$. Required details concerning the computing of $\mathbf{K}_{\mathrm{F}}$ are given in [17].

In the frame of the same concept, the manipulator stiffness model for the loaded mode was proposed in [14, 15], where numerous factors were taken into account (conventional external loading, gravity forces, antagonistic redundant actuation, etc.). The final results for the stiffness matrix were presented as

$$\mathbf{K}_{\mathrm{C}} = \mathbf{J}_{\theta}^{-\mathrm{T}} \, \mathbf{K}_{\mathrm{u}} \, \mathbf{J}_{\theta}^{-1} \tag{2.8}$$

where $\mathbf{K}_{\mathrm{u}}$ is a solution of a relevant non-linear matrix equation, which includes the joint stiffness and the external/internal loadings as parameters. However, this approach is rather hard from the computational point of view. Besides, in this work, the Jacobians and all their derivatives have been computed not in a "true" equilibrium configuration (it was unfairly replaced by the unloaded one). For this reason, since the equilibrium obviously depends on the loading magnitude, some essential issues were omitted. As a result, no nonlinear effects have been detected in the stiffness behaviour of the examined manipulators.

Another significant result, which should be mentioned here, is related to the stiffness modelling of the parallel manipulators with the cross-linkage. Firstly this problem was considered in [14, 15] where all coordinates were decomposed into two groups (dependent and independent ones). But no clear rule for such coordinate splitting has been proposed; besides, the developed technique involved very non-trivial computations. Further, the cross-linkage was in the focus of [13, 62] where a rather compact expression for the stiffness matrix has been proposed

$$\mathbf{K}_C = \mathbf{J}_\theta^{-T} (\mathbf{K}_\theta - \mathbf{K}_F + \mathbf{K}_I) \mathbf{J}_\theta^{-1} \quad (2.9)$$

which, compared to (2.7), includes an additional matrix $\mathbf{K}_I$ that is induced by the geometrical constraints (cross-linkage). However, there are still a number of open questions concerning the coordinate splitting rule (i.e. dividing them into dependent and independent ones) and computing the equilibrium configuration corresponding to the applied loading.

Among recent contributions, special attention deserves the works [18, 19, 63], where the authors addressed the stiffness modelling of parallel manipulators for the loaded mode. Here, principal advances are in the area of non-linear stiffness modelling and are related to the computing of the manipulator-loaded equilibrium as well as taking into account both gravity forces and internal preloading in the joints. The corresponding expression for the Cartesian stiffness matrix takes into account the first and second-order derivatives of direct kinematic equations and is able to estimate the impact of passive joints. In compact form, the result is presented as

$$\mathbf{K}_C = \mathbf{J}_\theta^{-T} \cdot (\mathbf{K}_\theta - \mathbf{K}_F) \cdot \mathbf{J}_\theta^{-1} \cdot \left(\mathbf{I} - \mathbf{K}_q\right) \quad (2.10)$$

where the first term $\mathbf{J}_\theta^{-T} \cdot (\mathbf{K}_\theta - \mathbf{K}_F) \cdot \mathbf{J}_\theta^{-1}$ corresponds to the conventional Cartesian stiffness matrix for the manipulator without passive joints, and the matrix $\mathbf{K}_q$ describes the passive joints' impact.

It should be noted that the problem of computing the loaded equilibrium has been omitted in most of the related works. At the same time, it is clear that the changes in the manipulator configuration directly influence the Jacobian (and high-order derivatives) as well as the end-effector location. For this reason, computing the Jacobians and Hessians in a traditional way (i.e. for the unloaded configuration) may lead to excessively rough simplification of the manipulator stiffness model. In particular, some non-linear phenomena in manipulator stiffness behaviour can be hardly detected, while from the point of view of a general theory of elastic structures, they should obviously exist. A short summary of the stiffness modelling results for robotic manipulators is given in Table 2.2.

Nonlinear behaviour of the manipulator under loading. In mechanics, it has been known for a long time that elastic structures may suddenly change their configuration if the loading exceeds some critical value. A classical example is the so-called Euler column that retains its straight shape until the loading is relatively small [64]. However, in robotics, this problem (so-called buckling) has attracted less attention, mainly due to the high rigidity of commercially available robots. On the other hand, current trends in the mechanical design of manipulators that are targeted at the essential reduction of moving masses lead to an increase in manipulator compliance. For this reason, non-linear stiffness analysis becomes a critical issue for modern robotics.

As follows from mechanics, in a general case the stiffness behaviour of a structure (or a mechanism) may be essentially non-linear as illustrated by typical force–deflection curves presented in Fig. 2.3 [64, 65]. In particular, in the simplest case, this curve is smooth and monotonic, but its inclination progressively reduces while

Table 2.2 Summary of the related works and expressions for the Cartesian stiffness matrix [18]

Publications	Model and assumptions	Stiffness matrix
Salisbury [11]	Serial manipulator, elasticity in actuators	$\mathbf{K}_\mathrm{C} = \mathbf{J}_\theta^{-\mathrm{T}}\,\mathbf{K}_\theta\,\mathbf{J}_\theta^{-1}$
Gosselin [12], Ciblak and Lipkin [30, 34] Pigoski et al. [23]	Parallel manipulator, elasticity in actuators, non-over-constrained	$\mathbf{K}_\mathrm{C} = \mathbf{J}_\theta^{-\mathrm{T}}\,\mathbf{K}_\theta\,\mathbf{J}_\theta^{-1}$
Zhang et al. [31, 32]	Serial kinematic chain without passive joints, elasticity in virtual joints	$\mathbf{K}_\mathrm{C} = \left(\sum_i \mathbf{J}_{\theta i}\,\mathbf{K}_{\theta i}^{-1}\,\mathbf{J}_{\theta i}^\mathrm{T}\right)^{-1}$
Pashkevich et al. [25]	Serial kinematic chain with passive joints, elasticity in virtual joints	$\begin{bmatrix}\mathbf{K}_\mathrm{C} & * \\ * & *\end{bmatrix} = \begin{bmatrix}\mathbf{J}_\theta\mathbf{K}_\theta^{-1}\mathbf{J}_\theta^\mathrm{T} & \mathbf{J}_\mathrm{q} \\ \mathbf{J}_\mathrm{q}^\mathrm{T} & \mathbf{0}\end{bmatrix}^{-1}$
Zhang and Gosselin [33]	Parallel manipulator without cross-linking between kinematic chains	$\mathbf{K}_\mathrm{C} = \sum_i \mathbf{K}_\mathrm{C}^{(i)}$; $\mathbf{K}_\mathrm{C}^{(i)} = \left(\sum_j \mathbf{J}_{\theta j}\,\mathbf{K}_{\theta j}^{-1}\,\mathbf{J}_{\theta j}^\mathrm{T}\right)^{-1}$
Pashkevich et al. [24, 48]	Parallel manipulator without cross-linking between kinematic chains	$\mathbf{K}_C = \sum_i \mathbf{K}_\mathrm{C}^{(i)}$; $\begin{bmatrix}\mathbf{K}_\mathrm{C}^{(i)} & * \\ * & *\end{bmatrix} = \begin{bmatrix}\mathbf{J}_{\theta i}\mathbf{K}_{\theta i}^{-1}\mathbf{J}_{\theta i}^T & \mathbf{J}_{\mathrm{q}i} \\ \mathbf{J}_{\mathrm{q}i}^\mathrm{T} & \mathbf{0}\end{bmatrix}^{-1}$
Alici and Shirinzadeh [16] Chen and Kao [27] Marlet and Gosselin [61]	Serial or parallel manipulator with external loading (non-over-constrained)	$\mathbf{K}_\mathrm{C} = \mathbf{J}_\theta^{-\mathrm{T}}\,(\mathbf{K}_\theta - \mathbf{K}_\mathrm{F})\,\mathbf{J}_\theta^{-1}$
Quennouelle and Gosselin [13, 62]	Parallel manipulator with external loading and supplementary geometric constraints (cross-linkings)	$\mathbf{K}_\mathrm{C} = \mathbf{J}_\theta^{-\mathrm{T}}\,(\mathbf{K}_\theta - \mathbf{K}_\mathrm{F} + \mathbf{K}_\mathrm{I})\,\mathbf{J}_\theta^{-1}$
Yi and Freeman [14, 15]	Parallel manipulator with external loading, inertia and gravity loads, joint stiffness, actuation redundancy	$\mathbf{K}_\mathrm{C} = \mathbf{J}_\theta^{-\mathrm{T}}\,\mathbf{K}_\mathrm{u}\,\mathbf{J}_\theta^{-1}$ $\mathbf{K}_\mathrm{u}$- solution of non-linear matrix equation
Klimchik et al. [19]	Parallel manipulator under end-effectors and auxiliary loadings	$\mathbf{K}_C = \sum_i \mathbf{K}_\mathrm{C}^{(i)}$; $\mathbf{K}_\mathrm{C} = \mathbf{K}_C^{0(i)} \cdot (\mathbf{I} - \mathbf{K}_\mathrm{q})$ $\mathbf{K}_C^{0(i)} = \mathbf{J}_\theta^{-\mathrm{T}} \cdot (\mathbf{K}_\theta - \mathbf{K}_F) \cdot \mathbf{J}_\theta^{-1}$

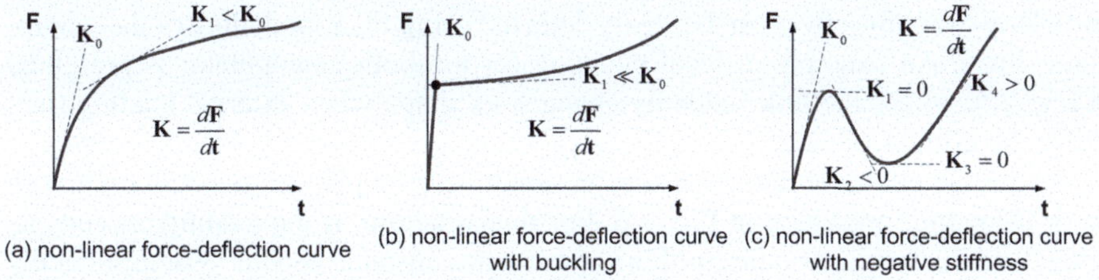

Fig. 2.3 Examples of non-linear force–deflection curves

the force is increasing (see Fig. 2.3a). Such a non-linear shape yields the stiffness coefficient, which depends on the applied force and monotonically decreases. The second example (see Fig. 2.3b) also shows a monotonic force–deflection curve but it contains a salient point that corresponds to an abrupt reduction of the stiffness coefficient (leading to buckling). The third example (see Fig. 2.3c) presents a non-monotonic (but smooth) force–deflection curve with local maximum and minimum. In this case, the curve includes a region with negative inclination (i.e. negative stiffness coefficient) and the system behaviour under the loading is highly non-linear. In the beginning, while the loading is gradually growing, the deflection is continuously increasing. Further, after reaching the critical value, the deflection changes abruptly. In structural mechanics, this phenomenon is usually called buckling. Analyzing these examples, we can conclude that in robotic systems, similar phenomena may exist and create undesired manipulator behaviour under loading.

In the following sections, we will pay special attention to the buckling phenomena since they are considered as potentially very dangerous for mechanical structures and mechanisms and may provoke failures. In the general theory of elasticity, several types of buckling are usually distinguished:

- *local buckling* of a structural element (rod, beam, etc.), observed as a sudden change of the element shape and significant reduction of its stiffness;
- *geometric buckling* of a mechanism or structure (arches, shells, etc.), observed as a sudden change of its geometric configuration;
- *contact buckling* associated with boundary conditions that can be observed similar to either local or geometric buckling.

For the local buckling, a classical example is usually related to the *axially compressed column* [65], which is also referred to as "Euler buckling". As shown in Fig. 2.4, depending on end-point constraints, the loading may provoke different changes in the column shape that appear after achieving a critical value for the compressing force. Besides, here the column end-point can abruptly be moved either in axial or lateral direction. It is obvious that in application to the robotic manipulator, similar behaviour may be observed both for a separate link and for a serial chain of links. It is unlikely that the VJM model can detect the local buckling for a separate link due to its simplified approximation (rigid rod and virtual spring at the end). However, this effect can be easily detected using FEA modelling of the manipulator links. An example of local buckling for a robotic manipulator is presented in Fig. 2.5, where the link suddenly changes its shape when external loading rises above the critical value.

For the geometrical buckling, a representative example is a *two-rod arch* with vertical loading presented in Fig. 2.6. For this structure, if the loading exceeds its critical value, the elastostatic equilibrium becomes unstable and the system suddenly changes its configuration. Making the analogy with robotic manipulators, it is clear that similar behaviour can be observed in parallel architectures with end-point loading and in serial chains with auxiliary loading applied in the intermediate nodes. Besides, in the case of serial chains with passive joints, it is possible to get buckling inside of the chain. A relevant example is given in Fig. 2.7, where the chain configuration suddenly changes from the straight-line to the trapezoidal profile if the loading is higher than the critical one. In addition, after buckling the chain stiffness becomes essentially lower.

It is worth mentioning that in previous works devoted to the manipulator stiffness analysis, the buckling phenomena have received rather limited attention. The authors

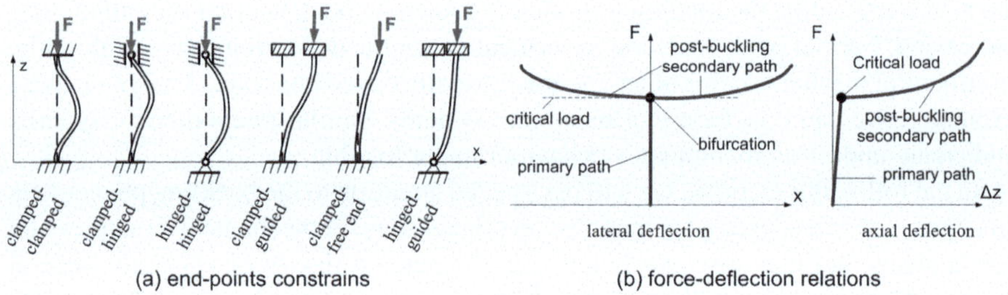

Fig. 2.4 Local buckling in axially compressed column

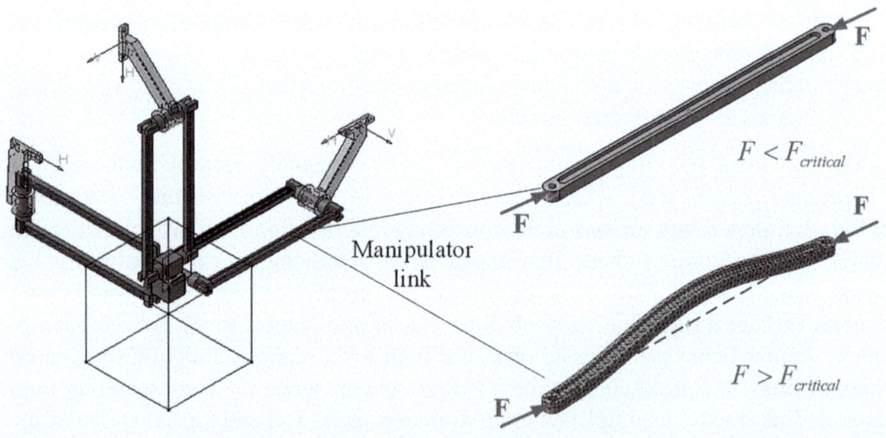

Fig. 2.5 Buckling of the manipulator link (Orthoglide parallel manipulator, critical load 15.2 kN)

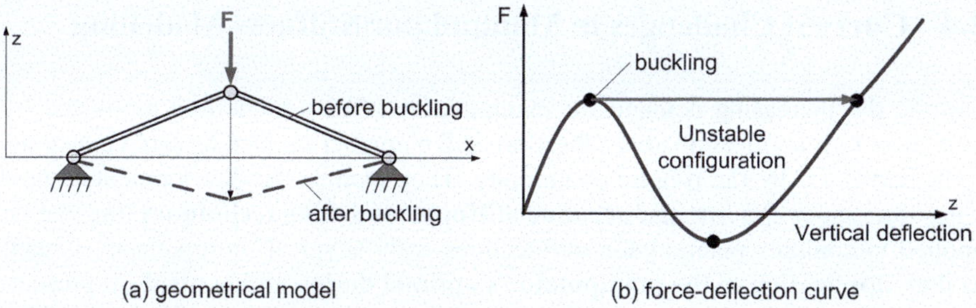

Fig. 2.6 Geometrical buckling in vertically loaded arch

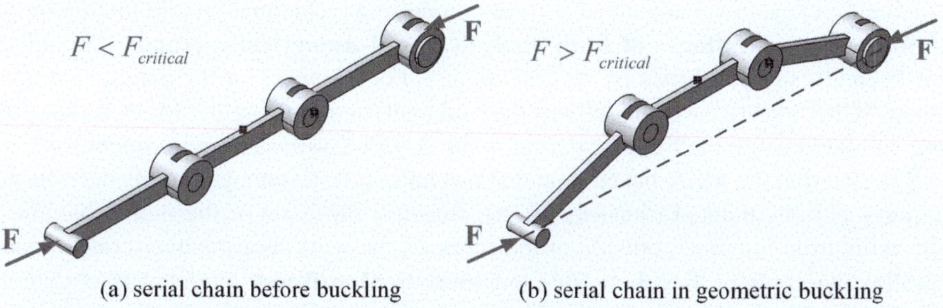

Fig. 2.7 Buckling of the serial chain (geometric buckling, critical load 4 kN for actuator compliances 10^{-4} rad/Nm, link lengths 0.4 m)

mainly concentrated on computing the stiffness matrices implicitly assuming that the force–deflection relation is monotonic and quasi-linear. Obviously, this approach does not allow us to detect salient points or local maximums/minimums that are essential for full-scale non-linear force–deflection analysis.

In practice, it is impossible to detect non-linear effects without finding the *loaded equilibrium,* while this question was also omitted in previous works. Besides, the loading may potentially lead to multiple equilibriums, bifurcations of the equilibriums and static instability of the manipulator configurations. These effects are essentially dangerous for parallel manipulators which impose numerous passive joints. Some aspects of the multiple-equilibrium problem for robotic manipulators have been examined in the works [66, 67] who applied the Catastrophe theory for the stability analysis of the planar parallel manipulators with several flexural elements under external loading. However, they did not propose a general approach for stability analysis of the manipulator configurations. More recent contributions in this area are in the works of the authors [50] that will be presented in detail in the following chapters.

2.4 Current Challenges in Manipulator Stiffness Modelling

Despite the increasing demand for manipulator stiffness modelling motivated by numerous new application areas, there are still a number of open research questions to be addressed by the robotic community. They include the development of new *computationally efficient and accurate* **stiffness modelling techniques** that can be applied to complex manipulator architectures with "atypical" components, as well as their **application to the manipulator's optimal design and control**. At present, these issues are in the focus of many research papers, but existing results do not allow us to overcome all difficulties arising in practice.

To meet to the current engineering requirements and ensure expected computational efficiency, novel advanced stiffness modelling techniques should intelligently **combine the advantages of both analytical and numerical methods**. This idea has been already developed by several authors. For instance, for the parallel manipulator 3CRS the stiffness model was derived analytically using the MSA technique and validated using the FEA modeling in the ANSYS software environment [68]. It was shown that the MSA-based model is fast and accurate enough, but its derivation requires tedious manual transformations. Besides, the issue of the model parameters estimation remained outside of the scope of the work. For the over-constrained parallel manipulator Exechon [69], an **analytical stiffness** model was obtained by applying the substructure decomposition approach. Here, the manipulator was decomposed into a moving platform and three limbs. Each limb was modelled as a rigid spatial beam with two sets of lumped springs at both ends, which provided simplified 6×6 diagonal stiffness matrices. The global kineto-static equations were derived using a conventional aggregation technique from the MSA. The obtained model was validated via the FEA modelling and was accurate enough, in spite of the doubtful simplification in the beam stiffness model. A semi-analytical approach was also applied in [70] where the Cartesian stiffness matrix of a 2PUS–2PRS parallel mechanism was derived using the screw theory. The model parameters were estimated through FEA modelling and numerical fitting. The obtained model allowed us to evaluate the stiffness distribution within the workspace, perform the sensitivity analysis and identify the key parameters affecting the manipulator stiffness.

An important issue directly influencing the stiffness model accuracy is related to the **precision of the model parameters**. In early works, the stiffness properties of the manipulator components were described by one-dimensional springs whose coefficients were estimated very roughly, approximating links by simple flexible structures described by a single stiffness coefficient or by a diagonal stiffness matrix. Currently, the manipulator stiffness parameters are estimated using either FEA modelling or physical experiments. Both approaches allow users to obtain rather accurate stiffness matrices of general (but symmetrical) structures for the links, however, estimation of the stiffness coefficients for the joints is still an open issue requiring additional efforts from the researchers.

Another important question in stiffness modelling, which is often missed out in the research literature, is computing of the manipulator **equilibrium configuration** corresponding to given external and internal loadings. In the simplest case, this equilibrium is obtained using direct and inverse kinematic equations, which explicitly assume that the desired Cartesian stiffness matrix is computed via linearization in the vicinity of the unloaded manipulator configuration. However, to detect some specific manipulator behaviour under the loading (buckling, etc.), the abovementioned linearization should be done for the loaded configuration. The latter can be computed from the non-linear static equilibrium equations that usually do not have analytical solutions and can be solved employing any available numerical technique. The main difficulties here are related to the multiplicity of solutions and the stability of relevant numerical routines. Some aspects of these issues were investigated in [18, 19, 36, 63, 71], where there were discovered the configuration instability and bulking phenomenon in the manipulator stiffness behaviour under the loading.

In the application area, the manipulator stiffness modelling is mainly used for **design optimization** of the robotic mechanisms to improve the positioning accuracy under the loading while ensuring desired dynamic properties. In literature, there are a number of such examples that are targeted at the link geometry optimization (shape, cross-section) in order to find a reasonable compromise between the link mass/inertia and the compliance errors [72]. One of the key issues in stiffness-based design optimization is the definition of an appropriate scalar **performance measure**, which has a clear physical meaning and can be easily computed using the manipulator stiffness model. To evaluate the manipulator resistance with respect to the external wrench applied to the end-effector, it is usually estimated the maximum compliance errors corresponding to the force/torque of the limited magnitude. These errors are straightforwardly computed by applying the singular value decomposition of relevant blocks of the Cartesian compliance matrix [73]. An alternative approach was proposed in [74] where the technology-oriented performance measure was used allowing us to evaluate the accuracy of the milling process.

Other useful applications of the considered techniques deal with stiffness modelling and optimization of robotic manipulators with specific components, such as single and double passive parallelograms [75–77], clutching mechanisms [78, 79], tensegrity-based mechanism [80], etc. For the **parallelogram-type chains**, there were derived analytical expressions [45, 81] that can be straightforwardly integrated into the manipulator VJM model. Another work dealing with kinematic parallelograms used the energy-based technique leading to equations similar to the MSA-based model [82]. In modern manipulator architectures, it is also used a new type of actuation where an actuated link can be switched through a special **clutching mechanism**. A typical example is the NaVARo robot which has eight actuation modes that are switched from one to another in order to avoid kinematic singularities [83, 84]. It is clear that such switching directly influences the manipulator stiffness behaviour which should be also investigated. Another interesting type of new manipulator is based on the **tensegrity-based** mechanisms that include several components in continuous compression and tension [85–88]. The main advantages of this design are low manipulator mass and high-efficiency material utilization as well as some useful properties in the stiffness behaviour. However, the classical stiffness modelling concepts can be hardly applied here because of non-linearity caused by possible transitions between the link state (from prestressed to non-stressed mode and vice versa).

References

1. Corradini C, Fauroux JC, Krut S Evaluation of a 4-degree of freedom parallel manipulator stiffness. In: Proceedings of the 11th world congress in mechanisms and machine science, Tianjin (China)
2. Akin JE (2005) Finite element analysis with error estimators: an introduction to the FEM and adaptive error analysis for engineering students. Butterworth-Heinemann
3. Bouzgarrou B, Fauroux J, Gogu G, Heerah Y Rigidity analysis of T3R1 parallel robot with uncoupled kinematics. In: Procceding of the 35th international symposium on robotics (ISR), Paris, France
4. Klimchik A, Pashkevich A, Chablat D (2013) CAD-based approach for identification of elasto-static parameters of robotic manipulators. Finite Elem Anal Des 75:19–30
5. Martin HC (1966) Introduction to matrix methods of structural analysis. McGraw-Hill
6. Marie S, Courteille E, Maurine P (2013) Elasto-geometrical modeling and calibration of robot manipulators: application to machining and forming applications. Mech Mach Theory 69:13–43
7. Nagai K, Liu Z A systematic approach to stiffness analysis of parallel mechanisms. In: Robotics and automation, 2008. ICRA 2008. IEEE international conference on. IEEE, pp 1543–1548
8. Deblaise D, Hernot X, Maurine P A systematic analytical method for PKM stiffness matrix calculation. In: IEEE international conference on robotics and automation (ICRA 2006). IEEE, pp 4213–4219
9. Deblaise D (2006) Contribution à la modélisation et à l'étalonnage élasto-géométriques des manipulateurs à structure parallèle. INSA de Rennes
10. Pashkevich A, Klimchik A, Chablat D, Wenger P Accuracy improvement for stiffness modeling of parallel manipulators. In: 42nd CIRP conference on manufacturing systems, pp 8
11. Salisbury JK Active stiffness control of a manipulator in Cartesian coordinates. In: Decision and control including the symposium on adaptive processes, 1980 19th IEEE Conference on. IEEE, pp 95–100
12. Gosselin C (1990) Stiffness mapping for parallel manipulators. Robot Autom IEEE Trans 6:377–382
13. Quennouelle C, Gosselin CÁ (2008) Stiffness matrix of compliant parallel mechanisms. In: Advances in robot kinematics: analysis and design. Springer, pp 331–341
14. Yi B-J, Freeman RA (1993) Geometric analysis of antagonistic stiffness in redundantly actuated parallel mechanisms. J Robot Syst 10:581–603
15. Yi B-J, Freeman RA (1992) Synthesis of actively adjustable springs by antagonistic redundant actuation. ASME Trans J Dynam Syst Measure Control B 114:454–461
16. Alici G, Shirinzadeh B (2005) Enhanced stiffness modeling, identification and characterization for robot manipulators. Robot IEEE Trans 21:554–564
17. Shih-Feng C, Imin K (2002) Geometrical approach to the conservative congruence transformation (CCT) for robotic stiffness control. In: Robotics and automation, proceedings. ICRA '02. IEEE international conference on, vol 541, pp 544–549
18. Pashkevich A, Klimchik A, Chablat D (2011) Enhanced stiffness modeling of manipulators with passive joints. Mech Mach Theory 46:662–679
19. Klimchik A, Chablat D, Pashkevich A (2014) Stiffness modeling for perfect and non-perfect parallel manipulators under internal and external loadings. Mech Mach Theory 79:1–28
20. Majou F, Gosselin C, Wenger P, Chablat D (2007) Parametric stiffness analysis of the Orthoglide. Mech Mach Theory 42:296–311

21. Görgülü İ, Carbone G, Dede MİC (2020) Time efficient stiffness model computation for a parallel haptic mechanism via the virtual joint method. Mech Mach Theory 143:103614
22. Gosselin C, Zhang D (2002) Stiffness analysis of parallel mechanisms using a lumped model. Int J Robot Autom 17:17–27
23. Pigoski T, Griffis M, Duffy J (1998) Stiffness mappings employing different frames of reference. Mech Mach Theory 33:825–838
24. Yang C, Li Q, Chen Q (2020) Analytical elastostatic stiffness modeling of parallel manipulators considering the compliance of the link and joint. Appl Math Model 78:322–349
25. Pashkevich A, Chablat D, Wenger P (2009) Stiffness analysis of overconstrained parallel manipulators. Mech Mach Theory 44:966–982
26. Lin L, Yang C, Lu J, Ying LEW (2011) A fast parallel algorithm for selected inversion of structured sparse matrices with application to 2D electronic structure calculations. SIAM J Sci Comput 33:1329–1351
27. Chen S-F, Kao I (2000) Conservative congruence transformation for joint and Cartesian stiffness matrices of robotic hands and fingers. Int J Robot Res 19:835–847
28. Li Y, Kao I On the stiffness control and congruence transformation using the conservative congruence transformation (CCT). In: Robotics and automation, 2001. proceedings 2001 ICRA. IEEE international conference on. IEEE, pp 3937–3942
29. Chen SF, Kao I (2000) Simulation of conservative congruence transformation. Conservative properties in the joint and Cartesian spaces. In: Robotics and automation, proceedings. ICRA'00. IEEE international conference on. IEEE, pp 1283–1288
30. Ciblak N, Lipkin H (1999) Synthesis of Cartesian stiffness for robotic applications. In: Robotics and automation, proceedings. IEEE international conference on. IEEE, pp 2147–2152
31. Zhang D, Xi F, Mechefske CM, Lang SYT (2004) Analysis of parallel kinematic machine with kinetostatic modelling method. Robot Comput Integrat Manufact 20:151–165
32. Xi F, Zhang D, Mechefske CM, Lang SY (2004) Global kinetostatic modelling of tripod-based parallel kinematic machine. Mech Mach Theory 39:357–377
33. Zhang D, Gosselin CM (2002) Kinetostatic modeling of parallel mechanisms with a passive constraining leg and revolute actuators. Mech Mach Theory 37:599–617
34. Ciblak N, Lipkin H (1998) Application of stiffness decompositions to synthesis by springs. In: 1998 ASME design engineering technical conferences, pp. DETC98/MECH-5880. ASME
35. Ciblak N, Lipkin H Asymmetric cartesian stiffness for the modeling of compliant robotic systems. In: Procceding 23rd biennial ASME mechanisms conference, Minneapolis, MN
36. Pashkevich A, Klimchik A, Chablat D (2009) Nonlinear effects in stiffness modeling of robotic manipulators. World Acad Sci Eng Technol 58:168–173
37. Company O, Pierrot F, Fauroux JC (2005) A method for modeling analytical stiffness of a lower mobility parallel manipulator. In: Robotics and automation, ICRA 2005. proceedings of the 2005 IEEE international conference on, pp 3232–3237
38. Arumugam HK, Voyles RM, Bapat S (2004) Stiffness analysis of a class of parallel mechanisms for micro-positioning applications. In: Intelligent robots and systems, 2004.(IROS 2004). Proceedings IEEE/RSJ International conference on. IEEE, pp 1826–1831
39. Company O, Krut S, Pierrot F (2002) Modelling and preliminary design issues of a four-axis parallel machine for heavy parts handling. Proceed Inst Mech Eng Part K: J Multi-body Dynam 216:1–12
40. Vertechy R, Parenti-Castelli V (2007) Static and stiffness analyses of a class of over-constrained parallel manipulators with legs of type US and UPS. In: Robotics and automation, 2007 IEEE international conference on. IEEE, pp 561–567
41. Abele E, Weigold M, Rothenbücher S (2007) Modeling and identification of an industrial robot for machining applications. CIRP Ann Manuf Technol 56:387–390
42. Ceccarelli M, Carbone G (2002) A stiffness analysis for CaPaMan (Cassino parallel manipulator). Mech Mach Theory 37:427–439
43. Tyapin I, Hovland G (2011) The Gantry-Tau parallel kinematic machine—kinematic and elastodynamic design optimisation. Meccanica 46:113–129

44. Li Y, Xu Q (2008) Stiffness analysis for a 3-PUU parallel kinematic machine. Mech Mach Theory 43:186–200
45. Pashkevich A, Chablat D, Wenger P (2008) Stiffness analysis of 3-d.o.f. overconstrained translational parallel manipulators. In: Robotics and automation, ICRA 2008. IEEE international conference on, pp 1562–1567
46. Chablat D, Wenger P (2003) Architecture optimization of a 3-DOF translational parallel mechanism for machining applications, the orthoglide. Robot Autom IEEE Trans 19:403–410
47. Quennouelle C (2009) Modelisation geometrico-statique des mecanismes paralleles compliants. Université Laval
48. Pashkevich A, Klimchik A, Chablat D, Wenger P (2009) Stiffness analysis of multi-chain parallel robotic systems. J Autom Mob Robot Intell Syst 3:75–82
49. Klimchik A, Pashkevich A, Caro S, Chablat D (2012) Stiffness matrix of manipulators with passive joints: computational aspects. Robot IEEE Trans 28:955–958
50. Rizk R, Andreff N, Fauroux JC, Lavest JM, Gogu G (2007) Precision study of a decoupled four degrees of freedom parallel robot including manufacturing and assembling errors. In: Tichkiewitch S, Tollenaere M, Ray P (eds) Advances in integrated design and manufacturing in mechanical engineering II. Springer, Netherlands, pp 111–127
51. Klimchik A, Pashkevich A, Caro S, Chablat D Stiffness modeling of robotic-manipulators under auxiliary loadings. In: ASME 2012 international design engineering technical conferences and computers and information in engineering conference. American Society of Mechanical Engineers, pp 469–476
52. Wei W, Simaan N (2010) Design of planar parallel robots with preloaded flexures for guaranteed backlash prevention. ASME J Mech Robot 2
53. Pashkevich A, Klimchik A, Chablat D (2010) Stiffness analysis of parallel manipulators with preloaded passive joints. In: Lenarcic J, Stanisic MM (eds) Advances in robot kinematics: motion in man and machine. Springer, Netherlands, pp 465–474
54. Klimchik A, Caro S, Wu Y, Chablat D, Furet B, Pashkevich A (2014) Stiffness modeling of robotic manipulator with gravity compensator. In: Thomas F, Perez Gracia A (eds) Computational kinematics, vol 15. Springer, Netherlands, pp 185–192
55. Klimchik A, Wu Y, Dumas C, Caro S, Furet B, Pashkevich A Identification of geometrical and elastostatic parameters of heavy industrial robots. In: IEEE international conference on robotics and automation (ICRA), pp 3707–3714
56. Chakarov D (2004) Study of the antagonistic stiffness of parallel manipulators with actuation redundancy. Mech Mach Theory 39:583–601
57. Chakarov D (1999) Study of the passive compliance of parallel manipulators. Mech Mach Theory 34:373–389
58. Koganezawa K, Nakazawa T, Inaba T (2006) Antagonistic control of multi-DOF joint by using the actuator with non-linear elasticity. In: Robotics and automation, ICRA 2006. Proceedings 2006 IEEE international conference on, pp 2201–2207
59. Koganezawa K (2013) Antagonistic control of multi-DOF joint. In: Desai JP, Dudek G, Khatib O, Kumar V (eds) Experimental robotics. Springer International Publishing, vol 88, pp 667–681
60. Li Y, Chen SF, Kao I (2002) Stiffness control and transformation for robotic systems with coordinate and non-coordinate bases. In: Robotics and automation, proceedings. ICRA '02. IEEE international conference on, pp 550–555
61. Merlet J-P, Gosselin C (2008) Parallel mechanisms and robots. In: Siciliano B, Khatib O (eds) Springer handbook of robotics. Springer, Berlin Heidelberg, pp 269–285
62. Quennouelle C, Gosselin CM (2008) Instantaneous kinemato-static model of planar compliant parallel mechanisms. In: ASME 2008 international design engineering technical conferences and computers and information in engineering conference. American Society of Mechanical Engineers, pp 163–173
63. Klimchik A (2011) Enhanced stiffness modeling of serial and parallel manipulators for robotic-based processing of high performance materials. Ecole Centrale de Nantes
64. Timoshenko S, Goodier JN (1970) Theory of elasticity. McGraw-Hill, Singapore
65. Jones RM (2006) Buckling of bars, plates, and shells. Bull Ridge Corporation

66. Carricato M, Duffy J, Parenti-Castelli V (2002) Catastrophe analysis of a planar system with flexural pivots. Mech Mach Theory 37:693–716
67. Hines R, Marsh D, Duffy J (1998) Catastrophe analysis of the planar two-spring mechanism. Int J Robot Res 17:89–101
68. Nguyen AV, Bouzgarrou BC, Charlet K, Béakou A (2015) Static and dynamic characterization of the 6-Dofs parallel robot 3CRS. Mech Mach Theory 93:65–82
69. Zhang J, Zhao Y, Jin Y Kinetostatic-model-based stiffness analysis of Exechon PKM. Robot Comput Integrat Manufac
70. Li Y, Zhang E, Song Y, Feng Z (2013) Stiffness modeling and analysis of a novel 4-DOF PKM for manufacturing large components. Chin J Aeronaut 26:1577–1585
71. Klimchik A, Chablat D, Pashkevich A (2015) Static stability of manipulator configuration: influence of the external loading. Eur J Mech A Solids 51:193–203
72. Janez G, Timi K, Karl G, Miran B (2020) Accuracy improvement of robotic machining based on robot's structural properties. Int J Adv Manufact Technol 108:1309–1329
73. Klimchik A, Pashkevich A (2017) Serial vs. quasi-serial manipulators: comparison analysis of elasto-static behaviors. Mechan Mach Theory 107:46–70
74. Klimchik A, Ambiehl A, Garnier S, Furet B, Pashkevich A (2017) Efficiency evaluation of robots in machining applications using industrial performance measure. Robot Comput Integrat Manufact 48:12–29
75. Kuo CH, Dai JS Robotics for minimally invasive surgery: a historical review from the perspective of kinematics. In: International symposium on history of machines and mechanisms. Springer Netherlands, pp 337–354
76. Christensen S, Bai S (2018) Kinematic analysis and design of a novel shoulder exoskeleton using a double parallelogram linkage. J Mech Robot 10
77. Kong K, Li J, Zhang H, Li J, Wang S (2016) Kinematic design of a generalized double parallelogram based remote center-of-motion mechanism for minimally invasive surgical robot. J Med Dev 10
78. Plooij M, Wisse M, Vallery H (2016) Reducing the energy consumption of robots using the bidirectional clutched parallel elastic actuator. IEEE Trans Rob 32:1512–1523
79. DeBoon B, Nokleby S, Delfa NL, Rossa C (2019) Differentially-clutched series elastic actuator for robot-aided musculoskeletal rehabilitation. In: 2019 international conference on robotics and automation (ICRA), pp 1507–1513
80. Moon Y, Crane Iii CD, Roberts RG (2013) Reverse kinetostatic analysis and stiffness synthesis of a spatial tensegrity-based compliant mechanism. Mech Mach Theory 70:320–337
81. Pashkevich A, Klimchik A, Caro S, Chablat D (2010) Stiffness modelling of parallelogram-based parallel manipulators. In: Pisla D, Ceccarelli M, Husty M, Corves B (eds) New trends in mechanism science, vol 5. Springer, Netherlands, pp 675–682
82. Yan SJ, Ong SK, Nee AYC (2016) Stiffness analysis of parallelogram-type parallel manipulators using a strain energy method. Robot Comput Integrat Manufact 37:13–22
83. Chablat D, Rolland L (2018) NaVARo II: a new parallel robot with eight actuation modes. V05AT07A043
84. Caro S, Chablat D, Hu Y (2014) Algorithm for the actuation mode selection of the parallel manipulator NAVARO. V05BT08A063
85. Zhao W, Pashkevich A, Klimchik A, Chablat D (2022) Elastostatic modeling of multi-link flexible manipulator based on two-dimensional dual-triangle tensegrity mechanism. J Mechan Robot 14
86. Zhakatayev A, Abdikadirova B, Sarmonov S, Varol HA (2020) Dynamics of tensegrity robots with negative stiffness elements. IEEE Access 8:187114–187125
87. Shintake J, Zappetti D, Peter T, Ikemoto Y, Floreano D (2020) Bio-inspired tensegrity fish robot. In: 2020 IEEE international conference on robotics and automation (ICRA), pp 2887–2892
88. Wenger P, Chablat D (2019) Kinetostatic analysis and solution classification of a class of planar tensegrity mechanisms. Robotica 37:1214–1224

Chapter 3
Finite Element Analysis in Manipulator Stiffness Modelling

3.1 Fundamentals of FEA Modelling Technique

Finite element analysis (FEA) is a numerical method for computing displacements, strains, and stresses in deformable structures and mechanisms under internal and external loads [1]. The basic idea of the FEA is to split an object into a number of small standard elements (triangles, prisms, tetrahedrons, hexahedrons, etc.) with a certain number of nodes connected by virtual springs describing the elasticity of the object. Mathematically, each element is described by a small matrix giving a relation between a vector of nodal displacements and a vector of applied nodal forces. It should be stressed that the complete system can be very complex, and its components may be irregularly shaped (Fig. 3.1), but each finite element is trivial and easy to analyse. After the element matrices are calculated, they are aggregated into a single large matrix describing the entire system, allowing to compute responses at the nodes to the applied loads. Basic assumptions here are that the elements are connected at the nodes only and the nodes are shared by several adjacent finite elements. Using the obtained large matrix, it is possible to evaluate the desired elastic deformations and stresses of the examined structure.

One of the most important advantages of the FEA method is that it may easily take into account the influence of non-homogeneous material properties and distributed forces. On the other hand, the FEA accuracy highly depends on discretization, i.e. on the type and size of finite elements. It is clear that increasing the number of elements leads to an essential increase in computational time and the impact of round-offs. For this reason, it is always required to find a reasonable trade-off between the size of finite elements and computational time/accuracy. Besides, in the frame of this method, it is rather individually and the FEA analysis should be repeated for each particular case.

For robotics, the FEA-based stiffness analysis is very attractive, since the links/joints are modeled with their true dimension and shape [2–4]. However, high computation expenses for the repeated re-meshing and re-computing do not allow it to be

A. Klimchik et al., *Stiffness Modeling of Parallel Robots*, Parallel Robots: Theory and Applications, https://doi.org/10.1007/978-3-031-76320-5_3

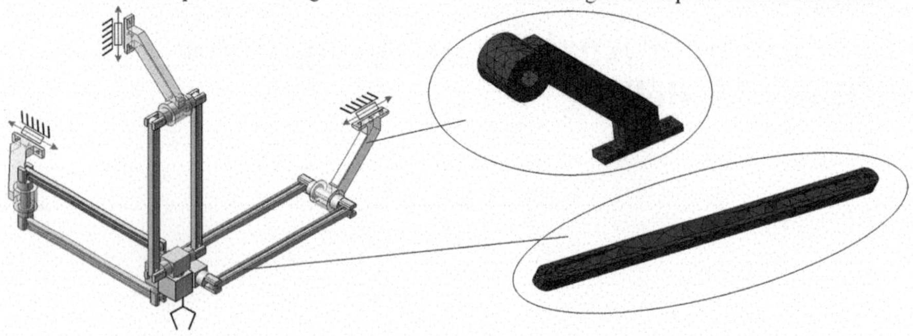

Fig. 3.1 Example of FEA-based presentation of Orthoglide manipulator links

used for online evaluation, so in robotics, the FEA method is usually applied at the final design stage only [2, 5–7].

To present the fundamentals of the FEA method, let us consider two simple examples that deal with the analysis of conventional spring systems. This allows us to show the main principles and equations using the simplest 1D finite elements.

The first example (Fig. 3.2) deals with a single spring of stiffness K. For this case, for each node, Hooke's law can be written in the following form

$$\begin{aligned} F_1 &= K \cdot (\delta_1 - \delta_2) \\ F_2 &= K \cdot (\delta_2 - \delta_1) \end{aligned} \tag{3.1}$$

where $F_1,\ F_2$ are the external forces applied to the corresponding nodes and $\delta_1,\ \delta_2$ are the deflections in these nodes. So, in a matrix form, the desired relation between the nodal displacements and the nodal forces can be presented as

$$\begin{bmatrix} F_1 \\ F_2 \end{bmatrix} = \begin{bmatrix} K & -K \\ -K & K \end{bmatrix} \cdot \begin{bmatrix} \delta_1 \\ \delta_2 \end{bmatrix} \tag{3.2}$$

or, in a general form,

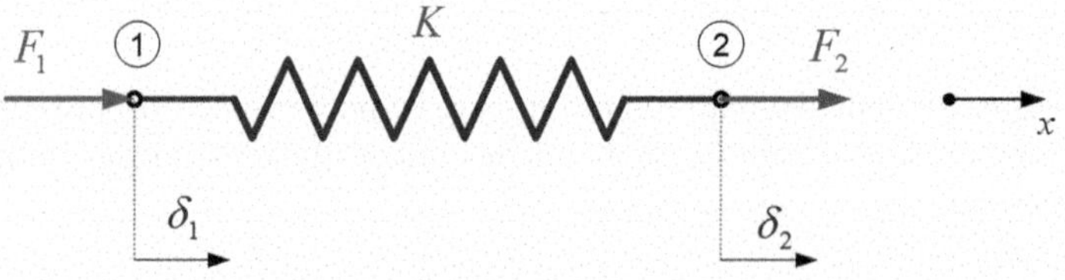

Fig. 3.2 FEA-based presentation of a linear spring (2 nodes)

$$\begin{bmatrix} F_1 \\ F_2 \end{bmatrix} = \begin{bmatrix} K_{11} & K_{12} \\ K_{21} & K_{22} \end{bmatrix} \cdot \begin{bmatrix} \delta_1 \\ \delta_2 \end{bmatrix} \tag{3.3}$$

where the stiffness matrix elements K_{ij} define the resistance of the node j to the external loading applied to the node i. For this trivial case, the obtained system of linear equations presents a stiffness model structurally similar to those used in the FEA method. It is clear that this model is redundant because only two variables among $\delta_1,\ \delta_2$ and $F_1,\ F_2$ should be treated as independent ones. In practice, to obtain valid results, it is necessary to take into account some constraints, for instance $\delta_1 = 0$ or $F_1, +F_2 = 0$, etc.

The second example (Fig. 3.3) deals with a multi-spring system, for which equations relating the nodal displacements/forces are derived. This system includes 4 nodes, where the node #1 is fixed to the base and the node #4 is subjected to the external load F. It is assumed that external loads in the intermediate nodes #2 and #3 are equal to zero. For this system, it is required to find nodal displacements $\delta_1,\ \delta_2,\ \delta_3,\ \delta_4$ corresponding to the external force F.

Using the above-presented expression (3.3), it is possible to write force–displacement relations for each finite element (i.e. spring) of the considered system separately

$$\begin{array}{ll} FE\ \#1\ |Nodes\ (\#1-\#2): & \begin{bmatrix} F_1^{(1)} \\ F_2^{(1)} \end{bmatrix} = \begin{bmatrix} K_{11}^{(1)} & K_{12}^{(1)} \\ K_{21}^{(1)} & K_{22}^{(1)} \end{bmatrix} \cdot \begin{bmatrix} \delta_1^{(1)} \\ \delta_2^{(1)} \end{bmatrix} \\ FE\ \#2\ |Nodes\ (\#2-\#3): & \begin{bmatrix} F_1^{(2)} \\ F_2^{(2)} \end{bmatrix} = \begin{bmatrix} K_{11}^{(2)} & K_{12}^{(2)} \\ K_{21}^{(2)} & K_{22}^{(2)} \end{bmatrix} \cdot \begin{bmatrix} \delta_1^{(2)} \\ \delta_2^{(2)} \end{bmatrix} \\ FE\ \#3\ |Nodes\ (\#1-\#3): & \begin{bmatrix} F_1^{(3)} \\ F_2^{(3)} \end{bmatrix} = \begin{bmatrix} K_{11}^{(3)} & K_{12}^{(3)} \\ K_{21}^{(3)} & K_{22}^{(3)} \end{bmatrix} \cdot \begin{bmatrix} \delta_1^{(3)} \\ \delta_2^{(3)} \end{bmatrix} \\ FE\ \#4\ |Nodes\ (\#3-\#4): & \begin{bmatrix} F_1^{(4)} \\ F_2^{(4)} \end{bmatrix} = \begin{bmatrix} K_{11}^{(4)} & K_{12}^{(4)} \\ K_{21}^{(4)} & K_{22}^{(4)} \end{bmatrix} \cdot \begin{bmatrix} \delta_1^{(4)} \\ \delta_2^{(4)} \end{bmatrix} \end{array} \tag{3.4}$$

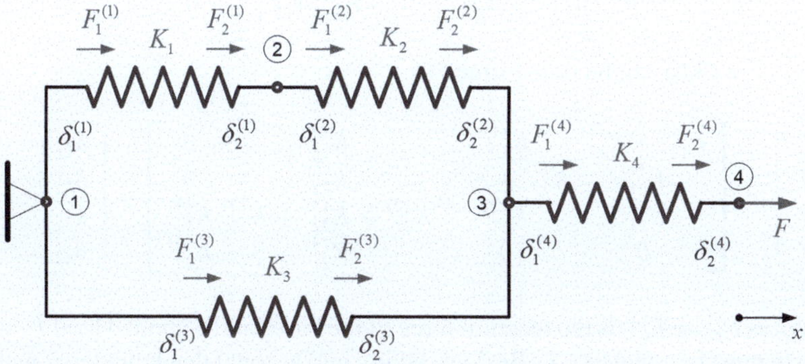

Fig. 3.3 FEA-based presentation of a multi-spring system (4 nodes)

where FE denotes the finite element and the superscript "(i)" defines the element number. Further, to derive the desired force–displacement relation for the entire system, let us rewrite each of the above matrix equations with respect to all node displacements $\delta_1,\ \delta_2,\ \delta_3,\ \delta_4$

$$
\begin{bmatrix} F_1^{(1)} \\ F_2^{(1)} \\ 0 \\ 0 \end{bmatrix} = \begin{bmatrix} K_{11}^{(1)} & K_{12}^{(1)} & 0 & 0 \\ K_{21}^{(1)} & K_{22}^{(1)} & 0 & 0 \\ 0 & 0 & 0 & 0 \\ 0 & 0 & 0 & 0 \end{bmatrix} \cdot \begin{bmatrix} \delta_1 \\ \delta_2 \\ \delta_3 \\ \delta_4 \end{bmatrix} \begin{bmatrix} 0 \\ F_1^{(2)} \\ F_2^{(2)} \\ 0 \end{bmatrix} = \begin{bmatrix} 0 & 0 & 0 & 0 \\ 0 & K_{11}^{(2)} & K_{12}^{(2)} & 0 \\ 0 & K_{21}^{(2)} & K_{22}^{(2)} & 0 \\ 0 & 0 & 0 & 0 \end{bmatrix} \cdot \begin{bmatrix} \delta_1 \\ \delta_2 \\ \delta_3 \\ \delta_4 \end{bmatrix}
$$

$$
\begin{bmatrix} F_1^{(3)} \\ 0 \\ F_2^{(3)} \\ 0 \end{bmatrix} = \begin{bmatrix} K_{11}^{(3)} & 0 & K_{12}^{(3)} & 0 \\ 0 & 0 & 0 & 0 \\ K_{21}^{(3)} & 0 & K_{22}^{(3)} & 0 \\ 0 & 0 & 0 & 0 \end{bmatrix} \cdot \begin{bmatrix} \delta_1 \\ \delta_2 \\ \delta_3 \\ \delta_4 \end{bmatrix} \begin{bmatrix} 0 \\ 0 \\ F_1^{(4)} \\ F_2^{(4)} \end{bmatrix} = \begin{bmatrix} 0 & 0 & 0 & 0 \\ 0 & 0 & 0 & 0 \\ 0 & 0 & K_{11}^{(4)} & K_{12}^{(4)} \\ 0 & 0 & K_{21}^{(4)} & K_{22}^{(4)} \end{bmatrix} \cdot \begin{bmatrix} \delta_1 \\ \delta_2 \\ \delta_3 \\ \delta_4 \end{bmatrix} \tag{3.5}
$$

This allows us to find resultant forces in all nodes as a simple sum, which yields a general matrix equation

$$
\begin{bmatrix} F_1^{(1)} + F_1^{(3)} \\ F_2^{(1)} + F_1^{(2)} \\ F_2^{(2)} + F_2^{(3)} + F_1^{(4)} \\ F_2^{(4)} \end{bmatrix} = \begin{bmatrix} K_{11}^{(1)} + K_{11}^{(3)} & K_{12}^{(1)} & K_{12}^{(3)} & 0 \\ K_{21}^{(1)} & K_{22}^{(1)} + K_{11}^{(2)} & K_{12}^{(2)} & 0 \\ K_{21}^{(3)} & K_{21}^{(2)} & K_{22}^{(2)} + K_{22}^{(3)} + K_{11}^{(4)} & K_{12}^{(4)} \\ 0 & 0 & K_{21}^{(4)} & K_{22}^{(4)} \end{bmatrix} \cdot \begin{bmatrix} \delta_1 \\ \delta_2 \\ \delta_3 \\ \delta_4 \end{bmatrix} \tag{3.6}
$$

defining the system reaction for any combination of external forces applied at the nodes.

To derive the desired expression for the considered system (see Fig. 3.3), it is necessary to take into account several constraints. In particular, taking into account that resultant forces in nodes #2 and #3 are equal to zero (since there are no external loads here), one can write that

$$
F_2^{(1)} + F_1^{(2)} = 0;\ F_2^{(2)} + F_2^{(3)} + F_1^{(4)} = 0 \tag{3.7}
$$

So, system (3.6) can be transformed to

$$
\begin{bmatrix} F_r \\ 0 \\ 0 \\ F \end{bmatrix} = \begin{bmatrix} K_{11}^{(1)} + K_{11}^{(3)} & K_{12}^{(1)} & K_{12}^{(3)} & 0 \\ K_{21}^{(1)} & K_{22}^{(1)} + K_{11}^{(2)} & k_{12}^{(2)} & 0 \\ K_{21}^{(3)} & K_{21}^{(2)} & K_{22}^{(2)} + K_{22}^{(3)} + K_{11}^{(4)} & K_{12}^{(4)} \\ 0 & 0 & K_{21}^{(4)} & K_{22}^{(4)} \end{bmatrix} \cdot \begin{bmatrix} \delta_1 \\ \delta_2 \\ \delta_3 \\ \delta_4 \end{bmatrix} \tag{3.8}
$$

where $F_r = F_1^{(1)} + F_1^{(3)}$ is the reaction force of the base for the applied load F, which obviously must be equal to $-F$. Besides, taking into account that boundary condition $\delta_1 = 0$ for node #1 (which is fixed to the base), by simple elimination of the first line and column in the corresponding matrix the final force–displacement relation is reduced to

$$\begin{bmatrix} 0 \\ 0 \\ F \end{bmatrix} = \begin{bmatrix} K_{22}^{(1)} + K_{11}^{(2)} & K_{12}^{(2)} & 0 \\ K_{21}^{(2)} & K_{22}^{(2)} + K_{22}^{(3)} + K_{11}^{(4)} & K_{12}^{(4)} \\ 0 & K_{21}^{(4)} & K_{22}^{(4)} \end{bmatrix} \cdot \begin{bmatrix} \delta_2 \\ \delta_3 \\ \delta_4 \end{bmatrix} \tag{3.9}$$

The obtained relations (3.8) and (3.9) show the typical solution of the stiffness analysis problem using the FEA-based technique. It allows the user to compute all node displacements for a given external load and, consequently, evaluate stresses in a mechanical system. Besides, the structure of the corresponding matrices is also typical here, it includes elements computed as the sums of the small-size matrices describing adjacent finite elements. Generalizing this idea for a complex system composed of a large number of 3D finite elements (prisms, tetrahedrons, hexahedrons, etc.), it is possible to obtain similar matrices for any mechanism or its element. However, the matrix size defined by the number of nodes is usually extremely high, which requires essential support of relevant CAD tools for the object splitting into the FE elements (meshing), building the global stiffness matrix, its inversion, stress analysis, etc.

In modern commercial software systems, FEA modelling is usually split into three sequential stages.

- *Preprocess*: creating a 3D solid model of the mechanism and its components, defining finite element types and material properties, assigning modelling parameters and boundary conditions;
- *Process*: discretization of all 3D objects into finite elements, building a system of linear equations describing nodal force-deflection relations, and solving this system for given boundary conditions;
- *Postprocess*: extraction results in the form suitable for the user, their transformation and visualization.

At the first stage (preprocess) user should define a final goal of the FEA modelling, which can be structural static analysis, buckling analysis, contact analysis, etc. Here it is required to specify the interpolation method and select the shape and size of finite elements (truss, beam, shell, solid, triangle, see Fig. 3.4). It should be noted that the type/size of finite elements may differ for different mechanism components. Besides, the user assigns material properties that include principal elastic parameters (Young's modulus, Poisson's ratio, Bulk modulus, material density, etc.), defines external forces and direction of gravity, connections between adjacent elements (rigid or elastic) as well as boundary conditions.

The second stage (process) is aimed at creating of true FEA model and its analysis. It begins with meshing, where all mechanical components are discretized into finite elements in accordance with user-defined settings (type, size, etc.). Each element is described by a system of linear equations relating the nodal forces and deflections. These systems are aggregated into a single one that is solved taking into account the boundary conditions.

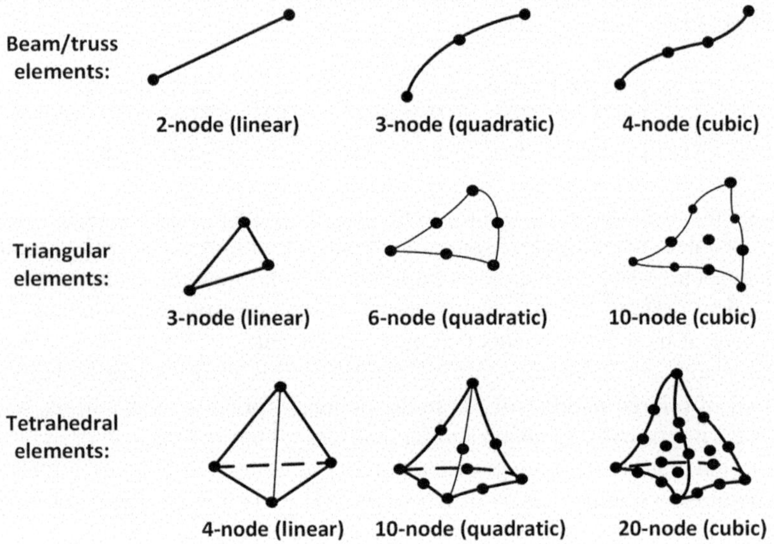

Fig. 3.4 Typical finite elements used in FEA modeling

At the final stage (post-process) computational results are presented in a user-friendly form and are visualized as node-points displacements or stresses, for instance (Fig. 3.5). In more detail, the FEA modelling process is illustrated in the following Section using an application example that deals with the stiffness modelling of the Orthoglide parallel manipulator.

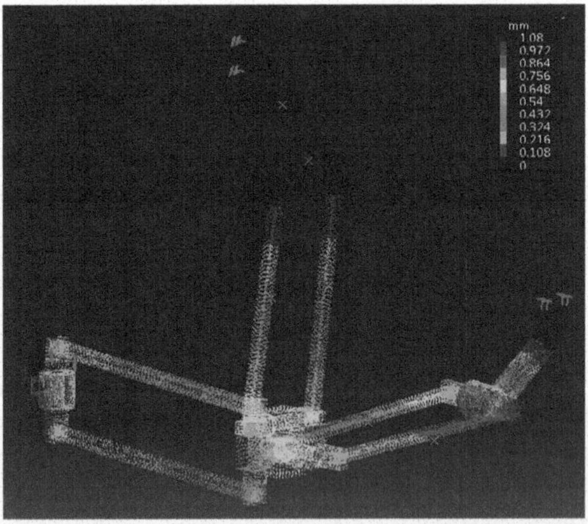

Fig. 3.5 Visualization of FEA results: node-point displacements under external loading

3.2 FEA in Robotics: Examples and Principal Difficulties

For robotic manipulators, the FEA technique provides the most detailed information on the stiffness properties, i.e. displacements, strains, and stresses of robot components under internal and external loads. There are many successful examples of FEA applications in this area that include stiffness analysis of robots CaPaMan [8], T3R1 [4], Tricept and TriVariant [9], Delta [10], DAEDALUS [2] and others. However, because of high computing expenses, the FEA method is mainly used as a supplementary tool allowing users to validate developed robot performances and optimize mechanical design. The principal difficulty here is the necessity of repetitive re-meshing and re-computing for each manipulator configuration. Nevertheless, the FEA method is very efficient for the estimation of manipulator stiffness coefficients and natural frequencies within the robot workspace.

Another effective FEA application in robotics is related to the comparison study of stiffness properties for complex parallel manipulators. For example, Wang et al. [9] compared 5-DOF reconfigurable hybrid robots Tricept and TriVariant. The authors studied FEA presentations of different types of elastic joints and developed a fast method for natural frequency estimation associated with different manipulator configurations.

Application of the FEA method in the manipulator design optimization allows the engineer to detect weak points and immediately adjust manipulator parameters in order to improve its elastostatic properties. For instance, the design of a closed-loop platform-type robot manipulator 6-SPS using the finite element method has been done by Ramachandran et al. [11]. Using FEA, the authors optimized the manipulator geometry and estimated the actuator parameters. Similarly, the FEA simulations were used by Kobel and Clavel [12] to analyze and optimize the structural proprieties and dynamical performance of micro robot ΦR for rotary desktop assembly lines. Their model takes into account the influence of the bearings via 6 idealized springs and the elasticity of the belt-driven pulleys with preloading of the belt and the idealized passive joints. Liang et al. [13] used FEA to estimate the stiffness of developed 6-dof. micro-manipulator based on the compliant parallel mechanism with an integrated force sensor. Huang and Wang [14] used the finite element approach to develop the generalized equations of motion for two- and three-dimension complex elastic mechanisms that include the nonlinear coupling terms of rigid body and elastic motions.

Some authors used FEA modelling for stiffness coefficient identification of manipulator components. It should be stressed that for links of complex shapes, it remains a unique way to get suitable values of stiffness parameters. For example, in [15] the stiffness coefficients of complicated components of the 2PUS–2PRS parallel mechanism were estimated by aggregating FEA and numerical fitting, which were further used in a semi-analytical model to predict stiffness variation within the robot workspace. Similarly, Aginaga et al. [16] used FEA modelling to identify stiffness coefficients of 6-RUS manipulator components with complex shapes, which were included in the developed analytical model. This approach was further developed by

Pashkevich et al. [17, 18] for the identification of 6×6 stiffness matrices of manipulator links and their integration in the VJM modelling technique. This idea allowed users to avoid repeating FEA modelling for all desired configurations since FEA was applied only once for each original manipulator component. This approach has been advanced in [19, 20], where the authors proposed dedicated numerical routines allowing accuracy improvement of the identified stiffness matrices. Besides, in these works the issues of optimal virtual experimental planning and selection of meshing types were analyzed. A relevant step-by-step procedure is presented in detail in the following section.

Some modifications of FEA methods are used in robotics for rather particular tasks where other methods are not well developed yet or cannot ensure high efficiency. For example, a planar fully parallel robot with flexible links was examined in [21] using dynamic finite element analysis. The authors stated the significant influence of high-speed motion and manipulator configuration on the natural frequencies. The generalized stochastic finite element method has been used in [22] for the stability analysis of elastic systems with random parameters.

However, the main application of FEA in robotics remains the validation of analytical models. In [23] the FEA was used to validate the stiffness model of a 6-dof. parallel robot 3CRS with 3 limbs derived using the matrix structural analysis (MSA). The FEA and MSA results were compared for 27 configurations regularly distributed within the robot workspace. They showed that the difference between the deflections computed using FEA and MSA methods was within 5%, which is usually acceptable for static analysis. Similarly, Deblaise [10] used FEA modelling to verify the accuracy of the MSA-based stiffness model for the parallel robot Delta. In [24] FEA modelling and experimental results were used to validate the stiffness model of a 6-dof. force/torque sensor based on a 3-RPS parallel mechanism. This approach was also used in [25] for a compliant micro-manipulator, for which the analytical stiffness model was validated through FEA and experimental study. In [26] the authors used FEA to verify the accuracy of the analytical stiffness model for a 3-UPU parallel manipulator; the obtained 25% difference in the force–deflection relation was accepted as a reasonable one for practical application. In [27] the FEA was used for specific parallelogram-type manipulators in order to show that the proposed algebraic method is a reasonable alternative for the pre-design stage: the maximum relative error between deflections computed using FEA and the algebraic method was less than 10%. The stiffness of a serial-parallel manipulator 2(SP + SPR + SPU) was analyzed in [28] using both analytical and FEA methods. In [29], a finite element model was used to validate the developed stiffness model of 6-dof high-accuracy UTI-arm. In [30], FEA was used to validate the developed methodology for the stiffness analysis of lower mobility parallel manipulators with decoupled motions (4-dof. parallel manipulator DAEDALUS I). A slightly different idea was implemented by Corradini et al. [2] who experimentally evaluated the accuracy of a simplified FEA-based stiffness model for an H4 parallel robot, which was constructed with 41 nodes and 44 beams of various cross sections. The validity of two analytical methods (distributed and lumped ones) for computing the Cartesian stiffness matrix of the 3-PSP parallel robot was verified using FEA in [31]. In the simulation study the manipulator links were approximated by elastic circular beams and the motor elasticities were presented by 1-dof virtual springs. The obtained results showed that the distributed stiffness model was very close to the FEA-based one, while the lumped model was rather inaccurate (since it was not able to take into account bending in the compliant elements). In [5], the FEA method and algebraic approach were applied to the stiffness modelling of a Stewart Platform, to get maps of minimum and maximum values of the stiffness coefficient in the arbitrary direction.

Despite the numerous advantages and good reputation of the FEA method, users should be prudent while applying it to the stiffness analysis of robotic manipulators since in reality FEA remains an approximation. This fact is confirmed by the comparison study of Theodore and Ghosal [32], who compared a finite element discretization model and an analytical representation of robot manipulator links. As follows from this work, the finite element model usually provides an overestimation of the stiffness matrix compared to its analytical counterpart, but this difference becomes negligible if the mashing step is small enough.

3.3 Illustrative Example: FEA-Based Analysis of Orthoglide Manipulator

Let us illustrate now application of the FEA technique with an example of stiffness modeling for parallel manipulator of the Orthoglide family [33, 34] in the CAD environment CATIA V5. This manipulator contains 3 identical chains with kinematic parallelograms connected to the end-platform (Fig. 3.1). The manipulator end-effector position is controlled by 3 linear motors (orthogonal to each other) installed in the actuated joints between the robot base and foot. The remaining joints (six for each chain) are passive ones. The presented example deals with FEA modelling for the manipulator configuration corresponding to the middle workspace point for which all kinematic parallelograms are orthogonal to each other. In the frame of this modelling, it is assumed that all manipulator links are made from isotropic aluminium alloy with Young modulus 7×10^{10} N/m^3 and Poison ratio 0.346. The density of the alloy is 2710 kg/m^3, the yield strength is 9.5×10^7 N/m^3 .

At the *preprocessing* stage, the 3D CAD models of all manipulator components were created and assembled in accordance with mechanism kinematics. Then, the physical properties of materials were assigned and finite element type/size was defined. The modelling environment of CATIA proposes user several options for the finite elements to select from. In particular, it is necessary to set the mesh step and to choose between linear and parabolic interpolation for tetrahedron 3D finite elements (Fig. 3.6), which provide either 4 or 10 nodes respectively. It is clear that these settings essentially influence the modelling accuracy and computing time. From our experience, the parabolic mesh is preferable and provides higher accuracy. For the mesh size, it is required to reduce essentially the default values in order to ensure that each cross-section is approximated by 3–5 finite elements at least that in our

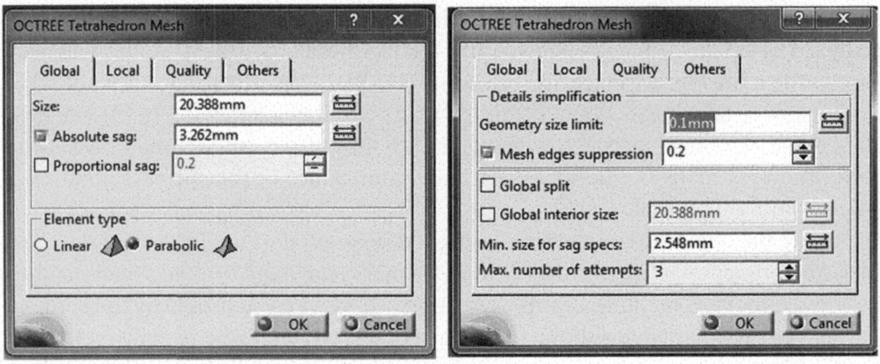

Fig. 3.6 Tuning meshing options for manipulator links

case corresponds to 2 mm size. However, for this example, further reduction of the element size does not give any benefit and leads to computing time increasing and accumulating round-offs only. It should be mentioned that the mesh parameters should be tuned separately for each manipulator component.

Also, at the preprocessing stage, there were assigned the boundary conditions describing the manipulator elements rigidly fixed to the robot base, defined connections between manipulator components (active and passive joints) and external actions, including gravity direction and disturbing forces applied to the end-effector. In this simulation, the actuating joints were assumed to be rigid. The passive joints were defined as a specific constraint: via a surface connection allowing their sliding without losing contact. In our experiments, the external loading was distributed and applied to all end-effector faces.

At the *process* stage, the model was meshed into a set of finite elements and a relevant system of linear equations (force–deflection relations for all nodes) was generated using CATIA built-in capabilities. Then, the system has been solved for the above-defined boundary conditions and the external loading of 1000 N is applied to the manipulator end-effector in the xy-plane.

Finally, at the *postprocessing* stage, the computational results were visualized in order to detect the weakest manipulator components. Here, CATIA provides the user with several options allowing them to show the nodal deflections, the stresses, computational precision, etc. (Fig. 3.7). Besides, it is possible to scale the nodal displacements in order to understand better directions for the link deformations and detect weak points of the mechanism (Fig. 3.8).

For comparison purposes, the above-described simulations were repeated for two other configurations, close to singular ones (so-called near-flat and near-bar configurations). Relevant results are presented in Fig. 3.9 and clearly show that in the singularity neighbourhood, the manipulator resistance against external loading essentially reduces and the deformations of manipulator components are essentially higher compared to the workspace mid-point.

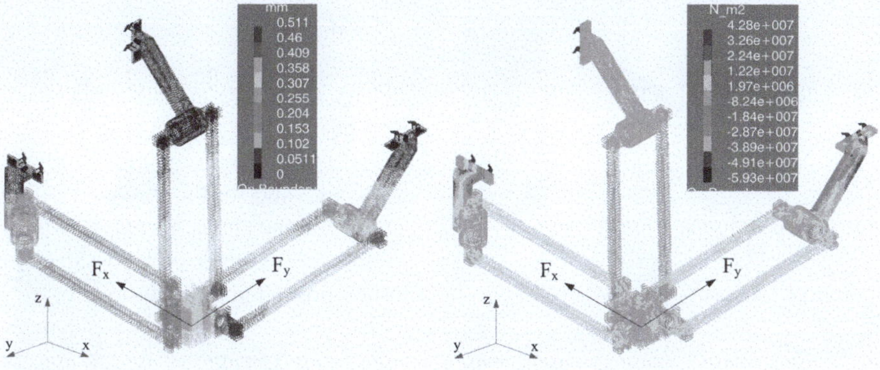

(a) Displacements caused by the loading 1 kN (b) Internal stresses caused by the loading 1 kN

Fig. 3.7 Visualization of FEA modeling results for the Orthoglide manipulator

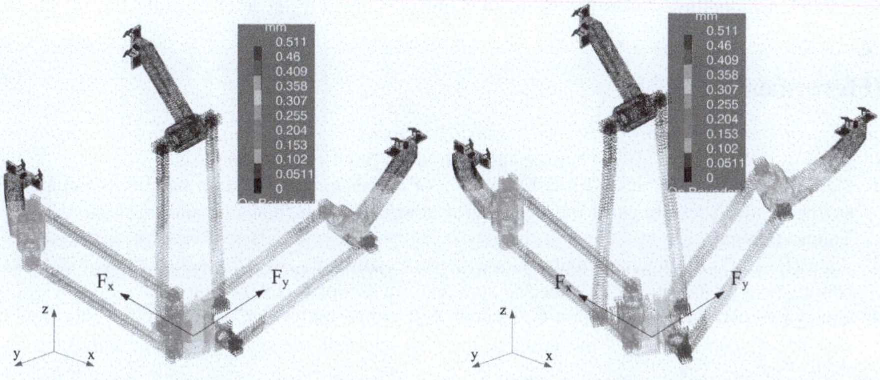

(a) Displacements with scaling factor 50 (b) Displacements with scaling factor 200

Fig. 3.8 Impact of scaling factor on the visualization of manipulator deformations

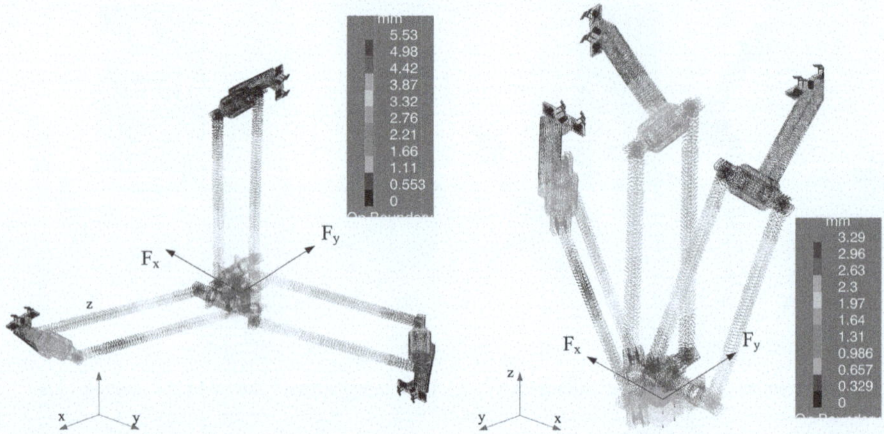

(a) FEA-modeling for the near-flat configuration (b) FEA-modeling for the near-bar configuration

Fig. 3.9 FEA modelling results for the quasi-singular configurations of Orthoglide

References

1. Bathe KJ, Wilson EL (1976) Numerical methods in finite element analysis
2. Corradini C, Fauroux JC, Krut S Evaluation of a 4-degree of freedom parallel manipulator stiffness. In: Proceedings of the 11th world congress in mechanisms and machine science, Tianjin (China)
3. Akin JE (2005) Finite element analysis with error estimators: an introduction to the FEM and adaptive error analysis for engineering students. Butterworth-Heinemann
4. Bouzgarrou B, Fauroux J, Gogu G, Heerah Y Rigidity analysis of T3R1 parallel robot with uncoupled kinematics. In: Procceding of the 35th international symposium on robotics (ISR), Paris, France
5. El-Khasawneh BS, Ferreira PM (1999) Computation of stiffness and stiffness bounds for parallel link manipulators. Int J Mach Tools Manuf 39:321–342
6. Rizk R, Fauroux JC, Mumteanu M, Gogu G (2006) A comparative stiffness analysis of a reconfigurable parallel machine with three or four degrees of mobility. J Mach Eng 6:45–55
7. Long CS, Snyman JA, Groenwold AA (2003) Optimal structural design of a planar parallel platform for machining. Appl Math Model 27:581–609
8. Pêgo JMT, Gonçalves RS, Carvalho JCM, Carbone G, Ceccarelli M Stiffness analysis of CaPaMan-2bis using finite element analysis. In: 22nd international congress of mechanical engineering (COBEM 2013), pp 3–7
9. Wang YY, Huang T, Zhao XM, Mei JP, Chetwynd DG, Hu SJ Finite element analysis and comparison of two hybrid robots-the Tricept and the TriVariant. In: Intelligent robots and systems, 2006 IEEE/RSJ international conference on, pp 490–495
10. Deblaise D (2006) Contribution à la modélisation et à l'étalonnage élasto-géométriques des manipulateurs à structure parallèle. INSA de Rennes
11. Ramachandran S, Nagarajan T, Siva Prasad N (1992) A finite element approach to the design and dynamic analysis of platform type robot manipulators. Finite Elem Anal Des 10:335–350
12. Kobel P, Clavel R Micro robot for rotary desktop assembly line. In: Assembly and manufacturing (ISAM), 2011 IEEE international symposium on, pp 1–6
13. Liang Q, Zhang D, Chi Z, Song Q, Ge Y, Ge Y (2011) Six-DOF micro-manipulator based on compliant parallel mechanism with integrated force sensor. Robot Comput Integrat Manufact 27:124–134

14. Huang S-J, Wang T-Y (1993) Structural dynamics analysis of spatial robots with finite element approach. Comput Struct 46:703–716
15. Li Y, Zhang E, Song Y, Feng Z (2013) Stiffness modeling and analysis of a novel 4-DOF PKM for manufacturing large components. Chin J Aeronaut 26:1577–1585
16. Aginaga J, Zabalza I, Altuzarra O, Nájera J (2012) Improving static stiffness of the parallel manipulator using inverse singularities. Robot Comput Integrat Manufact 28:458–471
17. Pashkevich A, Chablat D, Wenger P (2009) Stiffness analysis of overconstrained parallel manipulators. Mech Mach Theory 44:966–982
18. Pashkevich A, Chablat D, Wenger P Stiffness analysis of 3-d.o.f. overconstrained translational parallel manipulators. In: Robotics and automation, 2008. ICRA 2008. IEEE international conference on, pp 1562–1567
19. Klimchik A, Pashkevich A, Chablat D (2013) CAD-based approach for identification of elasto-static parameters of robotic manipulators. Finite Elem Anal Des 75:19–30
20. Pashkevich A, Klimchik A, Chablat D, Wenger P Accuracy improvement for stiffness modeling of parallel manipulators. In: 42nd CIRP conference on manufacturing systems, pp 8
21. Piras G, Cleghorn WL, Mills JK (2005) Dynamic finite-element analysis of a planar high-speed, high-precision parallel manipulator with flexible links. Mech Mach Theory 40:849–862
22. Kamiński MM, Świta P (2011) Generalized stochastic finite element method in elastic stability problems. Comput Struct 89:1241–1252
23. Nguyen AV, Bouzgarrou BC, Charlet K, Béakou A (2015) Static and dynamic characterization of the 6-Dofs parallel robot 3CRS. Mech Mach Theory 93:65–82
24. Lu Y, Chen L, Wang P, Zhang B, Zhao Y, Hu B (2013) Statics and stiffness analysis of a novel six-component force/torque sensor with 3-RPPS compliant parallel structure. Mech Mach Theory 62:99–111
25. Pham H-H, Chen IM (2005) Stiffness modeling of flexure parallel mechanism. Precis Eng 29:467–478
26. Hu B, Lu Y (2011) Solving stiffness and deformation of a 3-UPU parallel manipulator with one translation and two rotations. Robotica 29:815–822
27. Yan SJ, Ong SK, Nee AYC (2016) Stiffness analysis of parallelogram-type parallel manipulators using a strain energy method. Robot Comput Integrat Manufact 37:13–22
28. Hu B, Lu Y, Tan Q, Yu J, Han J (2011) Analysis of stiffness and elastic deformation of a 2(SP+SPR+SPU) serial–parallel manipulator. Robot Comput Integrat Manufact 27:418–425
29. Abdel-Malek K, Paul B (1998) Criteria for the design of manipulator arms for a high stiffness-to-weight ratio. J Manuf Syst 17:209–220
30. Pinto C, Corral J, Altuzarra O, Hernández A (2009) A methodology for static stiffness mapping in lower mobility parallel manipulators with decoupled motions. Robotica 28:719–735
31. Rezaei A, Akbarzadeh A, Akbarzadeh-T M-R (2012) An investigation on stiffness of a 3-PSP spatial parallel mechanism with flexible moving platform using invariant form. Mech Mach Theory 51:195–216
32. Theodore RJ, Ghosal A (1995) Comparison of the assumed modes and finite element models for flexible multilink manipulators. Int J Robot Res 14:91–111
33. Chablat D, Wenger P (2003) Architecture optimization of a 3-DOF translational parallel mechanism for machining applications, the orthoglide. Robot Autom IEEE Trans 19:403–410
34. Pashkevich A, Wenger P, Chablat D (2005) Design strategies for the geometric synthesis of Orthoglide-type mechanisms. Mech Mach Theory 40:907–930

Chapter 4
Matrix Structural Analysis and Its Application in Robotics

4.1 MSA Method Background and Stiffness Model Parameters

The MSA is considered a compromise technique, which incorporates the main ideas of the FEA but operates with rather large elements such as flexible links connected by the actuated and passive joints in the overall manipulator structure [1–3]. This obviously leads to the reduction of the computational expenses that are quite acceptable for robotics, but it requires some non-trivial actions for including passive and elastic joints in the related mathematical model. On the other side, the MSA is very convenient for a description of complex structures with numerous closed-loops and cross-linkage.

4.1.1 MSA-Based Stiffness Model of a Separate Link

In classical mechanics, the stiffness properties of the *cantilever beam* (fixed at one side, see Fig. 4.1a) are described by Hook's law which defines a linear relation between the applied external wrench (force/torque) $\mathbf{W}$ and corresponding deflection $\Delta\mathbf{t}$ at the free-end

$$\mathbf{W} = \mathbf{K} \cdot \Delta\mathbf{t} \tag{4.1}$$

where $\mathbf{K}$ is 6×6 stiffness matrix. It should be mentioned that here $\Delta\mathbf{t}$ is a 6-dimensional deflection vector that includes both translational $\Delta\mathbf{p} = [\Delta p_x, \Delta p_y, \Delta p_z]^T$ and rotation $\Delta\boldsymbol{\varphi} = [\Delta\varphi_x, \Delta\varphi_y, \Delta\varphi_z]^T$ components. Similarly, the wrench vector $\mathbf{W}$ is also 6-dimensional and contains both force $\mathbf{F} = [F_x, F_y, F_z]^T$ and torque $\mathbf{M} = [M_x, M_y, M_z]^T$ parts.

A. Klimchik et al., *Stiffness Modeling of Parallel Robots*, Parallel Robots: Theory and Applications, https://doi.org/10.1007/978-3-031-76320-5_4

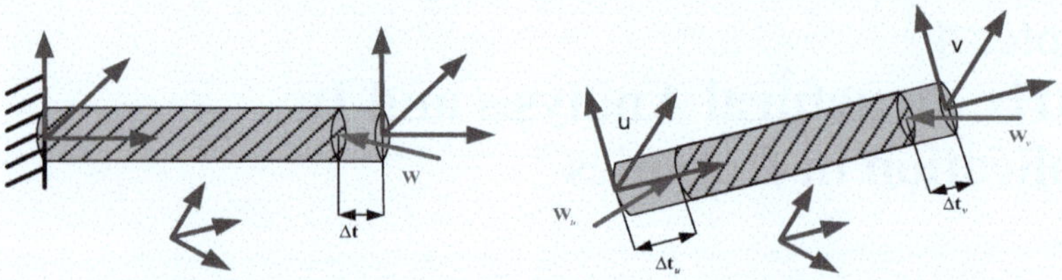

(a) Cantilever beam (b) Unsupported beam

Fig. 4.1 Cantilever and unsupported beams: definition of frames, deflections and wrenches [5]

In the general case, the stiffness matrix **K** from Hook's law is symmetric and positive definite but may include some off-diagonal elements [4]. For typical beams commonly used in practice (with regular cross-section) the stiffness matrix can be computed analytically as follows

$$\mathbf{K} = \begin{bmatrix} \frac{E\cdot S}{L} & 0 & 0 & 0 & 0 & 0 \\ 0 & \frac{12\cdot E\cdot Iz}{L^3} & 0 & 0 & 0 & \frac{-6\cdot E\cdot Iz}{L^2} \\ 0 & 0 & \frac{12\cdot E\cdot Iy}{L^3} & 0 & \frac{6\cdot E\cdot Iy}{L^2} & 0 \\ 0 & 0 & 0 & \frac{G\cdot J}{L} & 0 & 0 \\ 0 & 0 & \frac{6\cdot E\cdot Iy}{L^2} & 0 & \frac{4\cdot E\cdot Iy}{L} & 0 \\ 0 & \frac{-6\cdot E\cdot Iz}{L^2} & 0 & 0 & 0 & \frac{4\cdot E\cdot Iz}{L} \end{bmatrix} \tag{4.2}$$

where L the beam length, S is the beam cross-section area, Iy, Iz are the second moments, J is the polar moment, E and G are Young's module and Coulomb's modules of the beam material, respectively. The second and polar moment for typical cross-sections can be computed using expressions given in Table 4.1.

It should be also stressed that for the cantilever beam, both the wrench **W** and deflection $\Delta\mathbf{t}$ are usually expressed with respect to the coordinate system attached to the beam's fixed-end. However, in the general case, the vectors **W** and $\Delta\mathbf{t}$ may be presented in the global coordinate system, which requires some revision of (4.1) and (4.2). In fact, for small angular deflections, the deflection and wrench vectors in the global system can be presented as

$$\begin{bmatrix} \Delta\mathbf{p}_{\text{global}} \\ \Delta\boldsymbol{\varphi}_{\text{global}} \end{bmatrix} = \begin{bmatrix} \mathbf{R} & \mathbf{0}_{3\times3} \\ \mathbf{0}_{3\times3} & \mathbf{R} \end{bmatrix}_{6\times6} \cdot \begin{bmatrix} \Delta\mathbf{p} \\ \Delta\boldsymbol{\varphi} \end{bmatrix} \tag{4.3}$$

$$\begin{bmatrix} \mathbf{F}_{\text{global}} \\ \mathbf{M}_{\text{global}} \end{bmatrix} = \begin{bmatrix} \mathbf{R} & \mathbf{0}_{3\times3} \\ \mathbf{0}_{3\times3} & \mathbf{R} \end{bmatrix}_{6\times6} \cdot \begin{bmatrix} \mathbf{F} \\ \mathbf{M} \end{bmatrix} \tag{4.4}$$

where the orthogonal matrix **R** defines the orientation of the local coordinate system relative to the global one. The letter allows us to present the stiffness model of the cantilever beam in the global coordinate system

Table 4.1 Expressions for the moments for the beam with typical cross-sections

Type of the cross-section	Cross-section area	The second moments	The polar moment
	S	$Iy,\ Iz$	J
Cylinder 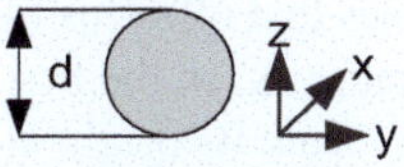	$S=\frac{\pi\cdot d^2}{4}$	$Iy=\dfrac{\pi\cdot d^4}{64};$ $Iz=\dfrac{\pi\cdot d^4}{64}$	$J=\frac{\pi\cdot d^4}{32}$
Quill cylinder 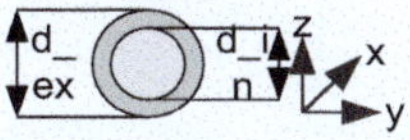	$S=\frac{\pi\cdot(d_{ex}^2-d_{in}^2)}{4}$	$Iy=\dfrac{\pi\cdot(d_{ex}^4-d_{in}^4)}{64};$ $Iz=\dfrac{\pi\cdot(d_{ex}^4-d_{in}^4)}{64}$	$J=\frac{\pi\cdot(d_{ex}^4-d_{in}^4)}{32}$
Square 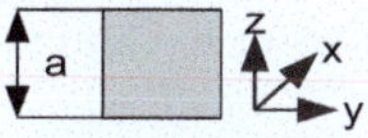	$S=a^2$	$Iy=\frac{a^4}{12};\ Iz=\frac{a^4}{12}$	$J=\frac{a^4}{6}$
Rectangle 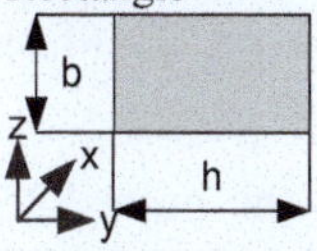	$S=b\cdot h$	$Iy=\frac{b\cdot h^3}{12};\ Iz=\frac{b^3\cdot h}{12}$	$J=\frac{h\cdot b}{12}\cdot(h^2+b^2)$
I-shape 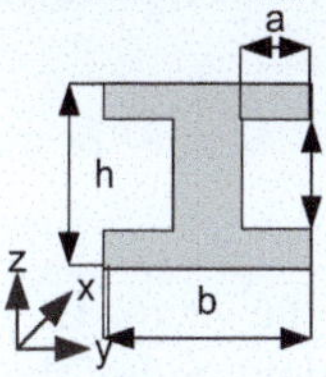	$S=b\cdot h-$ $-2\cdot a\cdot c$	$Iy=\dfrac{b\cdot h^3}{12}-2\cdot\dfrac{a\cdot c^3}{12}$ $Iz=\dfrac{b^3\cdot h}{12}-2\cdot\dfrac{a^3\cdot c}{12};$	$J=\dfrac{b\cdot h}{12}\cdot(b^2+h^2)-$ $-2\cdot\dfrac{a\cdot c}{12}\cdot(a^2+c^2)$

$$\begin{bmatrix}\mathbf{F}_{\text{global}}\\ \mathbf{M}_{\text{global}}\end{bmatrix}=\begin{bmatrix}\mathbf{R} & \mathbf{0}_{3\times3}\\ \mathbf{0}_{3\times3} & \mathbf{R}\end{bmatrix}_{6\times6}\cdot\mathbf{K}_{6\times6}\cdot\begin{bmatrix}\mathbf{R}^T & \mathbf{0}_{3\times3}\\ \mathbf{0}_{3\times3} & \mathbf{R}^T\end{bmatrix}_{6\times6}\cdot\begin{bmatrix}\Delta\mathbf{p}_{\text{global}}\\ \Delta\boldsymbol{\varphi}_{\text{global}}\end{bmatrix} \tag{4.5}$$

that gives a simple rule for transforming the local stiffness matrices of all mechanism components to a single coordinate frame.

For the *unsupported beams* (with two non-fixed ends) that are used in Matrix Structural Analysis (MSA) as principal components it is necessary to define the deflections and wrenches for both sides (see Fig. 4.1b). The latter will be further referred to as "*u*" and "*v*" or "1" and "2". In this case, the stiffness model is presented in an extended form

$$\begin{bmatrix}\mathbf{W}_1\\ \mathbf{W}_2\end{bmatrix}=\begin{bmatrix}\mathbf{K}_{11} & \mathbf{K}_{12}\\ \mathbf{K}_{21} & \mathbf{K}_{22}\end{bmatrix}_{12\times12}\cdot\begin{bmatrix}\Delta\mathbf{t}_1\\ \Delta\mathbf{t}_2\end{bmatrix} \tag{4.6}$$

that relates the deflections on both sides $\Delta\mathbf{t}_1,\ \Delta\mathbf{t}_2$ and corresponding wrenches $\mathbf{W}_1,\ \mathbf{W}_2$ through 12×12 extended stiffness matrix composed of four 6×6 blocks $\mathbf{K}_{11},\ \mathbf{K}_{12},\ \mathbf{K}_{21},\ \mathbf{K}_{22}$. It is clear that this 12×12 matrix is rank deficient since the wrenches $\mathbf{W}_1,\ \mathbf{W}_2$ should satisfy the static equilibrium equation that defines linear dependence between the matrix rows in Eq. (4.6).

To find the desired 6×6 blocks, let us consider two special cases that allow us to apply directly the cantilever beam Eq. (4.1) that includes 6×6 matrix $\mathbf{K}$. In the first case, let us orient the global system axes in the standard way and locate its origin at the left-end of the beam. This corresponds to $\Delta\mathbf{t}_1 = \mathbf{0},\ \mathbf{W}_2 = \mathbf{K} \cdot \Delta\mathbf{t}_2$ and leads to the simplification of Eq. (4.6)

$$\begin{aligned} \mathbf{W}_1 &= \mathbf{K}_{12} \cdot \Delta\mathbf{t}_2 \\ \mathbf{W}_2 &= \mathbf{K}_{22} \cdot \Delta\mathbf{t}_2 \end{aligned} \tag{4.7}$$

The relevant static equilibrium equation written at point "1" allows us to express the wrenches $\mathbf{W}_1,\ \mathbf{W}_2$ via the force $\mathbf{F}_2$ and torque $\mathbf{M}_2$ at the point "2" as

$$\mathbf{W}_1 = \begin{bmatrix} -\mathbf{F}_2 \\ -\mathbf{M}_2 - \mathbf{L} \times \mathbf{F}_2 \end{bmatrix};\ \mathbf{W}_2 = \begin{bmatrix} \mathbf{F}_2 \\ \mathbf{M}_2 \end{bmatrix} \tag{4.8}$$

where $\mathbf{L}$ is the beam length vector directed from the point "1" to the point "2" (i.e. along the beam principal axis). Applying further the cantilever beam standard equation $\mathbf{W}_2 = \mathbf{K} \cdot \Delta\mathbf{t}_2$ one can get explicit expressions for the matrices $\mathbf{K}_{12},\ \mathbf{K}_{22}$:

$$\mathbf{K}_{22} = \ \mathbf{K}; \qquad \mathbf{K}_{12} = -\begin{bmatrix} \mathbf{I}_{3\times3} & \mathbf{0}_{3\times3} \\ [\mathbf{L}\times] & \mathbf{I}_{3\times3} \end{bmatrix} \cdot \mathbf{K}_{22} \tag{4.9}$$

where $[\mathbf{L}\times]$ denotes the 3×3 skew-symmetric matrix derived from the vector $\mathbf{L}$.

Similarly, in the second case, the global system is located at the right-end of the beam (with the same orientation as in the previous case), which yields $\Delta\mathbf{t}_2 = \mathbf{0}$ and simplifies Eq. (4.6) to

$$\begin{aligned} \mathbf{W}_1 &= \mathbf{K}_{11} \cdot \Delta\mathbf{t}_1 \\ \mathbf{W}_2 &= \mathbf{K}_{21} \cdot \Delta\mathbf{t}_1 \end{aligned} \tag{4.10}$$

The static equilibrium equation leads to the following expressions

$$\mathbf{W}_1 = \begin{bmatrix} \mathbf{F}_1 \\ \mathbf{M}_1 \end{bmatrix}; \qquad \mathbf{W}_2 = \begin{bmatrix} -\mathbf{F}_1 \\ -\mathbf{M}_1 + \mathbf{L} \times \mathbf{F}_1 \end{bmatrix} \tag{4.11}$$

where $\mathbf{L}$ is the same beam length vector (directed from point "1" to point "2"). However, the cantilever beam equation should be slightly modified to take into account the standard directions of the coordinate axes

$$\begin{bmatrix} \mathbf{R}_{z\pi} & \mathbf{0}_{3\times3} \\ \mathbf{0}_{3\times3} & \mathbf{R}_{z\pi} \end{bmatrix} \mathbf{W}_1 = \mathbf{K} \cdot \begin{bmatrix} \mathbf{R}_{z\pi} & \mathbf{0}_{3\times3} \\ \mathbf{0}_{3\times3} & \mathbf{R}_{z\pi} \end{bmatrix} \cdot \Delta\mathbf{t}_1 \tag{4.12}$$

where $\mathbf{R}_{z\pi} = diag(-1, -1, 1)$ is 3×3 rotation matrix around the z-axis by the angle π, which changes the directions of the axes x, and y. The latter allows us to obtain explicit expressions for $\mathbf{K}_{11}, \ \mathbf{K}_{21}$

$$\mathbf{K}_{11} = \begin{bmatrix} \mathbf{R}_{z\pi} & \mathbf{0}_{3\times3} \\ \mathbf{0}_{3\times3} & \mathbf{R}_{z\pi} \end{bmatrix}^T \cdot \mathbf{K} \cdot \begin{bmatrix} \mathbf{R}_{z\pi} & \mathbf{0}_{3\times3} \\ \mathbf{0}_{3\times3} & \mathbf{R}_{z\pi} \end{bmatrix}; \qquad \mathbf{K}_{21} = -\begin{bmatrix} \mathbf{I}_{3\times3} & \mathbf{0}_{3\times3} \\ (\mathbf{L}\times) & \mathbf{I}_{3\times3} \end{bmatrix} \cdot \mathbf{K}_{11} \tag{4.13}$$

Physical interpretation of the matrices $\mathbf{K}_{11}, \ \mathbf{K}_{12}, \ \mathbf{K}_{21}, \ \mathbf{K}_{22}$ can be done in the following way. The stiffness matrix $\mathbf{K}_{11}$ describes the force/torque reaction at the beam's left-end point caused by the left-end deflection $\Delta\mathbf{t}_1$. The stiffness matrix $\mathbf{K}_{12}$ describes the force/torque reaction at the beam's left-end point caused by the right-end deflection $\Delta\mathbf{t}_2$. The stiffness matrix $\mathbf{K}_{21}$ describes the force/torque reaction at the beam's right-end point caused by the left-end deflection $\Delta\mathbf{t}_1$. The stiffness matrix $\mathbf{K}_{22}$ describes the force/torque reaction at the beam's right-end point caused by the right-end deflection $\Delta\mathbf{t}_2$. The physical meaning of $\mathbf{K}_{11}, \ \mathbf{K}_{12}, \ \mathbf{K}_{21}, \ \mathbf{K}_{22}$ is also illustrated for the simplest case of the 1D linear spring (Fig. 4.2), where 6×6 matrices are reduced to the scalar stiffness coefficients.

For the beam with a regular shape (with uniform cross-section), these stiffness matrices can be computed analytically using the following expressions

$$\mathbf{K}_{11} = \begin{bmatrix} \frac{E\cdot S}{L} & 0 & 0 & 0 & 0 & 0 \\ 0 & \frac{12\cdot E\cdot Iz}{L^3} & 0 & 0 & 0 & \frac{6\cdot E\cdot I}{L^2} \\ 0 & 0 & \frac{12\cdot E\cdot Iy}{L^3} & 0 & \frac{-6\cdot E\cdot Iy}{L^2} & 0 \\ 0 & 0 & 0 & \frac{G\cdot J}{L} & 0 & 0 \\ 0 & 0 & \frac{-6\cdot E\cdot Iy}{L^2} & 0 & \frac{4\cdot E\cdot Iy}{L} & 0 \\ 0 & \frac{6\cdot E\cdot Iz}{L^2} & 0 & 0 & 0 & \frac{4\cdot E\cdot Iz}{L} \end{bmatrix} \tag{4.14}$$

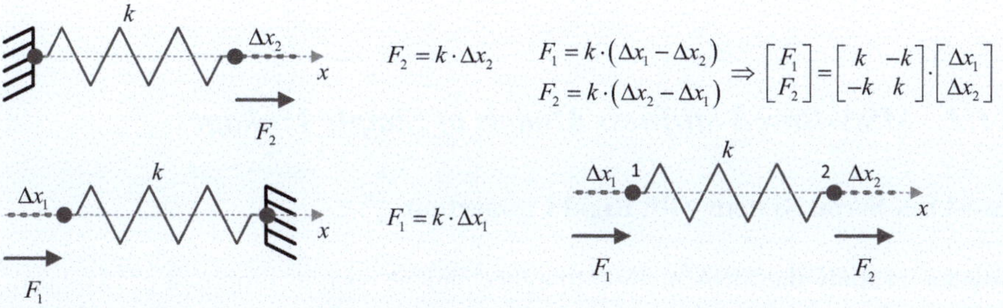

Fig. 4.2 Stiffness models of a simple spring with fixed and free ends [5]

$$\mathbf{K}_{12} = \begin{bmatrix} \frac{-E \cdot S}{L} & 0 & 0 & 0 & 0 & 0 \\ 0 & \frac{-12 \cdot E \cdot Iz}{L^3} & 0 & 0 & 0 & \frac{6 \cdot E \cdot Iz}{L^2} \\ 0 & 0 & \frac{-12 \cdot E \cdot Iy}{L^3} & 0 & \frac{-6 \cdot E \cdot Iy}{L^2} & 0 \\ 0 & 0 & 0 & \frac{-G \cdot J}{L} & 0 & 0 \\ 0 & 0 & \frac{6 \cdot E \cdot Iy}{L^2} & 0 & \frac{2 \cdot E \cdot Iy}{L} & 0 \\ 0 & \frac{-6 \cdot E \cdot Iz}{L^2} & 0 & 0 & 0 & \frac{2 \cdot E \cdot Iz}{L} \end{bmatrix} \tag{4.15}$$

$$\mathbf{K}_{21} = \begin{bmatrix} \frac{-E \cdot S}{L} & 0 & 0 & 0 & 0 & 0 \\ 0 & \frac{-12 \cdot E \cdot Iz}{L^3} & 0 & 0 & 0 & \frac{-6 \cdot E \cdot Iz}{L^2} \\ 0 & 0 & \frac{-12 \cdot E \cdot Iy}{L^3} & 0 & \frac{6 \cdot E \cdot Iy}{L^2} & 0 \\ 0 & 0 & 0 & \frac{-G \cdot J}{L} & 0 & 0 \\ 0 & 0 & \frac{-6 \cdot E \cdot Iy}{L^2} & 0 & \frac{2 \cdot E \cdot Iy}{L} & 0 \\ 0 & \frac{6 \cdot E \cdot Iz}{L^2} & 0 & 0 & 0 & \frac{2 \cdot E \cdot Iz}{L} \end{bmatrix} \tag{4.16}$$

$$\mathbf{K}_{22} = \begin{bmatrix} \frac{E \cdot S}{L} & 0 & 0 & 0 & 0 & 0 \\ 0 & \frac{12 \cdot E \cdot Iz}{L^3} & 0 & 0 & 0 & \frac{-6 \cdot E \cdot I}{L^2} \\ 0 & 0 & \frac{12 \cdot E \cdot Iy}{L^3} & 0 & \frac{6 \cdot E \cdot Iy}{L^2} & 0 \\ 0 & 0 & 0 & \frac{G \cdot J}{L} & 0 & 0 \\ 0 & 0 & \frac{6 \cdot E \cdot Iy}{L^2} & 0 & \frac{4 \cdot E \cdot Iy}{L} & 0 \\ 0 & \frac{-6 \cdot E \cdot Iz}{L^2} & 0 & 0 & 0 & \frac{4 \cdot E \cdot Iz}{L} \end{bmatrix} \tag{4.17}$$

where all notations have the same meaning as in the cantilever beam stiffness matrix (4.2).

Similar to the cantilever bean case, the stiffness matrix of the unsupported beam can be transformed to the global system using an extended version of Eq. (4.5)

$$\begin{bmatrix} \mathbf{K}_{11}^{\text{global}} & \mathbf{K}_{12}^{\text{global}} \\ \mathbf{K}_{21}^{\text{global}} & \mathbf{K}_{22}^{\text{global}} \end{bmatrix} = \begin{bmatrix} \mathbf{R} & \mathbf{0} & \mathbf{0} & \mathbf{0} \\ \mathbf{0} & \mathbf{R} & \mathbf{0} & \mathbf{0} \\ \mathbf{0} & \mathbf{0} & \mathbf{R} & \mathbf{0} \\ \mathbf{0} & \mathbf{0} & \mathbf{0} & \mathbf{R} \end{bmatrix} \cdot \begin{bmatrix} \mathbf{K}_{11}^{\text{local}} & \mathbf{K}_{12}^{\text{local}} \\ \mathbf{K}_{21}^{\text{local}} & \mathbf{K}_{22}^{\text{local}} \end{bmatrix} \cdot \begin{bmatrix} \mathbf{R}^T & \mathbf{0} & \mathbf{0} & \mathbf{0} \\ \mathbf{0} & \mathbf{R}^T & \mathbf{0} & \mathbf{0} \\ \mathbf{0} & \mathbf{0} & \mathbf{R}^T & \mathbf{0} \\ \mathbf{0} & \mathbf{0} & \mathbf{0} & \mathbf{R}^T \end{bmatrix} \tag{4.18}$$

where the orthogonal matrix **R** defines the orientation of the local coordinate system relative to the global one.

4.1.2 MSA-Based Stiffness Models of Simple Linkages

Two-Link Serial System with Rigid Connection

Now let us apply the MSA method to a mechanical system composed of two beam elements of lengths L_1 and L_2 with a rigid connection between them (Fig. 4.3). It is assumed that the beams are not aligned and the angle between them is equal to q.

Here, each beam has two local coordinate systems: 0, 1 for the first beam and 2, 3 for the second one, as shown in Fig. 4.3.

Let us assume that the global coordinate system is located at the left-end of the first beam (point 0) and the global coordinate axes are aligned with the local ones. Under this assumption, the force deflection relations for the first link can be written as

$$\begin{bmatrix} \mathbf{W}_0 \\ \mathbf{W}_1 \end{bmatrix} = \begin{bmatrix} \mathbf{K}_{11}^1 & \mathbf{K}_{12}^1 \\ \mathbf{K}_{21}^1 & \mathbf{K}_{22}^1 \end{bmatrix} \cdot \begin{bmatrix} \Delta\mathbf{t}_0 \\ \Delta\mathbf{t}_1 \end{bmatrix} \tag{4.19}$$

where the upper superscript denotes the link number and relevant matrices are computed using Eqs. (4.14–4.17). For the second beam, a similar relation should be written taking into account that the local axes (x_2, z_2) and (x_3, z_3) are turned with respect to the global ones. This yields the following force–deflection relation

$$\begin{bmatrix} \mathbf{W}_2 \\ \mathbf{W}_3 \end{bmatrix} = \begin{bmatrix} \mathbf{Q}\cdot\mathbf{K}_{11}^2\cdot\mathbf{Q}^T & \mathbf{Q}\cdot\mathbf{K}_{12}^2\cdot\mathbf{Q}^T \\ \mathbf{Q}\cdot\mathbf{K}_{21}^2\cdot\mathbf{Q}^T & \mathbf{Q}\cdot\mathbf{K}_{22}^2\cdot\mathbf{Q}^T \end{bmatrix} \cdot \begin{bmatrix} \Delta\mathbf{t}_2 \\ \Delta\mathbf{t}_3 \end{bmatrix} \tag{4.20}$$

where $\mathbf{Q} = diag(\mathbf{R}_y^q, \mathbf{R}_y^q)$ is the 6×6 matrix that is expressed via the orthogonal rotation matrix $\mathbf{R}_y^q$ describing rotation around the y-axis by the angle q. For further convenience let us denote $\mathbf{K}_{ij}^{2q} = \mathbf{Q}\cdot\mathbf{K}_{ij}^2\cdot\mathbf{Q}^T$ and rewrite the above equation in a more compact form

$$\begin{bmatrix} \mathbf{W}_2 \\ \mathbf{W}_3 \end{bmatrix} = \begin{bmatrix} \mathbf{K}_{11}^{2q} & \mathbf{K}_{12}^{2q} \\ \mathbf{K}_{21}^{2q} & \mathbf{K}_{22}^{2q} \end{bmatrix} \cdot \begin{bmatrix} \Delta\mathbf{t}_2 \\ \Delta\mathbf{t}_3 \end{bmatrix} \tag{4.21}$$

After collecting Eqs. (4.19) and (4.21) in a single system, one can get

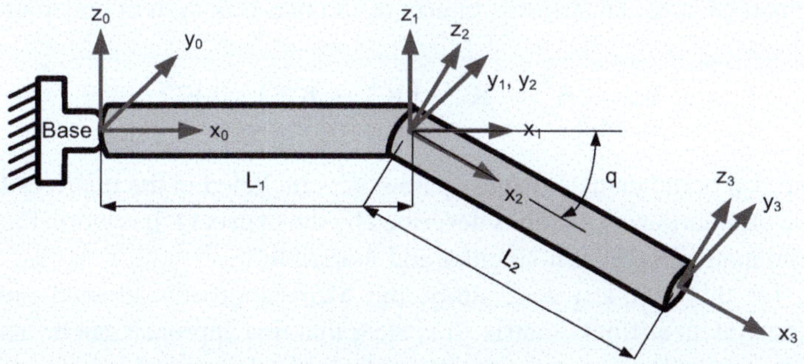

Fig. 4.3 Two-link serial system with rigid connection [5]

$$\begin{bmatrix} \mathbf{W}_0 \\ \mathbf{W}_1 \\ \mathbf{W}_2 \\ \mathbf{W}_3 \end{bmatrix}_{24\times 1} = \begin{bmatrix} \mathbf{K}^1_{11} & \mathbf{K}^1_{12} & \mathbf{0}_{6\times 6} & \mathbf{0}_{6\times 6} \\ \mathbf{K}^1_{21} & \mathbf{K}^1_{22} & \mathbf{0}_{6\times 6} & \mathbf{0}_{6\times 6} \\ \mathbf{0}_{6\times 6} & \mathbf{0}_{6\times 6} & \mathbf{K}^{2q}_{11} & \mathbf{K}^{2q}_{12} \\ \mathbf{0}_{6\times 6} & \mathbf{0}_{6\times 6} & \mathbf{K}^{2q}_{21} & \mathbf{K}^{2q}_{22} \end{bmatrix}_{24\times 24} \cdot \begin{bmatrix} \Delta\mathbf{t}_0 \\ \Delta\mathbf{t}_1 \\ \Delta\mathbf{t}_2 \\ \Delta\mathbf{t}_3 \end{bmatrix}_{24\times 1} \tag{4.22}$$

Further, taking into account the boundary condition $\Delta\mathbf{t}_0 = \mathbf{0}$ the above system can be reduced down to

$$\begin{bmatrix} \mathbf{W}_1 \\ \mathbf{W}_2 \\ \mathbf{W}_3 \end{bmatrix}_{18\times 1} = \begin{bmatrix} \mathbf{K}^1_{22} & \mathbf{0}_{6\times 6} & \mathbf{0}_{6\times 6} \\ \mathbf{0}_{6\times 6} & \mathbf{K}^{2q}_{11} & \mathbf{K}^{2q}_{12} \\ \mathbf{0}_{6\times 6} & \mathbf{K}^{2q}_{21} & \mathbf{K}^{2q}_{22} \end{bmatrix}_{18\times 18} \cdot \begin{bmatrix} \Delta\mathbf{t}_1 \\ \Delta\mathbf{t}_2 \\ \Delta\mathbf{t}_3 \end{bmatrix}_{18\times 1} \tag{4.23}$$

Furthermore, the rigid connection between the beams yields relations $\Delta\mathbf{t}_1 = \Delta\mathbf{t}_2$ and $\mathbf{W}_1 = -\mathbf{W}_2$, where the last one directly follows from Newton's third law. The latter allows us to sum up the first and second lines in the system (4.23) and obtain the final expression allowing us to compute deflections at the beam ends caused by external loading $\mathbf{W}_3$ applied to the free end of the considered 2-beam system

$$\begin{bmatrix} \mathbf{0} \\ \mathbf{W}_3 \end{bmatrix}_{12\times 1} = \begin{bmatrix} \mathbf{K}^1_{22} + \mathbf{K}^{2q}_{11} & \mathbf{K}^{2q}_{12} \\ \mathbf{K}^{2q}_{21} & \mathbf{K}^{2q}_{22} \end{bmatrix}_{12\times 12} \cdot \begin{bmatrix} \Delta\mathbf{t}_2 \\ \Delta\mathbf{t}_3 \end{bmatrix}_{12\times 1} \tag{4.24}$$

It is worth mentioning that here the 12×12 matrix is invertible, and equations can be easily solved with respect to the deflections $\Delta\mathbf{t}_2, \ \Delta\mathbf{t}_3$. In particular, an analytical solution for the free-end deflection $\Delta\mathbf{t}_3$ caused by $\mathbf{W}_3$ may be presented in the following form

$$\mathbf{W}_3 = \mathbf{K}_C \cdot \Delta\mathbf{t}_3 \tag{4.25}$$

where $\mathbf{K}_C$ is the Cartesian stiffness matrix of the two-link system under study

$$\mathbf{K}_C = \mathbf{K}^{2q}_{22} - \mathbf{K}^{2q}_{21} \cdot \left(\mathbf{K}^{2q}_{11} + \mathbf{K}^1_{22}\right)^{-1} \cdot \mathbf{K}^{2q}_{12} \tag{4.26}$$

that depends on both beams' stiffness parameters included in the matrices $\mathbf{K}^1_{ij}, \ \mathbf{K}^{2q}_{ij}$ and relative orientations of the links described by the matrix $\mathbf{Q}$. It is worth mentioning that here the matrix $\mathbf{K}_C$ is non-singular and invertible.

Hence, for this simplest case study, the MSA approach allowed rather easy compute the system stiffness matrix. It is clear that this approach can be also generalized for the multi-beam serial systems with rigid connections, however, the final analytical expression will include the matrix inversion of higher dimension. Besides, it is worth mentioning that the above-presented example implements the general MSA assembling technique that usually produces matrices similar to (4.24), which includes several matrix sums.

Two-Link Serial System With a Passive Joint

Let us apply the MSA method to a mechanical system composed of two beam elements of lengths L_1 and L_2 with a revolute passive joint between them (Fig. 4.4). In structural mechanics, this type of connection is also called as a pin joint. It is assumed that the passive joint axis is directed along y_1 and y_2, the beams are not aligned and the angle between them is equal to the passive joint coordinate q. It is worth mentioning that this problem cannot be solved straightforwardly using the standard MSA method, which was primarily designed for rigid and elastic connections. However, it is possible to apply some special techniques to adjust MSA for the passive joint case.

It is clear that the basic MSA equations for this 2-beam system have the same structure as above

$$\begin{bmatrix} \mathbf{W}_1 \\ \mathbf{W}_2 \\ \mathbf{W}_3 \end{bmatrix}_{18\times1} = \begin{bmatrix} \mathbf{K}_{22}^{1} & \mathbf{0}_{6\times6} & \mathbf{0}_{6\times6} \\ \mathbf{0}_{6\times6} & \mathbf{K}_{11}^{2q} & \mathbf{K}_{12}^{2q} \\ \mathbf{0}_{6\times6} & \mathbf{K}_{21}^{2q} & \mathbf{K}_{22}^{2q} \end{bmatrix}_{18\times18} \cdot \begin{bmatrix} \Delta\mathbf{t}_1 \\ \Delta\mathbf{t}_2 \\ \Delta\mathbf{t}_3 \end{bmatrix}_{18\times1} \tag{4.27}$$

but here the matrices $\mathbf{K}_{ij}^{2q}$ are not constant and depend on the passive joint coordinate q, which may vary under the influence of the external loading $\mathbf{W}_3$. However, in contrast to the previous case, where the connection is described by equations $\Delta\mathbf{t}_1 = \Delta\mathbf{t}_2$ and $\mathbf{W}_1 = -\mathbf{W}_2$, here these relations must be replaced by

$$\begin{aligned} &\mathbf{\Lambda}^r \cdot (\Delta\mathbf{t}_1 - \Delta\mathbf{t}_2) = \mathbf{0} \\ &\mathbf{\Lambda}^r \cdot (\mathbf{W}_2 + \mathbf{W}_1) = \mathbf{0}; \qquad \mathbf{\Lambda}^p \cdot \mathbf{W}_1 = \mathbf{0}; \qquad \mathbf{\Lambda}^p \cdot \mathbf{W}_2 = \mathbf{0} \end{aligned} \tag{4.28}$$

where the matrices $\mathbf{\Lambda}^r = diag(1, 1, 1, 1, 0, 1)$ and $\mathbf{\Lambda}^p = diag(0, 0, 0, 0, 1, 0)$ describe the passive joint geometry, and the superscripts 'r' and 'p' are referred to as the "rigid" and "passive" connections. The first of these relations takes into account that the connection 1 and 2 ensures equality of all components of $\Delta\mathbf{t}_1$, $\Delta\mathbf{t}_2$

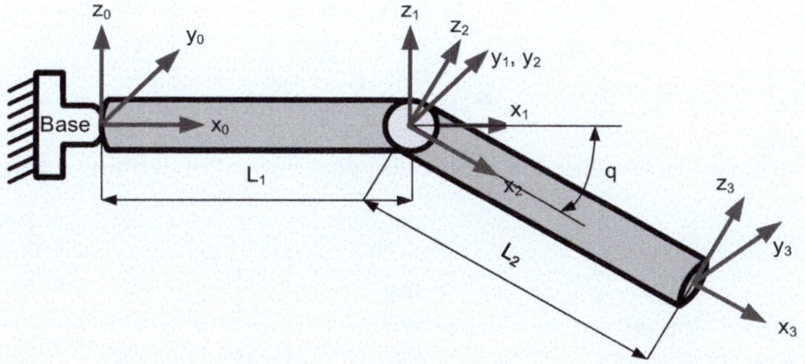

Fig. 4.4 Two-link serial system with a passive joint [5]

except φ_y^1, φ_y^2 (since the passive joint axis is directed along y). The second group of relations shows that the passive joint ensures the validity of the 3rd Newton law for five wrench components only (except M_y), but the torques M_y^1 and M_y^2 are equal to zero. It should be noted that in the general case the matrices $\mathbf{\Lambda}^r$ and $\mathbf{\Lambda}^p$ may be non-diagonal, but their rank remains the same and is equal to 5 and 1, respectively (as in the case of diagonal matrices).

Combining Eqs. (4.27) and (4.28) one can get the following linear system

$$\begin{bmatrix} \mathbf{0} \\ \mathbf{0} \\ \mathbf{0} \\ \mathbf{0} \\ \mathbf{W}_3 \end{bmatrix}_{30\times 1} = \begin{bmatrix} \mathbf{\Lambda}^r & -\mathbf{\Lambda}^r & \mathbf{0}_{6\times 6} \\ \mathbf{\Lambda}^p \cdot \mathbf{K}_{22}^1 & \mathbf{0}_{6\times 6} & \mathbf{0}_{6\times 6} \\ \mathbf{0}_{6\times 6} & \mathbf{\Lambda}^p \cdot \mathbf{K}_{11}^{2q} & \mathbf{\Lambda}^p \cdot \mathbf{K}_{12}^{2q} \\ \mathbf{\Lambda}^r \cdot \mathbf{K}_{22}^1 & \mathbf{\Lambda}^r \cdot \mathbf{K}_{11}^{2q} & \mathbf{\Lambda}^r \cdot \mathbf{K}_{12}^{2q} \\ \mathbf{0}_{6\times 6} & \mathbf{K}_{21}^{2q} & \mathbf{K}_{22}^{2q} \end{bmatrix}_{30\times 18} \cdot \begin{bmatrix} \Delta\mathbf{t}_1 \\ \Delta\mathbf{t}_2 \\ \Delta\mathbf{t}_3 \end{bmatrix}_{18\times 1} \tag{4.29}$$

that should be solved for $\Delta\mathbf{t}_3$. It should be mentioned that here the system matrix is non-square (of the size 30 × 18), but it contains 12 zero rows that are caused by zero diagonal components in $\mathbf{\Lambda}^r$ and $\mathbf{\Lambda}^p$ (each $\mathbf{\Lambda}^r$ produces one zero-row and each $\mathbf{\Lambda}^p$ produces five zero-rows). To eliminate the zero-rows that are useless here, let us introduce the modified non-singular matrices

$$\mathbf{\Lambda}_*^r = \begin{bmatrix} 1\,0\,0\,0\,0\,0 \\ 0\,1\,0\,0\,0\,0 \\ 0\,0\,1\,0\,0\,0 \\ 0\,0\,0\,1\,0\,0 \\ 0\,0\,0\,0\,0\,1 \end{bmatrix}; \qquad \mathbf{\Lambda}_*^p = \begin{bmatrix} 0\,0\,0\,0\,1\,0 \end{bmatrix} \tag{4.30}$$

and rewrite the system (4.29) in the reduced form (with a square matrix of size 18 × 18)

$$\begin{bmatrix} \mathbf{0}_{5\times 1} \\ \mathbf{0}_{1\times 1} \\ \mathbf{0}_{1\times 1} \\ \mathbf{0}_{5\times 1} \\ \hline \mathbf{W}_3 \end{bmatrix}_{18\times 1} = \left[\begin{array}{cc|c} \mathbf{\Lambda}_*^r & -\mathbf{\Lambda}_*^r & \mathbf{0}_{5\times 6} \\ \mathbf{\Lambda}_*^p \cdot \mathbf{K}_{22}^1 & \mathbf{0}_{1\times 6} & \mathbf{0}_{1\times 6} \\ \mathbf{0}_{1\times 6} & \mathbf{\Lambda}_*^p \cdot \mathbf{K}_{11}^{2q} & \mathbf{\Lambda}_*^p \cdot \mathbf{K}_{12}^{2q} \\ \mathbf{\Lambda}_*^r \cdot \mathbf{K}_{22}^1 & \mathbf{\Lambda}_*^r \cdot \mathbf{K}_{11}^{2q} & \mathbf{\Lambda}_*^r \cdot \mathbf{K}_{12}^{2q} \\ \hline \mathbf{0}_{6\times 6} & \mathbf{K}_{21}^{2q} & \mathbf{K}_{22}^{2q} \end{array}\right]_{18\times 18} \cdot \begin{bmatrix} \Delta\mathbf{t}_1 \\ \Delta\mathbf{t}_2 \\ \hline \Delta\mathbf{t}_3 \end{bmatrix}_{18\times 1} \tag{4.31}$$

This system can be presented in a more compact form after defining relevant block matrices

$$\begin{bmatrix} \mathbf{0}_{12\times 1} \\ \mathbf{W}_3 \end{bmatrix}_{18\times 1} = \begin{bmatrix} \mathbf{A}\ \mathbf{B} \\ \mathbf{C}\ \mathbf{D} \end{bmatrix}_{18\times 18} \cdot \begin{bmatrix} \Delta\mathbf{t}_1 \\ \Delta\mathbf{t}_2 \\ \Delta\mathbf{t}_3 \end{bmatrix}_{18\times 1} \tag{4.32}$$

where

$$\mathbf{A} = \begin{bmatrix} \mathbf{\Lambda}_*^r & -\mathbf{\Lambda}_*^r \\ \mathbf{\Lambda}_*^p \cdot \mathbf{K}_{22}^1 & \mathbf{0}_{1\times 6} \\ \mathbf{0}_{1\times 6} & \mathbf{\Lambda}_*^p \cdot \mathbf{K}_{11}^{2q} \\ \mathbf{\Lambda}_*^r \cdot \mathbf{K}_{22}^1 & \mathbf{\Lambda}_*^r \cdot \mathbf{K}_{11}^{2q} \end{bmatrix}_{12\times 12}; \qquad \mathbf{B} = \begin{bmatrix} \mathbf{0}_{5\times 6} \\ \mathbf{0}_{1\times 6} \\ \mathbf{\Lambda}_*^p \cdot \mathbf{K}_{12}^{2q} \\ \mathbf{\Lambda}_*^r \cdot \mathbf{K}_{12}^{2q} \end{bmatrix}_{12\times 6} \tag{4.33}$$

$$\mathbf{C} = \begin{bmatrix} \mathbf{0}_{6\times 6} & \mathbf{K}_{21}^{2q} \end{bmatrix}_{6\times 12}; \qquad \mathbf{D} = \mathbf{K}_{22}^{2q}$$

Further, after elimination $\Delta\mathbf{t}_1,\ \Delta\mathbf{t}_2$ one can get the following expression for the wrench applied to the system end-point

$$\mathbf{W}_3 = \mathbf{K}_C \cdot \Delta\mathbf{t}_3 \tag{4.34}$$

where

$$\mathbf{K}_C = \mathbf{D} - \mathbf{C} \cdot \mathbf{A}^{-1} \cdot \mathbf{B} \tag{4.35}$$

is the desired Cartesian stiffness matrix of the two-link system under study. It has a similar structure as (4.26) corresponding to the case of the rigid connection. However, it is easy to verify numerically that here the stiffness matrix $\mathbf{K}_C$ is rank-deficient, which is the result of the passive joint connection.

Hence, for this case study, some modifications of the MSA approach allowed us to compute the rank-deficient stiffness matrix. However, relevant transformations include a number of non-trivial steps, which will be generalized in the following Sections.

Two-Link Serial System with an Elastic Joint

Let us apply the MSA method to a mechanical system composed of two beam elements of lengths L_1 and L_2 with an elastic joint between them (Fig. 4.5). It is assumed that the elastic joint axis is directed along y_1 and y_2, the beams are not aligned and the angle between them is equal to the elastic joint coordinate q.

It is clear that the basic MSA equations for this 2-beam system have the same structure as in the cases (a) and (b)

$$\begin{bmatrix} \mathbf{W}_1 \\ \mathbf{W}_2 \\ \mathbf{W}_3 \end{bmatrix}_{18\times 1} = \begin{bmatrix} \mathbf{K}_{22}^1 & \mathbf{0}_{6\times 6} & \mathbf{0}_{6\times 6} \\ \mathbf{0}_{6\times 6} & \mathbf{K}_{11}^{2q} & \mathbf{K}_{12}^{2q} \\ \mathbf{0}_{6\times 6} & \mathbf{K}_{21}^{2q} & \mathbf{K}_{22}^{2q} \end{bmatrix}_{18\times 18} \cdot \begin{bmatrix} \Delta\mathbf{t}_1 \\ \Delta\mathbf{t}_2 \\ \Delta\mathbf{t}_3 \end{bmatrix}_{18\times 1} \tag{4.36}$$

Here the matrices $\mathbf{K}_{ij}^{2q}$ are not constant and depend on the elastic joint coordinate q, which may vary under the influence of the external loading $\mathbf{W}_3$. However, in contrast to the cases of the rigid and passive connections, here the joint static equations must

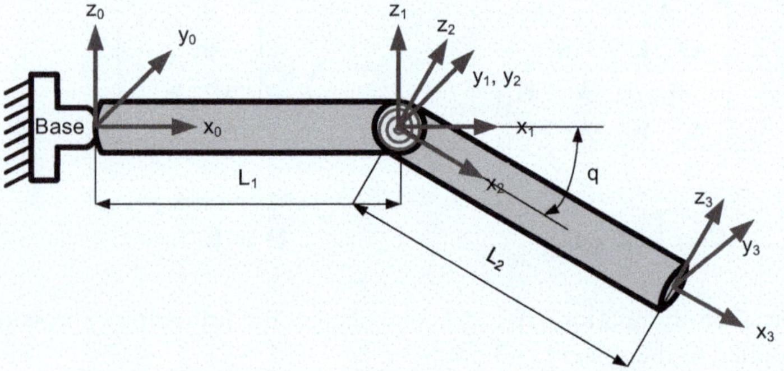

Fig. 4.5 Two-link serial system with an elastic joint [5]

be replaced by

$$\begin{aligned} &\mathbf{\Lambda}^r \cdot (\Delta\mathbf{t}_1 - \Delta\mathbf{t}_2) = \mathbf{0} \\ &\mathbf{W}_2 + \mathbf{W}_1 = \mathbf{0}; \\ &\mathbf{\Lambda}^e \cdot \mathbf{W}_2 = K_q \cdot \mathbf{\Lambda}^e \cdot (\Delta\mathbf{t}_2 - \Delta\mathbf{t}_1) \end{aligned} \tag{4.37}$$

where K_q is the stiffness coefficient of the elastic joint, the matrices $\mathbf{\Lambda}^r = diag(1, 1, 1, 1, 0, 1)$ and $\mathbf{\Lambda}^e = diag(0, 0, 0, 0, 1, 0)$ describe the passive joint geometry, and the superscripts 'r' and 'e' are referred to as "rigid" and "elastic" connections. The first of these relations takes into account that the connection 1 and 2 ensures equality of all components of $\Delta\mathbf{t}_1, \ \Delta\mathbf{t}_2$ except $\varphi_y^1, \ \varphi_y^2$ (since the elastic joint axis is directed along y). The second group of relations shows the elastic joint ensures the validity of the 3rd Newton law for all wrench components, and the third relation describes Hooke's law for the elastic joint.

After combining Eqs. (4.36) and (4.37) one can get the following linear system

$$\begin{bmatrix} \mathbf{0} \\ \mathbf{0} \\ \mathbf{0} \\ \mathbf{W}_3 \end{bmatrix}_{24\times 1} = \begin{bmatrix} \mathbf{\Lambda}^r & -\mathbf{\Lambda}^r & \mathbf{0}_{6\times 6} \\ K_q \cdot \mathbf{\Lambda}^e & \mathbf{\Lambda}^e \cdot \mathbf{K}_{11}^{2q} - K_q \cdot \mathbf{\Lambda}^e & \mathbf{\Lambda}^e \cdot \mathbf{K}_{12}^{2q} \\ \mathbf{K}_{22}^1 & \mathbf{K}_{11}^{2q} & \mathbf{K}_{12}^{2q} \\ \mathbf{0}_{6\times 6} & \mathbf{K}_{21}^{2q} & \mathbf{K}_{22}^{2q} \end{bmatrix}_{24\times 18} \cdot \begin{bmatrix} \Delta\mathbf{t}_1 \\ \Delta\mathbf{t}_2 \\ \Delta\mathbf{t}_3 \end{bmatrix}_{18\times 1} \tag{4.38}$$

that should be solved with respect to $\Delta\mathbf{t}_3$. It should be mentioned that here the system matrix is non-square of the size 24×18, but it contains 6 zero rows that are caused by zero diagonal components in $\mathbf{\Lambda}^r$ and $\mathbf{\Lambda}^e$ (each $\mathbf{\Lambda}^r$ produces one zero-row and each $\mathbf{\Lambda}^e$ produces five zero-rows). To eliminate the zero-rows that are useless here, let us introduce the modified non-singular matrices $\mathbf{\Lambda}_*^r$ of the size 5×6 (see Eq. (4.30)) and $\mathbf{\Lambda}_*^e = \begin{bmatrix} 0 \ 0 \ 0 \ 0 \ 1 \ 0 \end{bmatrix}$. This allows us to rewrite the system (4.38) in the reduced form (with a square matrix of size 18×18)

$$\begin{bmatrix} \mathbf{0}_{5\times1} \\ \mathbf{0}_{1\times1} \\ \mathbf{0}_{6\times1} \\ \hline \mathbf{W}_3 \end{bmatrix}_{18\times1} = \left[\begin{array}{cc|c} \boldsymbol{\Lambda}_*^r & -\boldsymbol{\Lambda}_*^r & \mathbf{0}_{6\times6} \\ K_q\cdot\boldsymbol{\Lambda}_*^e & \boldsymbol{\Lambda}_*^e\cdot\mathbf{K}_{11}^{2q} - K_q\cdot\boldsymbol{\Lambda}_*^e & \boldsymbol{\Lambda}_*^e\cdot\mathbf{K}_{12}^{2q} \\ \mathbf{K}_{22}^1 & \mathbf{K}_{11}^{2q} & \mathbf{K}_{12}^{2q} \\ \hline \mathbf{0}_{6\times6} & \mathbf{K}_{21}^{2q} & \mathbf{K}_{22}^{2q} \end{array}\right]_{18\times18} \cdot \begin{bmatrix} \Delta\mathbf{t}_1 \\ \Delta\mathbf{t}_2 \\ \hline \Delta\mathbf{t}_3 \end{bmatrix}_{18\times1} \tag{4.39}$$

This system can be simplified after defining relevant block matrices and presented as

$$\begin{bmatrix} \mathbf{0}_{12\times1} \\ \mathbf{W}_3 \end{bmatrix}_{18\times1} = \begin{bmatrix} \mathbf{A}\ \mathbf{B} \\ \mathbf{C}\ \mathbf{D} \end{bmatrix}_{18\times18} \cdot \begin{bmatrix} \Delta\mathbf{t}_1 \\ \Delta\mathbf{t}_2 \\ \Delta\mathbf{t}_3 \end{bmatrix}_{18\times1} \tag{4.40}$$

where

$$\mathbf{A} = \begin{bmatrix} \boldsymbol{\Lambda}_*^r & -\boldsymbol{\Lambda}_*^r \\ K_q\cdot\boldsymbol{\Lambda}_*^e & \boldsymbol{\Lambda}_*^e\cdot\mathbf{K}_{11}^{2q} - K_q\cdot\boldsymbol{\Lambda}_*^e \\ \mathbf{K}_{22}^1 & \mathbf{K}_{11}^{2q} \end{bmatrix}_{12\times12}; \qquad \mathbf{B} = \begin{bmatrix} \mathbf{0}_{6\times6} \\ \boldsymbol{\Lambda}_*^e\cdot\mathbf{K}_{12}^{2q} \\ \mathbf{K}_{12}^{2q} \end{bmatrix}_{12\times6} \tag{4.41}$$

$$\mathbf{C} = \begin{bmatrix} \mathbf{0}_{6\times6}\ \mathbf{K}_{21}^{2q} \end{bmatrix}_{6\times12}; \qquad \mathbf{D} = \mathbf{K}_{22}^{2q}$$

Further, after elimination $\Delta\mathbf{t}_1,\ \Delta\mathbf{t}_2$ one can get the following expression for the wrench applied to the system end-point

$$\mathbf{W}_3 = \mathbf{K}_C\cdot\Delta\mathbf{t}_3 \tag{4.42}$$

where

$$\mathbf{K}_C = \mathbf{D} - \mathbf{C}\cdot\mathbf{A}^{-1}\cdot\mathbf{B} \tag{4.43}$$

is the desired Cartesian stiffness matrix of the two-link system under study. It has a similar structure as (4.26) and (4.35) corresponding to the cases of rigid and passive connections. It is easy to verify numerically that here the stiffness matrix $\mathbf{K}_C$ is full-rank, which is in good agreement with the nature of the elastic joint.

Two-Link Parallel System with Passive Joints

Let us apply the MSA method to a closed-loop mechanical system composed of two beam elements with three revolute passive joints, which connect the beams to the rigid base at the left-hand side and the end-effector on the right-hand side (Fig. 4.6). It is assumed that the passive joint axes are directed along y and the angle between the beams is equal to q.

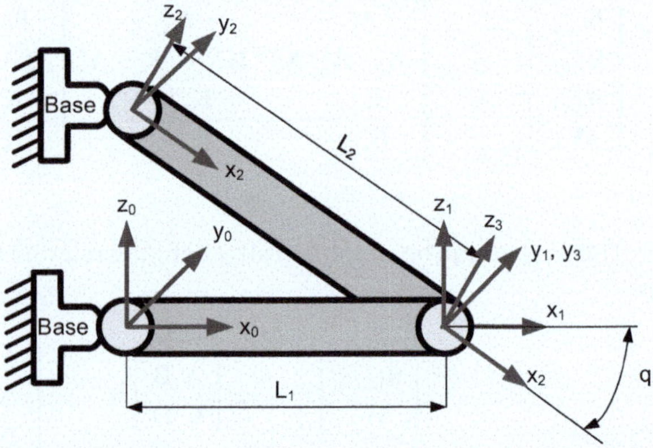

Fig. 4.6 Two-link parallel system with passive joints [5]

Similar to all previous cases the basic elasto-static equations are written as follows

$$\begin{bmatrix} \mathbf{W}_0 \\ \mathbf{W}_1 \\ \mathbf{W}_2 \\ \mathbf{W}_3 \end{bmatrix}_{24\times 1} = \begin{bmatrix} \mathbf{K}_{11}^1 & \mathbf{K}_{12}^1 & \mathbf{0}_{6\times 6} & \mathbf{0}_{6\times 6} \\ \mathbf{K}_{21}^1 & \mathbf{K}_{22}^1 & \mathbf{0}_{6\times 6} & \mathbf{0}_{6\times 6} \\ \mathbf{0}_{6\times 6} & \mathbf{0}_{6\times 6} & \mathbf{K}_{11}^{2q} & \mathbf{K}_{12}^{2q} \\ \mathbf{0}_{6\times 6} & \mathbf{0}_{6\times 6} & \mathbf{K}_{21}^{2q} & \mathbf{K}_{22}^{2q} \end{bmatrix}_{24\times 24} \cdot \begin{bmatrix} \Delta\mathbf{t}_0 \\ \Delta\mathbf{t}_1 \\ \Delta\mathbf{t}_2 \\ \Delta\mathbf{t}_3 \end{bmatrix}_{24\times 1} \tag{4.44}$$

where the matrices $\mathbf{K}_{ij}^{2q}$ depend on the system's geometric parameter q. However, because of the non-rigid connections between the beams and the base, here $\Delta\mathbf{t}_0 \neq 0$ and $\Delta\mathbf{t}_2 \neq 0$ that does not allow us to reduce the matrix dimension trivially (as it was done in all previous examples, where the lines corresponding to $\mathbf{W}_0$ were simply removed).

For the considered system, the left-hand side connections between the beam and rigid base (via the passive joints) produce the following equations

$$\begin{aligned} \mathbf{\Lambda}^r \cdot \Delta\mathbf{t}_0 = \mathbf{0}\,; \qquad \mathbf{\Lambda}^r \cdot \Delta\mathbf{t}_2 = \mathbf{0} \\ \mathbf{\Lambda}^p \cdot \mathbf{W}_0 = \mathbf{0}\,; \qquad \mathbf{\Lambda}^p \cdot \mathbf{W}_2 = \mathbf{0} \end{aligned} \tag{4.45}$$

where the first group describes the passive joint geometry and the second group follows from the 3rd Newton's law (the matrices $\mathbf{\Lambda}^r$ and $\mathbf{\Lambda}^p$ are defined above in the example (b)). For the right-hand side connections between the beams and end-effector, it is necessary to define additional variables $\Delta\mathbf{t}$ and $\mathbf{W}_e$ describing the end-effector deflection and wrench, respectively. Using these notations, the end-effector wrench may be expressed as follows

$$\mathbf{W}_e = \mathbf{\Lambda}^r \cdot (\mathbf{W}_3 + \mathbf{W}_1) \tag{4.46}$$

Besides, the passive joint geometry and the 3rd Newton's law at the right-hand side yield equations

$$\begin{aligned} \mathbf{\Lambda}^r \cdot (\Delta\mathbf{t}_1 - \Delta\mathbf{t}) = \mathbf{0}\,; \qquad & \mathbf{\Lambda}^r \cdot (\Delta\mathbf{t}_3 - \Delta\mathbf{t}) = \mathbf{0} \\ \mathbf{\Lambda}^p \cdot \mathbf{W}_1 = \mathbf{0}\,; \qquad & \mathbf{\Lambda}^p \cdot \mathbf{W}_3 = \mathbf{0} \end{aligned} \tag{4.47}$$

that allows us to take into account the particularities of the end-effector connection.

Combining Eqs. (4.44–4.47) yields the following linear system

$$\begin{bmatrix} \mathbf{0}_{6\times1} \\ \mathbf{0}_{6\times1} \\ \mathbf{0}_{6\times1} \\ \mathbf{0}_{6\times1} \\ \mathbf{0}_{6\times1} \\ \mathbf{0}_{6\times1} \\ \mathbf{0}_{6\times1} \\ \mathbf{0}_{6\times1} \\ \mathbf{0}_{6\times1} \\ \mathbf{W}_e \end{bmatrix}_{54\times1} = \begin{bmatrix} \mathbf{\Lambda}^r & \mathbf{0}_{6\times6} & \mathbf{0}_{6\times6} & \mathbf{0}_{6\times6} & \mathbf{0}_{6\times6} \\ \mathbf{0}_{6\times6} & \mathbf{0}_{6\times6} & \mathbf{\Lambda}^r & \mathbf{0}_{6\times6} & \mathbf{0}_{6\times6} \\ \mathbf{0}_{6\times6} & \mathbf{\Lambda}^r & \mathbf{0}_{6\times6} & \mathbf{0}_{6\times6} & -\mathbf{\Lambda}^r \\ \mathbf{0}_{6\times6} & \mathbf{0}_{6\times6} & \mathbf{0}_{6\times6} & \mathbf{\Lambda}^r & -\mathbf{\Lambda}^r \\ \mathbf{\Lambda}^p\mathbf{K}_{11}^1 & \mathbf{\Lambda}^p\mathbf{K}_{12}^1 & \mathbf{0}_{6\times6} & \mathbf{0}_{6\times6} & \mathbf{0}_{6\times6} \\ \mathbf{\Lambda}^p\mathbf{K}_{21}^1 & \mathbf{\Lambda}^p\mathbf{K}_{22}^1 & \mathbf{0}_{6\times6} & \mathbf{0}_{6\times6} & \mathbf{0}_{6\times6} \\ \mathbf{0}_{6\times6} & \mathbf{0}_{6\times6} & \mathbf{\Lambda}^p\mathbf{K}_{11}^{2q} & \mathbf{\Lambda}^p\mathbf{K}_{12}^{2q} & \mathbf{0}_{6\times6} \\ \mathbf{0}_{6\times6} & \mathbf{0}_{6\times6} & \mathbf{\Lambda}^p\mathbf{K}_{21}^{2q} & \mathbf{\Lambda}^p\mathbf{K}_{22}^{2q} & \mathbf{0}_{6\times6} \\ \mathbf{\Lambda}^r\mathbf{K}_{21}^1 & \mathbf{\Lambda}^r\mathbf{K}_{22}^1 & \mathbf{\Lambda}^r\mathbf{K}_{21}^{2q} & \mathbf{\Lambda}^r\mathbf{K}_{22}^{2q} & \mathbf{0}_{6\times6} \end{bmatrix}_{54\times30} \cdot \begin{bmatrix} \Delta\mathbf{t}_0 \\ \Delta\mathbf{t}_1 \\ \Delta\mathbf{t}_2 \\ \Delta\mathbf{t}_3 \\ \Delta\mathbf{t} \end{bmatrix}_{30\times1} \tag{4.48}$$

that should be solved for $\Delta\mathbf{t}$. It should be mentioned that here the system matrix is non-square of the size 54 × 30, but it contains 24 zero rows that are caused by zero diagonal components in $\mathbf{\Lambda}^r$ and $\mathbf{\Lambda}^p$ (each $\mathbf{\Lambda}^r$ produces one zero-row and each $\mathbf{\Lambda}^p$ produces five zero-rows). To eliminate the zero-rows, let us use the modified non-singular matrices $\mathbf{\Lambda}_*^r$ and $\mathbf{\Lambda}_*^p$ (defined in example (b)) and rewrite the system (4.48) in the reduced form

$$\begin{bmatrix} \mathbf{0}_{5\times1} \\ \mathbf{0}_{5\times1} \\ \mathbf{0}_{5\times1} \\ \mathbf{0}_{5\times1} \\ \mathbf{0}_{1\times1} \\ \mathbf{0}_{1\times1} \\ \mathbf{0}_{1\times1} \\ \mathbf{0}_{1\times1} \\ \hline \mathbf{W}_e \end{bmatrix}_{30\times1} = \left[\begin{array}{cccc|c} \mathbf{\Lambda}_*^r & \mathbf{0}_{5\times6} & \mathbf{0}_{5\times6} & \mathbf{0}_{5\times6} & \mathbf{0}_{5\times6} \\ \mathbf{0}_{5\times6} & \mathbf{0}_{5\times6} & \mathbf{\Lambda}_*^r & \mathbf{0}_{5\times6} & \mathbf{0}_{5\times6} \\ \mathbf{0}_{5\times6} & \mathbf{\Lambda}_*^r & \mathbf{0}_{5\times6} & \mathbf{0}_{5\times6} & -\mathbf{\Lambda}_*^r \\ \mathbf{0}_{5\times6} & \mathbf{0}_{5\times6} & \mathbf{0}_{5\times6} & \mathbf{\Lambda}_*^r & -\mathbf{\Lambda}_*^r \\ \mathbf{\Lambda}_*^p\mathbf{K}_{11}^1 & \mathbf{\Lambda}_*^p\mathbf{K}_{12}^1 & \mathbf{0}_{1\times6} & \mathbf{0}_{1\times6} & \mathbf{0}_{1\times6} \\ \mathbf{\Lambda}_*^p\mathbf{K}_{21}^1 & \mathbf{\Lambda}_*^p\mathbf{K}_{22}^1 & \mathbf{0}_{1\times6} & \mathbf{0}_{1\times6} & \mathbf{0}_{1\times6} \\ \mathbf{0}_{1\times6} & \mathbf{0}_{1\times6} & \mathbf{\Lambda}_*^p\mathbf{K}_{11}^{2q} & \mathbf{\Lambda}_*^p\mathbf{K}_{12}^{2q} & \mathbf{0}_{1\times6} \\ \mathbf{0}_{1\times6} & \mathbf{0}_{1\times6} & \mathbf{\Lambda}_*^p\mathbf{K}_{21}^{2q} & \mathbf{\Lambda}_*^p\mathbf{K}_{22}^{2q} & \mathbf{0}_{1\times6} \\ \hline \mathbf{\Lambda}^r\mathbf{K}_{21}^1 & \mathbf{\Lambda}^r\mathbf{K}_{22}^1 & \mathbf{\Lambda}^r\mathbf{K}_{21}^{2q} & \mathbf{\Lambda}^r\mathbf{K}_{22}^{2q} & \mathbf{0}_{6\times6} \end{array}\right]_{30\times30} \cdot \begin{bmatrix} \Delta\mathbf{t}_0 \\ \Delta\mathbf{t}_1 \\ \Delta\mathbf{t}_2 \\ \Delta\mathbf{t}_3 \\ \hline \Delta\mathbf{t} \end{bmatrix}_{30\times1} \tag{4.49}$$

where the main matrix of size 30 × 30 is obviously rank-deficient (since the 29th column is equal to zero). The latter is in good agreement with the physical nature of the system, where the end-effector is connected to the beams via the passive joints. To obtain the final expression, the reduced system can be presented in a compact form

$$\begin{bmatrix} \mathbf{0}_{24\times1} \\ \mathbf{W}_e \end{bmatrix}_{18\times1} = \begin{bmatrix} \mathbf{A} & \mathbf{B} \\ \mathbf{C} & \mathbf{D} \end{bmatrix}_{30\times30} \cdot \begin{bmatrix} \Delta\mathbf{t}_0 \\ \Delta\mathbf{t}_1 \\ \Delta\mathbf{t}_2 \\ \Delta\mathbf{t}_3 \\ \hline \Delta\mathbf{t} \end{bmatrix}_{30\times1} \tag{4.50}$$

where the block matrices $\mathbf{A},\ \mathbf{B},\ \mathbf{C},\ \mathbf{D}$ are defined as follows

$$\mathbf{A} = \begin{bmatrix} \mathbf{\Lambda}_*^r & \mathbf{0}_{5\times6} & \mathbf{0}_{5\times6} & \mathbf{0}_{5\times6} \\ \mathbf{0}_{5\times6} & \mathbf{0}_{5\times6} & \mathbf{\Lambda}_*^r & \mathbf{0}_{5\times6} \\ \mathbf{0}_{5\times6} & \mathbf{\Lambda}_*^r & \mathbf{0}_{5\times6} & \mathbf{0}_{5\times6} \\ \mathbf{0}_{5\times6} & \mathbf{0}_{5\times6} & \mathbf{0}_{5\times6} & \mathbf{\Lambda}_*^r \\ \mathbf{\Lambda}_*^p\mathbf{K}_{11}^1 & \mathbf{\Lambda}_*^p\mathbf{K}_{12}^1 & \mathbf{0}_{1\times6} & \mathbf{0}_{1\times6} \\ \mathbf{\Lambda}_*^p\mathbf{K}_{21}^1 & \mathbf{\Lambda}_*^p\mathbf{K}_{22}^1 & \mathbf{0}_{1\times6} & \mathbf{0}_{1\times6} \\ \mathbf{0}_{1\times6} & \mathbf{0}_{1\times6} & \mathbf{\Lambda}_*^p\mathbf{K}_{11}^{2q} & \mathbf{\Lambda}_*^p\mathbf{K}_{12}^{2q} \\ \mathbf{0}_{1\times6} & \mathbf{0}_{1\times6} & \mathbf{\Lambda}_*^p\mathbf{K}_{21}^{2q} & \mathbf{\Lambda}_*^p\mathbf{K}_{22}^{2q} \end{bmatrix}_{24\times24}; \quad \mathbf{B} = \begin{bmatrix} \mathbf{0}_{5\times6} \\ \mathbf{0}_{5\times6} \\ -\mathbf{\Lambda}_*^r \\ -\mathbf{\Lambda}_*^r \\ \mathbf{0}_{1\times6} \\ \mathbf{0}_{1\times6} \\ \mathbf{0}_{1\times6} \\ \mathbf{0}_{1\times6} \end{bmatrix}_{24\times6} \tag{4.51}$$

$$\mathbf{C} = \begin{bmatrix} \mathbf{\Lambda}^r\mathbf{K}_{21}^1 & \mathbf{\Lambda}^r\mathbf{K}_{22}^1 & \mathbf{\Lambda}^r\mathbf{K}_{21}^{2q} & \mathbf{\Lambda}^r\mathbf{K}_{22}^{2q} \end{bmatrix}_{6\times24}; \qquad \mathbf{D} = \mathbf{0}_{6\times6}$$

Further, after elimination $\Delta\mathbf{t}_0,\ \Delta\mathbf{t}_1,\ \Delta\mathbf{t}_2,\ \Delta\mathbf{t}_3$ one can express the end-effector wrench as

$$\mathbf{W}_e = \mathbf{K}_C \cdot \Delta\mathbf{t} \tag{4.52}$$

where

$$\mathbf{K}_C = -\mathbf{C} \cdot \mathbf{A}^{-1} \cdot \mathbf{B} \tag{4.53}$$

is the desired Cartesian stiffness matrix of the two-link closed-loop system under study. It has a similar structure as in the above examples. Also, it is easy to prove analytically that the stiffness matrix $\mathbf{K}_C$ is rank-deficient (here $rank(\mathbf{B}) = 5$ because of a zero column), which is the result of the passive joint connection to the end-effector.

Summarizing all the above-presented case studies, one can conclude that some modifications of the MSA approach allowed us to obtain the desired Cartesian stiffness matrix that may be either a full-rank or rank-deficient one. However, relevant techniques include several non-trivial steps, which are generalized in the following Section to be applied to more complex structures.

4.1.3 General Methodology of Classical MSA

As follows from the above case studies, basic ideas of the classical MSA can be successfully used in robotics but some enhancement is required in order to take into account numerous passive and actuated connections. Before presenting the proposed enhancement, let us remind briefly basic steps of the classical MSA that is perfectly suited for computer-aided analysis of complex structures such as trusses, bridges, high-voltage towers, etc. In general, the MSA-based stiffness analysis includes the following steps:

Step 1: Decoupling the original system into generic structural members such as beams, rods, etc. and presenting the system as a set of member elements connected at the nodes (MSA idealization).

Step 2: Creating "free-free" master stiffness models for each element using 12×12 stiffness matrices and presenting them in the global coordinate system (MSA members modelling).

Step 3: Creating the node-element connectivity matrix, that defines links between model members via rigid, elastic or passive connections (MSA members connecting).

Step 4: Merging the individual stiffness matrices with the global stiffness matrix for the entire structure using the node-element connectivity matrix and overlapping technique (MSA assembling).

Step 5: Defining nodal loads and system supports, incorporating them in the global stiffness model, application of the boundary conditions/constraints and reducing the global stiffness matrix dimension (MSA constraints).

Step 6: Solving the resulting set of the reduced equations for unknown nodal displacements corresponding to given external loads (MSA solving).

Step 7: Computing reaction forces at the supports, the nodal internal forces and stresses for all system members (MSA post-processing).

It should be noted that practical computer implementation of the MSA method significantly differs from the manual technique presented in the previous Section. There are two main differences: (i) at the assembling stage the individual stiffness matrices are not expanded but are directly merged through the use of a special "freedom pointer array"; (ii) both the individual and global stiffness matrices are stored using a special format that takes advantage of symmetry and sparseness. There are also some differences in the application of boundary conditions and the global stiffness matrix reduction. Nevertheless, relevant computer-oriented routines ensure the satisfaction of two basic rules of structural mechanics: compatibility of displacement and force equilibrium, which are equivalent to the simple merging of 6-dimensional rows and columns in the global stiffness matrix. Hence, some modifications of these routines are required to apply MSA in robotics where the size of the global stiffness matrix is not very high but the connections between the structural members (passive and actuated joints) should be treated differently.

4.2 MSA Application in Robotics

In contrast to structural mechanics where the main interest is in the area of the nodal displacements, reaction/internal forces and stresses, the stiffness analysis in robotics concentrates on computing the Cartesian stiffness matrix defining the force–deflection relation for the manipulator end-effector. Let us present the MSA-based technique adapted for robotic applications, which is an issue from previous works [3, 6] and also contains some new contributions and generalizations.

4.2.1 MSA Models Of Manipulator Links And Platforms

Modeling of a Flexible Link

If the link flexibility is non-negligible, the 2-node "free-free" stiffness model should be used, which can be obtained either from the approximation of the link by a beam or using the CAD-based technique [7] allowing evaluation of the stiffness parameters taking into account the link's real shape and geometry. In both cases the link is described by the linear matrix equation, containing 12 × 12 "free-free" stiffness matrix

$$\begin{bmatrix} \mathbf{W}_i \\ \mathbf{W}_j \end{bmatrix} = \begin{bmatrix} \mathbf{K}_{11}^{(ij)} & \mathbf{K}_{12}^{(ij)} \\ \mathbf{K}_{21}^{(ij)} & \mathbf{K}_{22}^{(ij)} \end{bmatrix}_{12\times 12} \cdot \begin{bmatrix} \Delta\mathbf{t}_i \\ \Delta\mathbf{t}_j \end{bmatrix} \tag{4.54}$$

where $\Delta\mathbf{t}_i,\ \Delta\mathbf{t}_j$ are the deflections at the link ends, $\mathbf{W}_i,\ \ \mathbf{W}_j$ are the link end wrenches, i and j are the node indices corresponding to the link ends, and $\mathbf{K}_{11}^{(ij)},\ \mathbf{K}_{12}^{(ij)},\ \mathbf{K}_{21}^{(ij)},\ \mathbf{K}_{22}^{(ij)}$ are 6×6 stiffness matrices. It should be noted that the system (4.54) ensures both the displacement and force equilibriums for the flexible link, it includes 12 scalar equations and 24 scalar variables contained in six-dimensional vectors $\Delta\mathbf{t}_i,\ \Delta\mathbf{t}_j$ and $\mathbf{W}_i,\ \mathbf{W}_j$. Besides, it can be proved the rank deficiency of this model is equal to 12.

It is clear that in general cases when the link is arbitrarily oriented with respect to the global coordinate system, the above equations should be slightly modified by simply rotating the local stiffness matrices

$$\begin{bmatrix} {}^g\mathbf{K}_{11}^{(ij)} & {}^g\mathbf{K}_{12}^{(ij)} \\ {}^g\mathbf{K}_{21}^{(ij)} & {}^g\mathbf{K}_{22}^{(ij)} \end{bmatrix} = \begin{bmatrix} \mathbf{Q}^{(ij)}\mathbf{K}_{11}^{(ij)} \cdot \mathbf{Q}^{(ij)T} & \mathbf{Q}^{(ij)} \cdot \mathbf{K}_{12}^{(ij)} \cdot \mathbf{Q}^{(ij)T} \\ \hline \mathbf{Q}^{(ij)}\mathbf{K}_{21}^{(ij)} \cdot \mathbf{Q}^{(ij)T} & \mathbf{Q}^{(ij)} \cdot \mathbf{K}_{22}^{(ij)} \cdot \mathbf{Q}^{(ij)T} \end{bmatrix} \tag{4.55}$$

where $\mathbf{Q}^{(ij)} = diag(\mathbf{R}^{(ij)}, \mathbf{R}^{(ij)})$ is composed of two similar orthogonal matrices $\mathbf{R}^{(ij)}$ defining the orientation of the local coordinate system of the link (ij) relative to the global one and left superscript "g" indicates that the matrix is presented in the global coordinate system.

Modeling of a Rigid Link

If the link flexibility is negligible, the above stiffness model should be replaced by two types of equations describing the displacement and force equilibriums. The first of them can be treated as a simple "rigidity constraint" that keeps constant the distance between the nodes i and j. Applying to this link the rigid body kinematic equations, one can get the following relations between the nodal displacements $\Delta\mathbf{t}_i = [\Delta\mathbf{p}_i;\ \Delta\boldsymbol{\varphi}_i]$, $\Delta\mathbf{t}_j = [\Delta\mathbf{p}_j;\ \Delta\boldsymbol{\varphi}_j]$ expressed via the linear and angular components

$$\begin{aligned} \Delta\boldsymbol{\varphi}_j &= \Delta\boldsymbol{\varphi}_i \\ \Delta\mathbf{p}_j &= \Delta\mathbf{p}_i + \Delta\boldsymbol{\varphi}_i \times \mathbf{d}^{(ij)} \end{aligned} \tag{4.56}$$

where the vector $\mathbf{d}^{(ij)} = [d_x^{(ij)}, d_y^{(ij)}, d_z^{(ij)}]^T$ describes the link geometry and is directed from the ith to the jth node. This constraint can be also rewritten in the matrix form

$$\begin{bmatrix} \mathbf{I}_{3\times3} & [\mathbf{d}^{(ij)}\times]^T & -\mathbf{I}_{3\times3} & \mathbf{0}_{3\times3} \\ \mathbf{0}_{3\times3} & \mathbf{I}_{3\times3} & \mathbf{0}_{3\times3} & -\mathbf{I}_{3\times3} \end{bmatrix} \cdot \begin{bmatrix} \Delta\mathbf{t}_i \\ \Delta\mathbf{t}_j \end{bmatrix} = \mathbf{0}_{6\times1} \tag{4.57}$$

where $[\mathbf{d}^{(ij)}\times]$ denotes the 3×3 skew-symmetric matrix derived from the vector $\mathbf{d}^{(ij)}$. Further, after definition 6×6 block matrix

$$\mathbf{D}^{(ij)} = \begin{bmatrix} \mathbf{I}_{3\times3} & [\mathbf{d}^{(ij)}\times]^T \\ \mathbf{0}_{3\times3} & \mathbf{I}_{3\times3} \end{bmatrix}_{6\times6} \tag{4.58}$$

the displacement constraint for the rigid-link can be presented in the following form

$$\begin{bmatrix} \mathbf{D}^{(ij)} & -\mathbf{I}_{6\times6} \end{bmatrix} \cdot \begin{bmatrix} \Delta\mathbf{t}_i \\ \Delta\mathbf{t}_j \end{bmatrix} = \mathbf{0}_{6\times1} \tag{4.59}$$

that is convenient for further aggregation of the stiffness model components.

The second group of equations describing the force equilibrium can be derived using the rigid body static equations, which yields the following relations

$$\begin{aligned} &\mathbf{M}_j + \mathbf{M}_i + \mathbf{d}^{(ij)} \times \mathbf{F}_j = 0 \\ &\mathbf{F}_j + \mathbf{F}_i = 0 \end{aligned} \tag{4.60}$$

where $\mathbf{F}_i,\ \mathbf{F}_j$ and $\mathbf{M}_i,\ \mathbf{M}_j$ denote the forces and torques applied at the nodes i and j respectively. In a matrix form, they can be presented as follows

$$\mathbf{W}_i + \begin{bmatrix} \mathbf{I} & 0 \\ [\mathbf{d}^{(ij)}\times] & \mathbf{I} \end{bmatrix} \mathbf{W}_j = 0 \tag{4.61}$$

Similarly, using the above-introduced definition for $\mathbf{D}^{(ij)}$, this equation can be rewritten in a more compact form

$$\mathrm{W}_i + \mathrm{D}^{(ij)T}\mathrm{W}_j = 0_{6\times 1} \tag{4.62}$$

Hence, the rigid link produces 12 scalar equations describing the displacement and force equilibriums

$$\begin{aligned} &\left[\mathbf{D}^{(ij)} \; -\mathbf{I}_{6\times 6}\right]\cdot\begin{bmatrix}\Delta\mathbf{t}_i\\ \Delta\mathbf{t}_j\end{bmatrix} = \mathbf{0}_{6\times 1} \\ &\left[\mathrm{I}_{6\times 6}\,\mathrm{D}^{(ij)T}\right]\cdot\begin{bmatrix}\mathbf{W}_i\\ \mathbf{W}_j\end{bmatrix} = \mathbf{0}_{6\times 1} \end{aligned} \tag{4.63}$$

which are written with respect to 24 scalar variables contained in six-dimensional vectors $\Delta\mathbf{t}_i$, $\Delta\mathbf{t}_j$ and $\mathbf{W}_i$, $\mathbf{W}_j$, similar to the flexible link case (the rank deficiency of the matrix is also equal to 12 here).

Modeling of a Rigid Mobile Platform

If the platform can be treated as a rigid body, it can be included in the global stiffness model employing several virtual rigid links connecting the nodes $i, j, k, \ldots$ of the manipulator leg clamping points and the virtual rigid node e corresponding to the manipulator end-effector reference point (see Fig. 4.7a). From the displacement equilibrium, one can derive several equations similar to (4.59) that may be aggregated in a common matrix equation

$$\begin{bmatrix} -\mathbf{I}_{6\times 6} & \mathbf{0}_{6\times 6} & \mathbf{0}_{6\times 6} & \mathbf{D}^{(ei)} \\ \mathbf{0}_{6\times 6} & -\mathbf{I}_{6\times 6} & \mathbf{0}_{6\times 6} & \mathbf{D}^{(ej)} \\ \mathbf{0}_{6\times 6} & \mathbf{0}_{6\times 6} & -\mathbf{I}_{6\times 6} & \mathbf{D}^{(ek)} \end{bmatrix} \cdot \begin{bmatrix}\Delta\mathbf{t}_i\\ \Delta\mathbf{t}_j\\ \Delta\mathbf{t}_k\\ \Delta\mathbf{t}_e\end{bmatrix} = \mathbf{0}_{18\times 1} \tag{4.64}$$

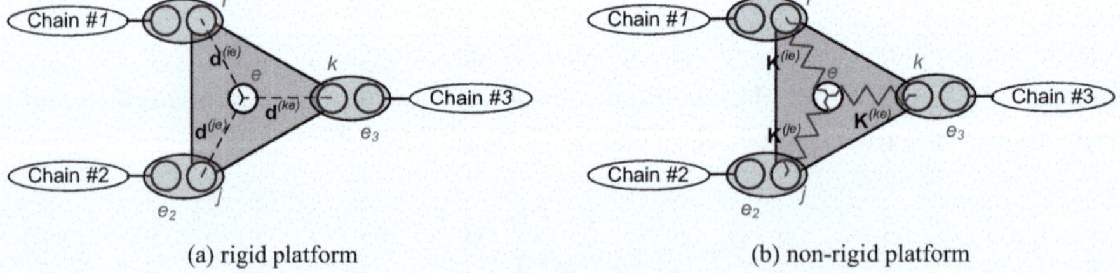

Fig. 4.7 Presentation of rigid (**a**) and non-rigid (**b**) platforms for the stiffness model of the parallel manipulator [5]

where $\mathbf{D}^{(ei)}$, $\mathbf{D}^{(ej)}$, $\mathbf{D}^{(ek)}$ describe the virtual links geometry in the same way as in (4.58). However, the force equilibrium produces here a single equation only

$$\mathbf{D}^{(ej)T}\,\mathbf{W}_i + \mathbf{D}^{(ej)T}\,\mathbf{W}_j + \mathbf{D}^{(ek)T}\,\mathbf{W}_k + \mathbf{W}_e = \mathbf{0}_{6\times 1} \tag{4.65}$$

which is an extended form of (4.62). In a matrix form, it can be presented as

$$\left[\mathbf{D}^{(ei)T}\ \mathbf{D}^{(ej)T}\mathbf{D}^{(ek)T}\mathbf{I}_{6\times 6}\right]\begin{bmatrix}\mathbf{W}_i\\ \mathbf{W}_j\\ \mathbf{W}_k\\ \mathbf{W}_e\end{bmatrix} = \mathbf{0}_{6\times 1} \tag{4.66}$$

Hence, for this example where the number of the leg clamping points is assumed to be equal to three, the model includes 24 scalar equations and 48 scalar variables. The rank deficiency of the model matrix is equal to 24, which agrees with the rigid body mechanics. It is clear that this model can be easily generalized to the case of 4, 5, … of attached legs by straightforward expanding the matrix sizes.

Modeling of a Non-rigid Mobile Platform

If the platform cannot be treated as a rigid body, it should be approximated by means of an equivalent structure consisting of several virtual flexible links connecting the nodes $i, j, k, \ldots$ of the manipulator leg clamping and the virtual rigid node e corresponding to the manipulator end-effector reference point (see Fig. 4.7b). These links are described by 12×12 stiffness matrix of the type (4.54) that should be aggregated in a common matrix equation.

$$\begin{bmatrix}\mathbf{W}_i\\ \mathbf{W}_j\\ \mathbf{W}_k\\ \mathbf{W}_e\end{bmatrix} = \begin{bmatrix} {}^g\mathbf{K}_{11}^{(ie)} & \mathbf{0}_{6\times 6} & \mathbf{0}_{6\times 6} & {}^g\mathbf{K}_{12}^{(ie)} \\ \mathbf{0}_{6\times 6} & {}^g\mathbf{K}_{11}^{(je)} & \mathbf{0}_{6\times 6} & {}^g\mathbf{K}_{12}^{(je)} \\ \mathbf{0}_{6\times 6} & \mathbf{0}_{6\times 6} & {}^g\mathbf{K}_{11}^{(ke)} & {}^g\mathbf{K}_{12}^{(ke)} \\ {}^g\mathbf{K}_{21}^{(ie)} & {}^g\mathbf{K}_{21}^{(je)} & {}^g\mathbf{K}_{21}^{(ke)} & {}^g\mathbf{K}_{22}^{(ie)} + {}^g\mathbf{K}_{22}^{(je)} + {}^g\mathbf{K}_{22}^{(ke)} \end{bmatrix}_{24\times 24} \cdot \begin{bmatrix}\Delta\mathbf{t}_i\\ \Delta\mathbf{t}_j\\ \Delta\mathbf{t}_k\\ \Delta\mathbf{t}_e\end{bmatrix} \tag{4.67}$$

where $\Delta\mathbf{t}_i$, $\Delta\mathbf{t}_j$, $\Delta\mathbf{t}_k$ are the deflections at the leg clamping points, $\Delta\mathbf{t}_e$ is the deflection at the end-effector reference point, $\mathbf{W}_i$, $\mathbf{W}_j$ $\mathbf{W}_k$ are the wrenches at the leg clamping points, $\mathbf{W}_e$ is the total wrench applied to the end-effector from the virtual links side. It is assumed here that all 6×6 stiffness matrices are presented in the global coordinate system. For this case of three attached legs, the model includes 24 scalar equations and 48 scalar variables, the rank deficiency of the model matrix is equal to 24, which is in good agreement with four free-points of the considered elastic mechanical system.

4.2.2 MSA Models of a Manipulator's Joints

In robotic manipulators, the robot base, legs/links and mobile platform are connected using either passive, actuated, elastic or rigid joints (Fig. 4.8). Let us consider in detail their contributions to the global stiffness model of the manipulator.

Modeling of a Rigid Joint

If adjacent links are connected by means of a rigid joint, the stiffness model must include relations describing two principal rules of structural mechanics: (a) displacement compatibility, and (b) force equilibrium. Assuming that the nodes corresponding to the adjacent link are denoted as $i, j, k, \ldots$ (Fig. 4.9), the first of these rules can be expressed as

$$\Delta \mathbf{t}_i = \Delta \mathbf{t}_j = \Delta \mathbf{t}_k = \ldots \tag{4.68}$$

that allows the user to merge corresponding columns in the global stiffness matrix at the assembling stage. It is clear that the above equations can be also included in the stiffness model as an additional constraint without the elimination of redundant variables. In a matrix form, this constraint can be expressed as

$$\begin{bmatrix} \mathbf{I}_{6\times 6} & -\mathbf{I}_{6\times 6} \end{bmatrix}_{6\times 12} \cdot \begin{bmatrix} \Delta \mathbf{t}_i \\ \Delta \mathbf{t}_j \end{bmatrix} = \mathbf{0}_{6\times 1} \tag{4.69}$$

in the case of two adjacent links with the rigid connection of the nodes i, j, and as

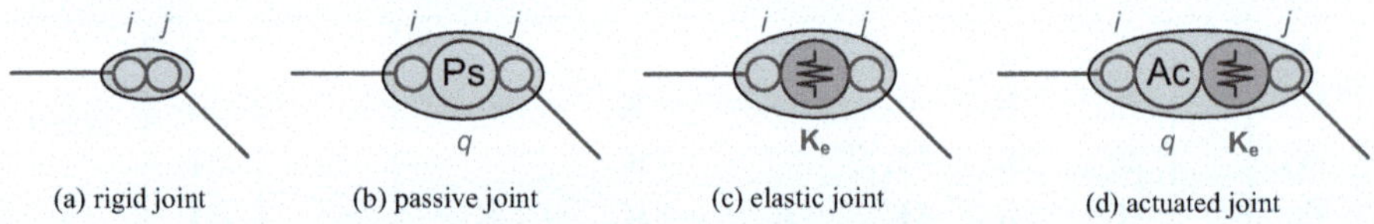

Fig. 4.8 Connections between elements of a robotic manipulator [5]

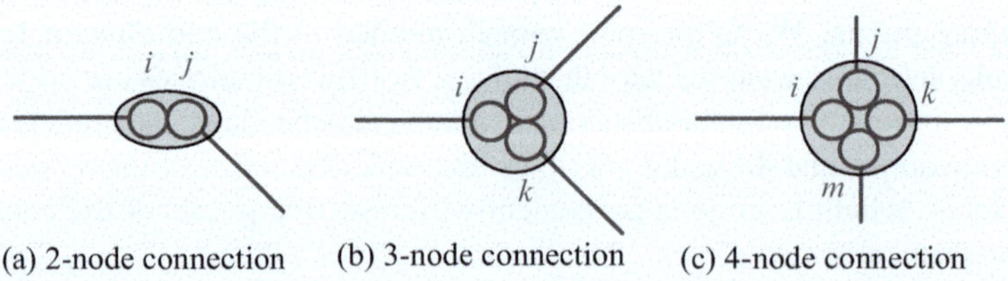

Fig. 4.9 Rigid linking of several joints [5]

$$\begin{bmatrix} \mathbf{I}_{6\times6} & \mathbf{0}_{6\times6} & -\mathbf{I}_{6\times6} \\ \mathbf{0}_{6\times6} & \mathbf{I}_{6\times6} & -\mathbf{I}_{6\times6} \end{bmatrix}_{12\times18} \cdot \begin{bmatrix} \Delta\mathbf{t}_i \\ \Delta\mathbf{t}_j \\ \Delta\mathbf{t}_k \end{bmatrix} = \mathbf{0}_{12\times1} \tag{4.70}$$

in the case of three adjacent links with the rigid connection of the nodes i, j, k. In the case of more adjacent links, the constraint is modified by a simple extension of the relevant matrix.

The second rule (the force equilibrium) produces the following relation

$$\mathbf{W}_i + \mathbf{W}_j + \mathbf{W}_k + \ldots = \mathbf{0} \tag{4.71}$$

that directly follows Newton's third law and allows the user to merge corresponding rows in the global stiffness matrix at the assembling stage. This relation can be also included in the stiffness model as an additional constraint without the elimination of redundant variables. In the case of two adjacent links with the rigid connection of the nodes i, j, the force equilibrium for the rigid connection of two links can be expressed as follows

$$\begin{bmatrix} \mathbf{I}_{6\times6} & \mathbf{I}_{6\times6} \end{bmatrix}_{6\times12} \cdot \begin{bmatrix} \mathbf{W}_i \\ \mathbf{W}_j \end{bmatrix} = \mathbf{0}_{6\times1} \tag{4.72}$$

Similarly, the equation can be written for the case of three or more adjacent links with a rigid connection. For example, for three adjacent links, it will have the form

$$\begin{bmatrix} \mathbf{I}_{6\times6} & \mathbf{I}_{6\times6} & \mathbf{I}_{6\times6} \end{bmatrix}_{12\times18} \cdot \begin{bmatrix} \mathbf{W}_i \\ \mathbf{W}_j \\ \mathbf{W}_k \end{bmatrix} = \mathbf{0}_{6\times1} \tag{4.73}$$

Hence, the rigid joint imposes some additional constraints on the model variables $\Delta\mathbf{t}_i$ and $\mathbf{W}_i$. The number of these constraints is equal to 12 for the case of two links, 18 for three links, etc.

Modeling of a Passive Joint

If adjacent links are connected by means of a passive joint, weaker constraints should be applied compared to the rigid joint case. In particular, for two adjacent links (ai), (jb) with common nodes i, j, the displacement compatibility condition of $\Delta\mathbf{t}_i$, $\Delta\mathbf{t}_j$ must be replaced by the matrix equation

$$\mathbf{\Lambda}_{ij}^{r} \cdot \left(\Delta\mathbf{t}_i - \Delta\mathbf{t}_j\right) = \mathbf{0} \tag{4.74}$$

where $\mathbf{\Lambda}_{ij}^{r}$ is a rank-deficient matrix that defines directions for the passive joint that does not admit free relative motions. For instance, for the revolute passive joint directed along the global y-axis, the matrix $\mathbf{\Lambda}^r$ is expressed as

$$\mathbf{\Lambda}^r_{Ry} = diag(1, 1, 1, 1, 0, 1) \tag{4.75}$$

For the spherical joint, it is presented in the form

$$\mathbf{\Lambda}^r_{Rxyz} = diag(1, 1, 1, 0, 0, 0) \tag{4.76}$$

In general case, to derive the matrix $\mathbf{\Lambda}^r$, let us define the orthonormal basis $\mathbf{u}_1, \mathbf{u}_2, ..., \mathbf{u}_6$ associated with the passive joint in such a way that the unit vectors $\mathbf{u}_1, ..., \mathbf{u}_r$ describe the directions of the rigid connection and the unit vector corresponding to the passive connection allowing free relative motions of the links. This definition allows us to present the matrix $\mathbf{\Lambda}^r$ as

$$\mathbf{\Lambda}^r = \left[\mathbf{u}_1, ..., \mathbf{u}_r, \mid \mathbf{0}_{6\times1}, .., \mathbf{0}_{6\times1}\right]^T_{6\times6} \tag{4.77}$$

that corresponds to the set of linear equations

$$\mathbf{u}_k^T \cdot (\Delta\mathbf{t}_i - \Delta\mathbf{t}_j) = 0, \qquad \forall k = \overline{1, r} \tag{4.78}$$

Using the matrix $\mathbf{\Lambda}^r_{ij}$, the deflection compatibility constraint may be directly written in the form

$$\begin{bmatrix} \mathbf{\Lambda}^r_{ij} & -\mathbf{\Lambda}^r_{ij} \end{bmatrix}_{6\times12} \cdot \begin{bmatrix} \Delta\mathbf{t}_i \\ \Delta\mathbf{t}_j \end{bmatrix} = \mathbf{0}_{6\times1} \tag{4.79}$$

which may be further reduced to a full-rank matrix equation

$$\begin{bmatrix} \mathbf{\Lambda}^r_{*ij} & -\mathbf{\Lambda}^r_{*ij} \end{bmatrix}_{r\times12} \cdot \begin{bmatrix} \Delta\mathbf{t}_i \\ \Delta\mathbf{t}_j \end{bmatrix} = \mathbf{0}_{r\times1} \tag{4.80}$$

that contains a rectangular matrix $\mathbf{\Lambda}^r_{*ij}$ of the rank r. In the simple cases, the matrix $\mathbf{\Lambda}^r_{*ij}$ can be easily derived from $\mathbf{\Lambda}^r_{ij}$ by simple elimination of zero rows (see an expression (4.30)). In the general case, the reduced matrix $\mathbf{\Lambda}^r_{*ij}$ is formed as follows

$$\mathbf{\Lambda}^r_{*} = [\mathbf{u}_1, ..., \mathbf{u}_r]^T_{r\times6} \tag{4.81}$$

The force equilibrium condition for the passive joint should be also presented in the weaker form as

$$\mathbf{\Lambda}^r_{ij} \cdot \left(\mathbf{W}_i + \mathbf{W}_j\right) = \mathbf{0} \tag{4.82}$$

that after elimination zero lines give the second constraint

$$\left[\boldsymbol{\Lambda}^r_{*ij}\ \boldsymbol{\Lambda}^r_{*ij} \right]_{r\times 12} \cdot \begin{bmatrix} \mathbf{W}_i \\ \mathbf{W}_j \end{bmatrix} = \mathbf{0}_{r\times 1} \tag{4.83}$$

In addition, it is necessary to take into account that passive joints do not transmit the force/torque in the direction corresponding to zero columns in the matrix $\boldsymbol{\Lambda}^r_{ij}$, which yields the following equations

$$\boldsymbol{\Lambda}^p_{ij} \cdot \mathbf{W}_i = \mathbf{0}\,; \qquad \boldsymbol{\Lambda}^p_{ij} \cdot \mathbf{W}_j = \mathbf{0} \tag{4.84}$$

Here, the matrix $\boldsymbol{\Lambda}^p_{ij}$ defines the passive joint axes and contains the remaining unit vectors of the basis $\mathbf{u}_1, \mathbf{u}_2, ..., \mathbf{u}_6$ that are not included in the $\boldsymbol{\Lambda}^r_{ij}$:

$$\boldsymbol{\Lambda}^p = \left[\, \mathbf{0}_{6\times 1}, .., \mathbf{0}_{6\times 1},\ |\ \mathbf{u}_{r+1}, ..., \mathbf{u}_6 \,\right]^T_{6\times 6} \tag{4.85}$$

Similar to the above, the latter equation can be easily reduced by simply replacing the singular 6×6 matrix $\boldsymbol{\Lambda}^p_{ij}$ by the full-rank matrix $\boldsymbol{\Lambda}^p_{*ij}$ of size4 $p \times 6$, $p = 6 - r$ that contains non-zero column only, i.e.

$$\boldsymbol{\Lambda}^p_* = \left[\, \mathbf{u}_{r+1}, ..., \mathbf{u}_6 \,\right]^T_{p\times 6} \tag{4.86}$$

Hence, for the passive joints, the force equilibrium produces two types of constraints (4.83), (4.84) that may be aggregated in the following matrix equation

$$\begin{bmatrix} \boldsymbol{\Lambda}^r_{*ij} & \boldsymbol{\Lambda}^r_{*ij} \\ \boldsymbol{\Lambda}^p_{*ij} & \mathbf{0} \\ \mathbf{0} & \boldsymbol{\Lambda}^p_{*ij} \end{bmatrix}_{(6+p)\times 12} \cdot \begin{bmatrix} \mathbf{W}_i \\ \mathbf{W}_j \end{bmatrix}_{12\times 1} = \mathbf{0}_{(6+p)\times 1} \tag{4.87}$$

It should be noted that the above equations deal with the case when the passive joint connects two links only. In the case when more than two links are connected by a single passive joint, similar equations should be derived for each pair of the links, i.e. all passive connections should be treated separately (in contrast to the multiple rigid connections presented in the previous Section). It is also worth mentioning that the passive joint imposes the same number of constraints on the model variables $\Delta\mathbf{t}_i$ and $\mathbf{W}_i$ as in the case of the rigid joint, but their physical nature is different. In particular, the trivial (one dof) passive connection of two links produces 12 constraints, where 5 ones are imposed on $\Delta\mathbf{t}_i$ and 7 others are imposed on $\mathbf{W}_i$ (in contrast to the rigid joint case, where there are 6 constraints on $\Delta\mathbf{t}_i$ and 6 constraints on $\mathbf{W}_i$).

Modeling of an Elastic Joint

In some manipulators, the adjacent links are connected by means of an elastic joint that can be also treated as a passive compliant joint with springs. In this case, the deflection compatibility condition (4.80) remains the same as above, it ensures equality for r components of deflection vectors $\Delta\mathbf{t}_i$, $\Delta\mathbf{t}_j$. However, the force equilibrium condition must be slightly revised to take into account that some wrench components may be expressed via Hooke's law:

$$\begin{aligned}&\mathbf{W}_i + \mathbf{W}_j = \mathbf{0}\,;\\&\mathbf{\Lambda}^e_{*ij}\cdot\mathbf{W}_i = {}^g\mathbf{K}^e_{ij}\cdot\mathbf{\Lambda}^e_{*ij}\cdot(\Delta\mathbf{t}_i - \Delta\mathbf{t}_j)\end{aligned} \tag{4.88}$$

where the matrix $\mathbf{\Lambda}^e_{*ij}$ of size $e\times 6$, $e = 6 - r$ corresponds to the non-rigid directions of the joint (similar to $\mathbf{\Lambda}^p_{*ij}$ in the passive joint case)

$$\mathbf{\Lambda}^e_{*} = \left[\,\mathbf{u}_{r+1},\,\ldots,\,\mathbf{u}_6\,\right]^T_{e\times 6}, \tag{4.89}$$

and ${}^g\mathbf{K}^e_{ij}$ is $e\times e$ stiffness matrix describing elastic properties of the joint. So, the force equilibrium condition can be presented as the following matrix constraint

$$\begin{bmatrix}\mathbf{0}_{6\times6} & \mathbf{0}_{6\times6} & \mathbf{I}_{6\times6} & \mathbf{I}_{6\times6}\\ \mathbf{K}^e_{ij}\mathbf{\Lambda}^e_{*ij} & -\mathbf{K}^e_{ij}\mathbf{\Lambda}^e_{*ij} & \mathbf{\Lambda}^e_{*ij} & \mathbf{0}_{e\times6}\end{bmatrix}_{(6+e)\times24}\cdot\begin{bmatrix}\Delta\mathbf{t}_i\\ \Delta\mathbf{t}_j\\ \mathbf{W}_i\\ \mathbf{W}_j\end{bmatrix}_{24\times1} = \mathbf{0}_{(6+e)\times1} \tag{4.90}$$

It should be stressed that the above expressions are valid for the so-called "non-preloaded case" when equal deflections do not generate the elastic forces, i.e. $\mathbf{\Lambda}^e_{*ij}\cdot\mathbf{W}_i = \mathbf{\Lambda}^e_{*ij}\cdot\mathbf{W}_j = \mathbf{0}$. In the case when the springs are initially preloaded by the wrench $\mathbf{W}^0_{ij}$, the force equilibrium equation must be presented in the form

$$\begin{aligned}&\mathbf{W}_i + \mathbf{W}_j = \mathbf{0}\,;\\&\mathbf{\Lambda}^e_{*ij}\cdot(\mathbf{W}_i - \mathbf{W}^0_{ij}) = \mathbf{K}^e_{ij}\cdot\mathbf{\Lambda}^e_{*ij}\cdot(\Delta\mathbf{t}_i - \Delta\mathbf{t}_j)\end{aligned} \tag{4.91}$$

that leads to the following linear matrix constraint

$$\begin{bmatrix}\mathbf{0}_{6\times6} & \mathbf{0}_{6\times6} & \mathbf{I}_{6\times6} & \mathbf{I}_{6\times6}\\ \mathbf{K}^e_{ij}\mathbf{\Lambda}^e_{*ij} & -\mathbf{K}^e_{ij}\mathbf{\Lambda}^e_{*ij} & \mathbf{\Lambda}^e_{*ij} & \mathbf{0}_{e\times6}\end{bmatrix}_{(6+e)\times24}\cdot\begin{bmatrix}\Delta\mathbf{t}_i\\ \Delta\mathbf{t}_j\\ \mathbf{W}_i\\ \mathbf{W}_j\end{bmatrix}_{24\times1} = \begin{bmatrix}\mathbf{0}_{6\times1}\\ \mathbf{\Lambda}^e_{*ij}\mathbf{W}^0_{ij}\end{bmatrix}_{(6+e)\times1} \tag{4.92}$$

Hence, the elastic joint may be included in the global stiffness model similarly to the passive one by minor modifications of the corresponding matrix constraints.

Modeling of an Actuated Joint

The actuated joint ensures transmission of the force/torque between the manipulator links. In the frame of the MSA technique, it can be presented either as a rigid, passive or elastic connection. In the first case, the actuating effort is transmitted in all direction $\mathbf{u}_1, ..., \mathbf{u}_6$. In the second case, it is transmitted in the direction corresponding to the vectors $\mathbf{u}_1, ..., \mathbf{u}_r$ while in the last case transmission is performed in the direction defined by the vectors $\mathbf{u}_{r+1}, ..., \mathbf{u}_6$ (see previous sections for details). Hence, the actuated joint may be included in the global stiffness model as a set of two linear matrix constraints describing the deflection compatibility and force equilibrium, which slightly differ depending on the accepted joint idealization.

It is also worth mentioning that in the frame of the global stiffness model the actuated joint influences the manipulator configuration, i.e. on the relative position and orientation of the links, which is obtained from the static equilibrium condition. Depending on the joint control type (position or torque-based control), the actuator influence in the above-presented constraints is included either via the relation between $\Delta\mathbf{t}_i$, $\Delta\mathbf{t}_j$ or via the preloading $\mathbf{W}_{ij}^0$.

4.2.3 *Including Boundary Conditions and External Forces*

To complete the global stiffness model, it is necessary to take into account the specificity of the manipulator links connections to the robot base as well as the influence of the external forces (including reactions at the supports). Let us consider in detail contributions to the global stiffness model of different types of supports that may be implemented using either fixed, passive or elastic connections (Fig. 4.10). In the frame of the MSA method, the equations produced by the supports are treated as the boundary conditions.

Modeling of a Rigid Support

The rigid connection of the link (*jb*) to the robot base can be presented as a special case of the rigid joint with an eliminated link (*ai*) and zero deflection $\Delta\mathbf{t}_j = \mathbf{0}$. This simplifies the deflection compatibility constraint (4.69) down to

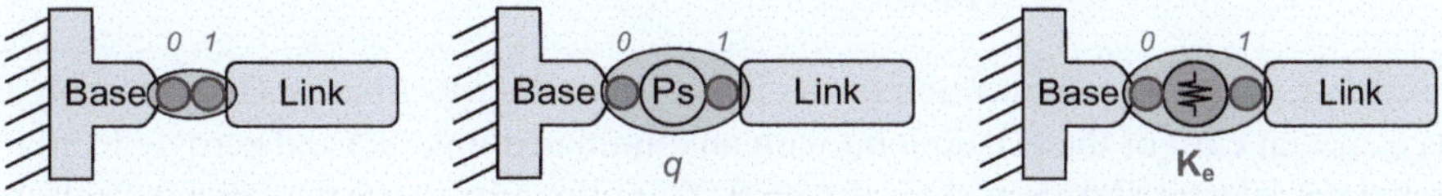

(a) via rigid joint (b) via passive joint (c) via elastic joint

Fig. 4.10 Connections between the manipulator links and base [5]

$$\left[\mathbf{I}_{6\times 6}\right]\cdot\left[\Delta\mathbf{t}_j\right]=\mathbf{0}_{6\times 1} \tag{4.93}$$

that contains 6 linear equations to be included in the global stiffness model. The corresponding reaction wrench at the rigid support may be computed as

$$\mathbf{W}_j={}^{g}\mathbf{K}_{12}^{(jb)}\cdot\Delta\mathbf{t}_b \tag{4.94}$$

where the deflection $\Delta\mathbf{t}_b$ is obtained from the solution of the global stiffness equations.

Modeling of a Passive Support

The passive connection of the link (jb) to the robot base can be presented as a special case of the passive joint with an eliminated link (ai) and zero deflection in the non-passive directions $\mathbf{\Lambda}_{*ij}^{r}\cdot\Delta\mathbf{t}_j=\mathbf{0}$. This simplifies the deflection compatibility constraint (4.80) down to

$$\left[\mathbf{\Lambda}_{*ij}^{r}\right]_{r\times 6}\cdot\left[\Delta\mathbf{t}_j\right]=\mathbf{0}_{r\times 1} \tag{4.95}$$

that contributes r linear equations to the global stiffness model. In addition, it is necessary to take into account that the wrench components for the passive direction are equal to zero, which leads to the simplified form of (4.84)

$$\mathbf{\Lambda}_{*ij}^{p}\cdot\mathbf{W}_j=\mathbf{0}_{p\times 1} \tag{4.96}$$

that contributes another p equations to the global stiffness model. Totally, this gives $r+p=6$ independent linear constraints. The corresponding reaction wrench at the passive support may be computed as

$$\mathbf{W}_j={}^{g}\mathbf{K}_{11}^{(jb)}\cdot\Delta\mathbf{t}_j+{}^{g}\mathbf{K}_{12}^{(jb)}\cdot\Delta\mathbf{t}_b \tag{4.97}$$

where the deflections $\Delta\mathbf{t}_j,\ \Delta\mathbf{t}_b$ are obtained from the solution of the global stiffness equations.

Modeling of an Elastic Support

Similarly, the elastic connection of the link (jb) to the robot base can be presented as a special case of the elastic joint with an eliminated link (ai) and zero deflection in the non-elastic directions $\mathbf{\Lambda}_{*ij}^{r}\cdot\Delta\mathbf{t}_j=\mathbf{0}$. This allows us to use the same deflection compatibility constraint (4.95) as above, which contributes r linear equations to the global stiffness model. In addition, it is necessary to take into account that the wrench components corresponding to the non-rigid directions are produced by the elastic forces satisfying Hooke's law, i.e.

$$\boldsymbol{\Lambda}^{e}_{*ij} \cdot \mathbf{W}_j = {}^{g}\mathbf{K}^{e}_{ij} \cdot \boldsymbol{\Lambda}^{e}_{*ij} \cdot \Delta\mathbf{t}_j \tag{4.98}$$

The latter leads to the linear equation

$$\left[-{}^{g}\mathbf{K}^{e}_{ij} \cdot \boldsymbol{\Lambda}^{e}_{*ij} \,\middle|\, \boldsymbol{\Lambda}^{e}_{*ij} \right]_{e\times 12} \cdot \left[\frac{\Delta\mathbf{t}_j}{\mathbf{W}_j} \right] = \left[\boldsymbol{\Lambda}^{e}_{*ij} \mathbf{W}^{0}_{ij} \right]_{e\times 1} \tag{4.99}$$

that contributes another e equations to the global stiffness model. Totally, this gives $r + e = 6$ independent linear constraints. To compute the reaction wrench at the elastic support the same expression (4.97) can be used.

Including an External Loading

In robotic manipulators, both serial and parallel, there is at least one node that is not connected directly to the robot base. It corresponds to the end-effector that interacts with the robot environment by applying the force/torque to the external objects. For the global stiffness model, the end-effector produces the boundary conditions that should include the vector of the external wrench $\mathbf{W}$, which is also necessary for the computation of the Cartesian stiffness matrix.

To take into account the external loading $\mathbf{W}_{ext}$, the global stiffness model must be completed by the linear constraint derived from the force equilibrium at the node e, i.e. $\mathbf{W}_i + \mathbf{W}_j = \mathbf{W}_{ext}$, which can be rewritten in the form

$$\left[\mathbf{I}_{6\times 6} \; \mathbf{I}_{6\times 6} \right]_{6\times 12} \cdot \begin{bmatrix} \mathbf{W}_i \\ \mathbf{W}_j \end{bmatrix}_{12\times 1} = \left[\mathbf{W}_{ext} \right]_{6\times 1} \tag{4.100}$$

Similarly, it is possible to derive the constraints for the case of three or more adjacent links, for instance

$$\left[\mathbf{I}_{6\times 6} \; \mathbf{I}_{6\times 6} \; \mathbf{I}_{6\times 6} \right]_{6\times 18} \cdot \begin{bmatrix} \mathbf{W}_i \\ \mathbf{W}_j \\ \mathbf{W}_k \end{bmatrix}_{18\times 1} = \left[\mathbf{W}_{ext} \right]_{6\times 1} \tag{4.101}$$

It is clear that this technique should be applied to all nodes with applied external wrenches.

4.2.4 Aggregation of MSA Model Components

In contrast to large mechanical structures consisting of a huge number of flexible components, a robotic manipulator is rather simple for the MSA analysis. Since the number of flexible links in the manipulator is relatively small, the assembling stage can be simplified and the columns/rows merging operations can be avoided and replaced by adding relevant constraints [3]. Let us present this technique in detail.

As it was shown above, the MSA equations for robotic manipulators are derived from three main sources: (i) link models, (ii) joint models and (iii) boundary conditions. The first of them describes the force–displacement relations for all links (both flexible and rigid) yielding 12 scalar equations per link relating 24 variables $\{\Delta\mathbf{t}_i, \Delta\mathbf{t}_j, \mathbf{W}_i, \mathbf{W}_j\}$. The second group of equations ensures the displacement compatibility and force/torque equilibrium for each internal connection (both rigid, passive and elastic), it includes two types of relations written for $\{\Delta\mathbf{t}_i, \Delta\mathbf{t}_j, ...\}$ and $\{\mathbf{W}_i, \mathbf{W}_j, ...\}$ separately. Independent of the connection type, each joint provides 12 scalar equations if it connects two links, 18 scalar equations for the connection of three links, etc. The third group of equations is issued from the manipulator connections to the environment, they are presented as the force/displacement constraints for certain nodes that give up to 6 scalar equations for $\{\Delta\mathbf{t}_i, \Delta\mathbf{t}_j, ...\}$ or at least 6 scalar equations for $\{\mathbf{W}_i, \mathbf{W}_j, ...\}$ depending on the connection type (rigid, elastic or passive).

In the general form, the aggregated stiffness model can be presented as follows

$$\begin{bmatrix} \mathbf{A}_{\mathrm{W}}^{(1)} & \mathbf{A}_{\Delta t}^{(1)} \\ \mathbf{A}_{\mathrm{W}}^{(2)} & \mathbf{A}_{\Delta t}^{(2)} \\ \cdots & \cdots \end{bmatrix} \cdot \begin{bmatrix} \{\mathbf{W}_i\} \\ \{\Delta\mathbf{t}_i\} \end{bmatrix} = \begin{bmatrix} \mathbf{b}_0^{(1)} \\ \mathbf{b}_0^{(2)} \\ \cdots \end{bmatrix} \tag{4.102}$$

where each line is issued from the equations/constraints presented in Sects. 3.1–3.3, the matrices $\mathbf{A}_{\mathrm{W}}^{(i)}$ and $\mathbf{A}_{\Delta t}^{(i)}$ correspond to the terms with $\mathbf{W}_i$ and $\Delta\mathbf{t}_i$ respectively, the vectors $\mathbf{b}_0^{(i)}$ represent the right-hand side of the corresponding equation/constraint. In this model, the contribution of the links can be expressed in the following way

$$\forall \textit{flexible links}: \begin{bmatrix} \{-\mathbf{I}_{12\times12}\} & \{\mathbf{K}_{12\times12}^{ij}\} \end{bmatrix} \cdot \begin{bmatrix} \{\mathbf{W}_i\} \\ \{\Delta\mathbf{t}_i\} \end{bmatrix} = \mathbf{0} \tag{4.103}$$

$$\forall \textit{rigid links}: \begin{bmatrix} \mathbf{0} & \{\mathbf{D}_{6\times6}^{ij}, -\mathbf{I}_{6\times6}\} \\ \{\mathbf{I}_{6\times6}, \mathbf{D}_{6\times6}^{ij\,T}\} & \mathbf{0} \end{bmatrix} \cdot \begin{bmatrix} \{\mathbf{W}_i\} \\ \{\Delta\mathbf{t}_i\} \end{bmatrix} = \begin{bmatrix} \mathbf{0} \\ \mathbf{0} \end{bmatrix} \tag{4.104}$$

which gives 12 scalar equations per link (both flexible and rigid). The platform is described by one of the equations

$$\textit{for flexible platform}: \begin{bmatrix} \{-\mathbf{I}_{6\times6}\} & \{\mathbf{K}_{6\times6}^{ij}\} \\ \mathbf{0} & \{\mathbf{K}_{6\times6}^{ij}\} \end{bmatrix} \cdot \begin{bmatrix} \{\mathbf{W}_i\} \\ \{\Delta\mathbf{t}_i\} \end{bmatrix} = \begin{bmatrix} \mathbf{0} \\ \mathbf{W}_{ext} \end{bmatrix} \tag{4.105}$$

$$\textit{for rigidplatform}: \left[\begin{array}{c|c} \mathbf{0} & \left\{-\mathbf{I}_{6\times 6}, \mathbf{D}^{\left(\frac{ei}{6\times 6}\right)}\right\} \\ \hline \left\{\mathbf{D}^{(ei)T}_{6\times 6}\right\} & 0 \end{array}\right] \cdot \begin{bmatrix} \{\mathbf{W}_i\} \\ \{\Delta \mathbf{t}_i\} \end{bmatrix} = \begin{bmatrix} \mathbf{0} \\ \mathbf{W}_{ext} \end{bmatrix} \tag{4.106}$$

whose dimension depends on the number of legs. The joints contribute to the aggregated model with the following constraints

$$\forall\, \textit{rigid joints}: \left[\begin{array}{c|c} 0 & \{\mathbf{I}_{6\times 6}, -\mathbf{I}_{6\times 6}\} \\ \hline \{\mathbf{I}_{6\times 6}, \mathbf{I}_{6\times 6}\} & 0 \end{array}\right] \cdot \begin{bmatrix} \{\mathbf{W}_i\} \\ \{\Delta \mathbf{t}_i\} \end{bmatrix} = \begin{bmatrix} \mathbf{0} \\ \{\mathbf{W}^0_{ij}\} \end{bmatrix} \tag{4.107}$$

$$\forall\, \textit{passivejoints}: \begin{bmatrix} \mathbf{0} & \{\mathbf{\Lambda}^r_{*ij}, -\mathbf{\Lambda}^r_{*ij}\} \\ \{\mathbf{\Lambda}^r_{*ij}, \mathbf{\Lambda}^r_{*ij}\} & \mathbf{0} \\ \{\mathbf{\Lambda}^p_{*ij}\} & \mathbf{0} \end{bmatrix} \cdot \begin{bmatrix} \{\mathbf{W}_i\} \\ \{\Delta \mathbf{t}_i\} \end{bmatrix} = \begin{bmatrix} \mathbf{0} \\ \{\mathbf{\Lambda}^r_{*ij} \mathbf{W}^0_{ij}\} \\ \mathbf{0} \end{bmatrix} \tag{4.108}$$

$$\forall\, \textit{elastic joints}: \begin{bmatrix} \mathbf{0} & \{\mathbf{\Lambda}^r_{*ij}, -\mathbf{\Lambda}^r_{*ij}\} \\ \{\mathbf{I}_{6\times 6}, \mathbf{I}_{6\times 6}\} & \mathbf{0} \\ \mathbf{\Lambda}^e_{*ij} & \{\mathbf{K}^e_{ij}\mathbf{\Lambda}^e_{*ij}, -\mathbf{K}^e_{ij}\mathbf{\Lambda}^e_{*ij}\} \end{bmatrix} \cdot \begin{bmatrix} \{\mathbf{W}_i\} \\ \{\Delta \mathbf{t}_i\} \end{bmatrix} = \begin{bmatrix} \mathbf{0} \\ \{\mathbf{W}^0_{ij}\} \\ \{\mathbf{\Lambda}^e_{*ij}\mathbf{W}^0_{ij}\} \end{bmatrix} \tag{4.109}$$

giving 12 scalar equations per internal joint (both rigid, passive and elastic). It should be noted that the external "joints" (connected to the base or end-effector) produce 6 scalar equations per connection only. So, their contribution to the aggregated model can be presented as

$$\forall\, \textit{rigid supports}: \begin{bmatrix} \mathbf{0} \; \{\mathbf{I}_{6\times 6}\} \end{bmatrix} \cdot \begin{bmatrix} \{\mathbf{W}_i\} \\ \{\Delta \mathbf{t}_i\} \end{bmatrix} = \mathbf{0} \tag{4.110}$$

$$\forall\, \textit{passive supports}: \begin{bmatrix} \mathbf{0} & \{\mathbf{\Lambda}^r_{*ij}\} \\ \{\mathbf{\Lambda}^p_{*ij}\} & \mathbf{0} \end{bmatrix} \cdot \begin{bmatrix} \{\mathbf{W}_i\} \\ \{\Delta \mathbf{t}_i\} \end{bmatrix} = \begin{bmatrix} \mathbf{0} \\ \mathbf{0} \end{bmatrix} \tag{4.111}$$

$$\forall\, \text{elastic supports}: \begin{bmatrix} \mathbf{0} & \{\mathbf{\Lambda}^r_{*ij}\} \\ \{\mathbf{\Lambda}^e_{*ij}\} & \{\mathbf{K}^e_{ij}\mathbf{\Lambda}^e_{*ij}\} \end{bmatrix} \cdot \begin{bmatrix} \{\mathbf{W}_i\} \\ \{\Delta \mathbf{t}_i\} \end{bmatrix} = \begin{bmatrix} \mathbf{0} \\ \{\mathbf{W}^0_{ij}\} \end{bmatrix} \tag{4.112}$$

Finally, the external loadings applied to some manipulator nodes are included in the model by means of the equations

$$\forall\, \textit{external loadings}: \begin{bmatrix} \{-\mathbf{I}_{6\times 6}\} \; \mathbf{0} \end{bmatrix} \cdot \begin{bmatrix} \{\mathbf{W}_i\} \\ \{\Delta \mathbf{t}_i\} \end{bmatrix} = \{\mathbf{W}^{ext}_i\} \tag{4.113}$$

It is worth mentioning that in practice the external loading applied to the manipulator end-effector is in the focus of stiffness analysis only.

For computational convenience, the Eqs. (4.103–4.113) can be arranged in the aggregated linear matrix equation of the following structure

$$\begin{bmatrix} \mathbf{S}_{agr} & \mathbf{K}_{agr} \\ \mathbf{E}_{agr} & \mathbf{F}_{agr} \end{bmatrix} \cdot \begin{bmatrix} \mathbf{W}_{agr} \\ \Delta\mathbf{t}_{agr} \end{bmatrix} = \begin{bmatrix} \mathbf{b}_{agr} \\ \mathbf{W}_{ext} \end{bmatrix} \tag{4.114}$$

where the matrices $\mathbf{S}_{agr}$, $\mathbf{K}_{agr}$, $\mathbf{E}_{agr}$, $\mathbf{F}_{agr}$, $\mathbf{b}_{agr}$, $\mathbf{W}_{ext}$ are generated using the relevant link/joint models or boundary conditions, while the vectors $\mathbf{W}_{agr}$ and $\Delta\mathbf{t}_{agr}$ contain all variables describing the wrenches and displacements, respectively. It can be proved that usually the system (4.114) cannot be solved straightforwardly because of the rank deficiency of the matrix to be inverted. Nevertheless, this system allows us to obtain the Cartesian stiffness matrix defining the relation between the end-effector deflection and applied force/torque (which obviously always exists but can be singular). To find the desired Cartesian stiffness matrix, let us divide the node displacement variables $\Delta\mathbf{t}_{agr}$ into two groups $\Delta\mathbf{t}_m$ and $\Delta\mathbf{t}_e$ corresponding to the manipulator's internal nodes and the end-effector node, where the external wrench $\mathbf{W}_e$ is applied. The latter allows us to rewrite the system (4.114) in the form

$$\left[\begin{array}{cc|c} \mathbf{S}_{agr} & \mathbf{K}_m & \mathbf{K}_e \\ \mathbf{E}_m & \mathbf{F}_m & \mathbf{C}_e \\ \hline \mathbf{E}_e & \mathbf{F}_e & \mathbf{D} \end{array}\right] \cdot \left[\begin{array}{c} \mathbf{W}_{agr} \\ \Delta\mathbf{t}_m \\ \hline \Delta\mathbf{t}_e \end{array}\right] = \left[\begin{array}{c} \mathbf{b}_{agr} \\ \mathbf{W}_m \\ \hline \mathbf{W}_e \end{array}\right] \tag{4.115}$$

and further, present it as

$$\begin{bmatrix} \mathbf{A} & \mathbf{B} \\ \mathbf{C} & \mathbf{D} \end{bmatrix} \cdot \left[\begin{array}{c} \boldsymbol{\mu} \\ \hline \Delta\mathbf{t}_e \end{array}\right] = \left[\begin{array}{c} \mathbf{b} \\ \hline \mathbf{W}_e \end{array}\right] \tag{4.116}$$

where all internal variables are included in the vector $\boldsymbol{\mu} = col(\mathbf{W}_{agr}, \Delta\mathbf{t}_m)$ and

$$\mathbf{A} = \begin{bmatrix} \mathbf{S}_{agr} & \mathbf{K}_m \\ \mathbf{E}_m & \mathbf{F}_m \end{bmatrix};\ \mathbf{B} = \begin{bmatrix} \mathbf{K}_e \\ \mathbf{C}_e \end{bmatrix};\ \mathbf{C} = \begin{bmatrix} \mathbf{E}_e & \mathbf{F}_e \end{bmatrix};\ \mathbf{b} = \begin{bmatrix} \mathbf{b}_{agr} \\ \mathbf{W}_m \end{bmatrix} \tag{4.117}$$

Using the obtained system, the relation between the displacement $\Delta\mathbf{t}_e$ and applied an external wrench $\mathbf{W}_e$ can be written as

$$\mathbf{W}_e = \mathbf{K}_C \cdot \Delta\mathbf{t}_e + \mathbf{W}_e^0 \tag{4.118}$$

that includes the desired Cartesian stiffness matrix

$$\mathbf{K}_C = \mathbf{D} - \mathbf{C} \cdot \mathbf{A}^{-1} \cdot \mathbf{B} \tag{4.119}$$

and the constant component

$$\mathbf{W}_e^0 = \mathbf{C} \cdot \mathbf{A}^{-1} \cdot \mathbf{b} \tag{4.120}$$

depending on the manipulator loadings and preloading included in the vector **b**.

It is worth mentioning that the matrix inversion in $\mathbf{A}^{-1}$ usually exists (if the manipulator does not include redundant passive joints), otherwise, the matrix inversion should be replaced by a pseudo-inverse that distributes deflections between the redundant joints. Besides, the Cartesian stiffness matrix $\mathbf{K}_C$ may be rank-deficient even for the non-singular matrix **A**, if the examined manipulator is under-constrained. Another important comment is related to the manipulator equilibrium configuration corresponding to given loadings and preloadings, which is implicitly assumed to be known here (i.e. computed before applying the MSA). However, for robotic manipulators, it is often not a trivial task [8]. It should be also mentioned that the final expression (4.119) presents the manipulator stiffness model in the form of the classical Cartesian stiffness matrix $\mathbf{K}_C$ of a size 6×6 corresponding to the fixed base and free end-effector. However, it can be easily transformed into the 'free-free' form that operates with 12×12 matrices using expressions similar to (4.9) and (4.13) taking into account that $\mathbf{K}_C$ corresponds to $\mathbf{K}_{22}$.

4.3 Illustrative Examples: MSA-Based Analysis of NAVARO Manipulator

To demonstrate the utility of the MSA-based technique for robotics, let us apply it to the stiffness analysis of the NaVaRo robot (Fig. 4.11a), a three-degree-of-freedom planar parallel manipulator with variable actuation schemes. It is composed of three identical legs and a moving platform formed of three segments rigidly linked at the central point. Each leg consists of four non-rigid links connected by five revolute joints to create a parallelogram linkage. Among them, there are four passive joints and one actuated joint connected to the motor via a double-clutching mechanism allowing us to actuate one of two links adjacent to the motor axis (Fig. 4.11b). The latter determines the main particularity of the NaVARo robot which has eight actuation modes that are switched from one to another in order to avoid kinematic singularities.

To show the application of the MSA technique step-by-step, let us split the manipulator mechanism into four parts: three kinematically identical legs and a mobile platform. For convenience, let us divide the longest link of the leg into two rigidly connected parts. This allows us to present each leg (Fig. 4.12) as a set of five flexible links (1,2), (3,4), (5,6), (7,8), (9,e) with passive connections of the nodes <2,3>, <4,5>, <6,7> and rigid connection of the nodes <6,9>. Besides, depending on the actuation mode, the boundary conditions induced by the motor/clutch mechanics can be specified as the passive connection of the nodes <0,8> and elastic connection of the nodes <0,1>, or vice versa.

The stiffness parameters of the NAVARO manipulator links were estimated using the CAD-based technique proposed in [7] and will be described in detail in Chap. 8. To find them, the virtual forces/torques were applied to all links separately and

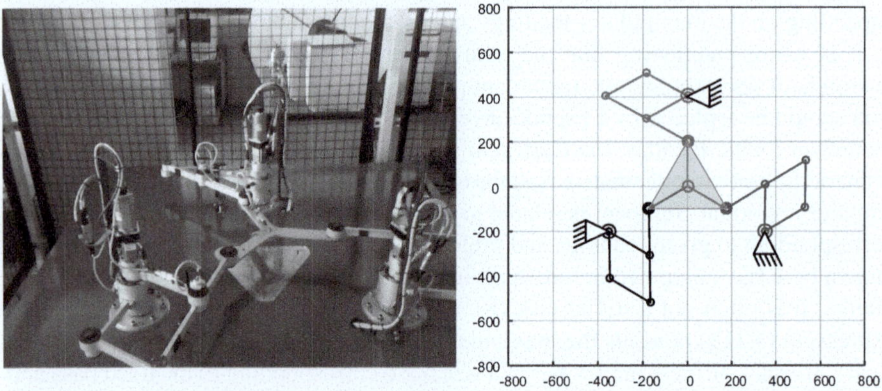

(a) robot NaVaRo (b) geometry of NaVaRo robot

Fig. 4.11 Parallel robot NaVaRo and its geometry [5]

Fig. 4.12 MSA-based representation of the NaVaRo leg (actuation of the link (1,2)) [5]

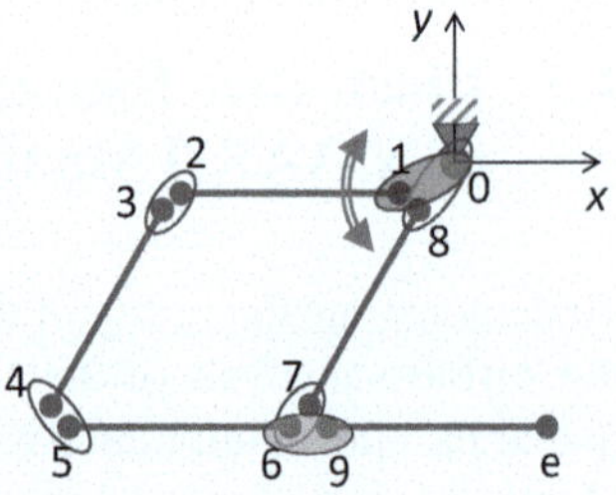

corresponding deflections were obtained via the FEA modelling. These data were used for the identification of the link compliance coefficients describing the relation between the external force/torque and the linear/angular deflections. The obtained values of the compliance matrices are given in Table 4.2, where the $\mathbf{C}_{22}^{(ij)}$ corresponds to the inverse of the stiffness matrix $\mathbf{K}_{22}^{(ij)}$ in Eq. (4.54). The stiffness coefficient of the actuator was estimated as 10^4Nm/rad.

Using the notation adopted in this paper, the "free-free" stiffness models of the links (4.54) can be presented in the form

$$\begin{bmatrix} \mathbf{W}_i \\ \mathbf{W}_j \end{bmatrix} = \begin{bmatrix} \mathbf{K}_{11}^{(ij)} & \mathbf{K}_{12}^{(ij)} \\ \mathbf{K}_{21}^{(ij)} & \mathbf{K}_{22}^{(ij)} \end{bmatrix}_{12\times 12} \cdot \begin{bmatrix} \Delta\mathbf{t}_i \\ \Delta\mathbf{t}_j \end{bmatrix} \tag{4.121}$$
$$(i,j) \in \{(1,2),(3,4),(5,6),(7,8),(9,e)\}$$

where all matrices are assumed to be already presented in the global coordinate system (here and below, the left superscript 'g' denoting the global system is omitted to improve readability). After assembling (4.121) into a single matrix, one can get the following aggregated stiffness matrix

Table 4.2 Compliance matrices of the NAVARO manipulator links [5]

Link	Compliance matrix $\mathbf{C}_{22}^{(ij)}$
Link (1,2) 	$\begin{bmatrix} 1.16\cdot10^{-8} & 0 & 0 & 0 & 0 & 0 \\ 0 & 9.21\cdot10^{-6} & 0 & 0 & 0 & 8.66\cdot10^{-5} \\ 0 & 0 & 2.32\cdot10^{-6} & 0 & -1.90\cdot10^{-5} & 0 \\ 0 & 0 & 0 & 8.67\cdot10^{-4} & 0 & 0 \\ 0 & 0 & -1.90\cdot10^{-5} & 0 & 2.00\cdot10^{-4} & 0 \\ 0 & 8.66\cdot10^{-5} & 0 & 0 & 0 & 9.90\cdot10^{-4} \end{bmatrix}$
Link (3.4) 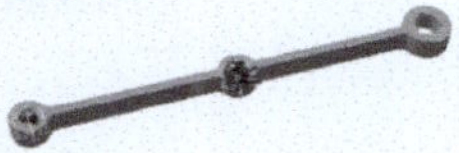	$\begin{bmatrix} 1.10\cdot10^{-8} & 0 & 0 & 0 & 0 & 0 \\ 0 & 1.68\cdot10^{-5} & 0 & 0 & 0 & 1.30\cdot10^{-4} \\ 0 & 0 & 3.20\cdot10^{-6} & 0 & -2.40\cdot10^{-5} & 0 \\ 0 & 0 & 0 & 1.10\cdot10^{-3} & 0 & 0 \\ 0 & 0 & -2.40\cdot10^{-5} & 0 & 2.28\cdot10^{-4} & 0 \\ 0 & 1.30\cdot10^{-4} & 0 & 0 & 0 & 1.24\cdot10^{-3} \end{bmatrix}$
Link (5,6) 	$\begin{bmatrix} 1.00\cdot10^{-8} & 0 & 0 & 0 & 0 & 0 \\ 0 & 1.78\cdot10^{-5} & 0 & 0 & 0 & 1.18\cdot10^{-4} \\ 0 & 0 & 2.85\cdot10^{-6} & 0 & -2.20\cdot10^{-5} & 0 \\ 0 & 0 & 0 & 9.93\cdot10^{-4} & 0 & 0 \\ 0 & 0 & -2.20\cdot10^{-5} & 0 & 2.10\cdot10^{-4} & 0 \\ 0 & 1.18\cdot10^{-4} & 0 & 0 & 0 & 1.13\cdot10^{-3} \end{bmatrix}$

(continued)

Table 4.2 (continued)

Link	Compliance matrix $\mathbf{C}_{22}^{(ij)}$
Link (7,8) 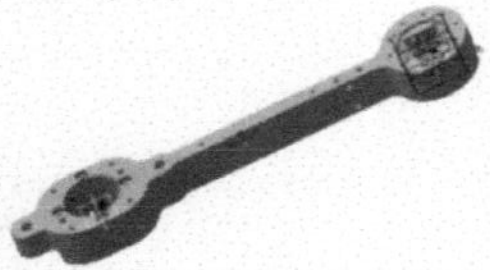	$\begin{bmatrix} 1.01\cdot10^{-8} & 0 & 0 & 0 & 0 & 0 \\ 0 & 1.09\cdot10^{-5} & 0 & 0 & 0 & 9.61\cdot10^{-5} \\ 0 & 0 & 2.65\cdot10^{-6} & 0 & -2.12\cdot10^{-5} & 0 \\ 0 & 0 & 0 & 8.99\cdot10^{-4} & 0 & 0 \\ 0 & 0 & -2.12\cdot10^{-5} & 0 & 2.10\cdot10^{-4} & 0 \\ 0 & 9.61\cdot10^{-5} & 0 & 0 & 0 & 1.01\cdot10^{-3} \end{bmatrix}$
Link (9,*e*) 	$\begin{bmatrix} 1.05\cdot10^{-8} & 0 & 0 & 0 & 0 & 0 \\ 0 & 1.68\cdot10^{-5} & 0 & 0 & 0 & 1.29\cdot10^{-4} \\ 0 & 0 & 3.21\cdot10^{-6} & 0 & -2.39\cdot10^{-5} & 0 \\ 0 & 0 & 0 & 1.05\cdot10^{-3} & 0 & 0 \\ 0 & 0 & -2.39\cdot10^{-5} & 0 & 2.20\cdot10^{-4} & 0 \\ 0 & 1.29\cdot10^{-4} & 0 & 0 & 0 & 1.18\cdot10^{-3} \end{bmatrix}$
Link (*i*,*e*) 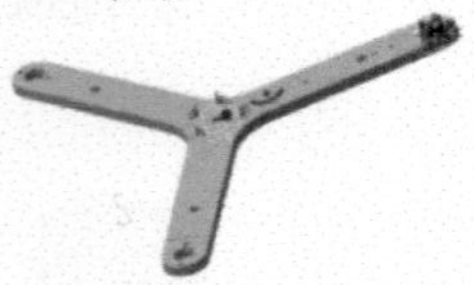	$\begin{bmatrix} 7.66\cdot10^{-9} & 0 & 0 & 0 & 0 & 0 \\ 0 & 1.36\cdot10^{-6} & 0 & 0 & 0 & 8.42\cdot10^{-6} \\ 0 & 0 & 1.12\cdot10^{-5} & 0 & -8.48\cdot10^{-5} & 0 \\ 0 & 0 & 0 & 6.58\cdot10^{-4} & 0 & 0 \\ 0 & 0 & -8.48\cdot10^{-5} & 0 & 8.51\cdot10^{-4} & 0 \\ 0 & 8.42\cdot10^{-6} & 0 & 0 & 0 & 7.09\cdot10^{-5} \end{bmatrix}$

$$\mathbf{K}_{links} = \begin{bmatrix} \begin{matrix}\mathbf{K}_{11}^{(12)} & \mathbf{K}_{12}^{(12)} \\ \mathbf{K}_{21}^{(12)} & \mathbf{K}_{22}^{(12)}\end{matrix} & & & & \\ & \begin{matrix}\mathbf{K}_{11}^{(34)} & \mathbf{K}_{12}^{(34)} \\ \mathbf{K}_{21}^{(34)} & \mathbf{K}_{22}^{(34)}\end{matrix} & & & \\ & & \begin{matrix}\mathbf{K}_{11}^{(56)} & \mathbf{K}_{12}^{(56)} \\ \mathbf{K}_{21}^{(56)} & \mathbf{K}_{22}^{(56)}\end{matrix} & & \\ & & & \begin{matrix}\mathbf{K}_{11}^{(78)} & \mathbf{K}_{12}^{(78)} \\ \mathbf{K}_{21}^{(78)} & \mathbf{K}_{22}^{(78)}\end{matrix} & \\ & & & & \begin{matrix}\mathbf{K}_{11}^{(9e)} & \mathbf{K}_{12}^{(9e)} \\ \mathbf{K}_{21}^{(9e)} & \mathbf{K}_{22}^{(9e)}\end{matrix} \end{bmatrix}_{60\times 60} \quad (4.122)$$

allowing to present the first component of the manipulator leg stiffness model in the following form

$$\left[-\mathbf{I}_{60\times 60}\ \mathbf{K}_{links}\right]_{60\times 120} \cdot \begin{bmatrix}\mathbf{W}_{agr} \\ \Delta\mathbf{t}_{agr}\end{bmatrix}_{120\times 1} = \mathbf{0}_{60\times 1} \quad (4.123)$$

where $\Delta\mathbf{t}_{agr}$ and $\mathbf{W}_{agr}$ are the aggregated displacement and wrenches defined as

$$\Delta\mathbf{t}_{agr} = col(\Delta\mathbf{t}_1, ..., \Delta\mathbf{t}_9, \Delta\mathbf{t}_e);\ \mathbf{W}_{agr} = col(\mathbf{W}_1, ..., \mathbf{W}_9, \mathbf{W}_e) \quad (4.124)$$

Further, the stiffness model of the leg should be complemented by the second component that includes an equation issued from the inter-link connections. For the considered manipulator, the links are connected by means of the rigid joint <6,9> , passive joints <2,3>, <4,5>, <6,7> , and elastic revolute joint <0,1>. The passive and elastic joints provide rotation around the z-axis, which corresponds to the constraint matrices of the following form

$$\mathbf{\Lambda}_*^r = \begin{bmatrix} 1\,0\,0\,0\,0\,0 \\ 0\,1\,0\,0\,0\,0 \\ 0\,0\,1\,0\,0\,0 \\ 0\,0\,0\,1\,0\,0 \\ 0\,0\,0\,0\,1\,0 \end{bmatrix};\ \mathbf{\Lambda}_*^p = \mathbf{\Lambda}_*^e = \left[0\,0\,0\,0\,0\,1\right] \quad (4.125)$$

Using this notation, the constraints imposed by the passive joints <2,3>, <4,5> can be presented as

$$
\begin{gathered}
\left[\mathbf{\Lambda}_*^r \; -\mathbf{\Lambda}_*^r\right] \cdot \begin{bmatrix} \Delta\mathbf{t}_i \\ \Delta\mathbf{t}_j \end{bmatrix} = \mathbf{0}_{5\times1}; \qquad \left[\mathbf{\Lambda}_*^r \; \mathbf{\Lambda}_*^r\right] \cdot \begin{bmatrix} \mathbf{W}_i \\ \mathbf{W}_j \end{bmatrix} = \mathbf{0}_{5\times1}; \\
\begin{bmatrix} \mathbf{\Lambda}^p & \mathbf{0}_{1\times6} \\ \mathbf{0}_{1\times6} & \mathbf{\Lambda}^p \end{bmatrix} \cdot \begin{bmatrix} \mathbf{W}_i \\ \mathbf{W}_j \end{bmatrix} = \mathbf{0}_{2\times1} \\
(i,j) \in \{(2,3),(4,5)\}
\end{gathered} \tag{4.126}
$$

The rigid joint constraint <6,9> , which ensures a stiff connection of the links (5,6) and (9,e), should be combined with the passive joint constraint <6,7> that yields the following equations

$$
\begin{gathered}
\left[\mathbf{\Lambda}_*^r \; -\mathbf{\Lambda}_*^r\right] \cdot \begin{bmatrix} \Delta\mathbf{t}_6 \\ \Delta\mathbf{t}_7 \end{bmatrix} = \mathbf{0}_{5\times1}; \qquad \left[\mathbf{I}_{6\times6} \; -\mathbf{I}_{6\times6}\right] \cdot \begin{bmatrix} \Delta\mathbf{t}_6 \\ \Delta\mathbf{t}_9 \end{bmatrix} = \mathbf{0}_{6\times1} \\
\left[\mathbf{\Lambda}_*^r \; \mathbf{\Lambda}_*^r \; \mathbf{\Lambda}_*^r\right] \cdot \begin{bmatrix} \mathbf{W}_6 \\ \mathbf{W}_7 \\ \mathbf{W}_9 \end{bmatrix} = \mathbf{0}_{5\times1}; \qquad \left[\mathbf{\Lambda}_*^p \; \mathbf{\Lambda}_*^p\right] \cdot \begin{bmatrix} \mathbf{W}_6 \\ \mathbf{W}_9 \end{bmatrix} = \mathbf{0}_{1\times1} \\
\mathbf{\Lambda}^p \cdot \mathbf{W}_7 = \mathbf{0}_{1\times1}
\end{gathered} \tag{4.127}
$$

The third model component includes the boundary conditions. For the adopted actuation mode when the motor is connected to the link (1,2) via an elastic transmission and the link (7,8) is passively connected to the base, the boundary condition at the node <0,1,8> is presented in the form

$$
\begin{gathered}
\mathbf{\Lambda}_*^r \Delta\mathbf{t}_1 = \mathbf{0}_{5\times1}; \qquad \mathbf{\Lambda}_*^r \Delta\mathbf{t}_8 = \mathbf{0}_{5\times1} \\
\mathbf{\Lambda}^p \cdot \mathbf{W}_8 = \mathbf{0}_{1\times1}; \qquad K_e \cdot \mathbf{\Lambda}_*^e \cdot \Delta\mathbf{t}_1 - \mathbf{\Lambda}_*^e \cdot \mathbf{W}_1 = 0
\end{gathered} \tag{4.128}
$$

where $\Delta\mathbf{t}_0$ is assumed to be equal to zero.

The external loading can be introduced into the system using the following equation

$$
\mathbf{W}_e = \mathbf{W}_{ext} \tag{4.129}
$$

After assembling all equations describing the constraints and boundary conditions the aggregated stiffness model of a single manipulator leg is presented as follows

$$
\begin{bmatrix} -\mathbf{I}_{60\times60} & \mathbf{K}_{links} \\ \mathbf{0}_{31\times60} & \mathbf{A}_{agr} \\ \mathbf{B}_{agr} & \mathbf{0}_{22\times60} \\ \mathbf{C}_{agr} & \mathbf{D}_{agr} \\ \mathbf{E}_{agr} & \mathbf{0}_{6\times60} \end{bmatrix}_{120\times120} \cdot \begin{bmatrix} \mathbf{W}_{agr} \\ \Delta\mathbf{t}_{agr} \end{bmatrix}_{120\times1} = \begin{bmatrix} \mathbf{0}_{114\times1} \\ \mathbf{W}_e \end{bmatrix}_{120\times1} \tag{4.130}
$$

where

$$\mathbf{A}_{agr} = \left[\begin{array}{cc|cc|cc|cc|cc} 0 & \mathbf{\Lambda}_*^r & -\mathbf{\Lambda}_*^r & 0 & 0 & 0 & 0 & 0 & 0 & 0 \\ \hline 0 & 0 & 0 & \mathbf{\Lambda}_*^r & -\mathbf{\Lambda}_*^r & 0 & 0 & 0 & 0 & 0 \\ \hline 0 & 0 & 0 & 0 & 0 & \mathbf{\Lambda}_*^r & -\mathbf{\Lambda}_*^r & 0 & 0 & 0 \\ 0 & 0 & 0 & 0 & 0 & \mathbf{I} & 0 & 0 & -\mathbf{I} & 0 \\ 0 & 0 & 0 & 0 & 0 & 0 & 0 & \mathbf{\Lambda}_*^r & 0 & 0 \\ \hline \mathbf{\Lambda}_*^r & 0 & 0 & 0 & 0 & 0 & 0 & 0 & 0 & 0 \end{array}\right]_{31\times 60} \tag{4.131}$$

$$\mathbf{B}_{agr} = \left[\begin{array}{cc|cc|cc|cc|cc} 0 & \mathbf{\Lambda}_*^r & \mathbf{\Lambda}_*^r & 0 & 0 & 0 & 0 & 0 & 0 & 0 \\ 0 & \mathbf{\Lambda}_*^p & 0 & 0 & 0 & 0 & 0 & 0 & 0 & 0 \\ 0 & 0 & \mathbf{\Lambda}_*^p & 0 & 0 & 0 & 0 & 0 & 0 & 0 \\ \hline 0 & 0 & 0 & \mathbf{\Lambda}_*^r & \mathbf{\Lambda}_*^r & 0 & 0 & 0 & 0 & 0 \\ 0 & 0 & 0 & \mathbf{\Lambda}_*^p & 0 & 0 & 0 & 0 & 0 & 0 \\ 0 & 0 & 0 & 0 & \mathbf{\Lambda}_*^p & 0 & 0 & 0 & 0 & 0 \\ \hline 0 & 0 & 0 & 0 & 0 & \mathbf{\Lambda}_*^r & \mathbf{\Lambda}_*^r & 0 & \mathbf{\Lambda}_*^r & 0 \\ 0 & 0 & 0 & 0 & 0 & \mathbf{\Lambda}_*^p & 0 & 0 & \mathbf{\Lambda}_*^p & 0 \\ 0 & 0 & 0 & 0 & 0 & 0 & \mathbf{\Lambda}_*^p & 0 & 0 & 0 \\ \hline 0 & 0 & 0 & 0 & 0 & 0 & 0 & \mathbf{\Lambda}_*^p & 0 & 0 \end{array}\right]_{22\times 60} \tag{4.132}$$

$$\mathbf{C}_{agr} = \left[-\mathbf{\Lambda}_*^e \; 0 \,|\, 0\; 0 \,|\, 0\; 0 \,|\, 0\; 0 \,|\, 0\; 0\right]_{1\times 60} \tag{4.133}$$

$$\mathbf{D}_{agr} = \left[K_e \mathbf{\Lambda}_*^e \; 0 \,|\, 0\; 0 \,|\, 0\; 0 \,|\, 0\; 0 \,|\, 0\; 0\right]_{1\times 60} \tag{4.134}$$

$$\mathbf{E}_{agr} = \left[0\; 0 \,|\, 0\; 0 \,|\, 0\; 0 \,|\, 0\; 0 \,|\, 0\; \mathbf{I}\right]_{6\times 60} \tag{4.135}$$

$$\mathbf{F}_{agr} = \left[0\; 0 \,|\, 0\; 0 \,|\, 0\; 0 \,|\, 0\; 0 \,|\, 0\; 0\right]_{6\times 60} \tag{4.136}$$

For computational convenience, this system can be also presented in the form (4.114), where

$$\mathbf{K}_{agr} = \begin{bmatrix} \mathbf{K}_{links} \\ \mathbf{A}_{agr} \\ \mathbf{0}_{22\times 60} \\ \mathbf{D}_{agr} \end{bmatrix}_{114\times 60} \quad ; \quad \mathbf{S}_{agr} = \begin{bmatrix} -\mathbf{I}_{60\times 60} \\ \mathbf{0}_{31\times 60} \\ \mathbf{B}_{agr} \\ \mathbf{C}_{agr} \end{bmatrix}_{114\times 60} \tag{4.137}$$

Further, after separating the node variables $\{\Delta\mathbf{t}_i\}$ in two groups corresponding to the internal nodes $\Delta\mathbf{t}_m$ and to the end effector node $\Delta\mathbf{t}_e$ one can get the linear systems (4.115) and (4.116) with $\mathbf{b} = \mathbf{0}_{114\times 1}$, which allows us to compute the desired stiffness matrices for all manipulator legs that will be further denoted as $\mathbf{K}_C^{(1)}$, $\mathbf{K}_C^{(2)}$, $\mathbf{K}_C^{(3)}$. It should be also noted that here the Cartesian stiffness matrices of the legs are rank-deficient (their ranks are equal to 5), i.e. the presented enhanced MSA technique is able to deal with under-constrained structures. Nevertheless, the assembling of three legs and the mobile platform in the NaVaRo robot provides the full-rank stiffness matrix (for non-singular configurations).

To find the Cartesian stiffness matrix of the entire manipulator, let us consider the moving platform as a mechanical structure composed of three flexible links, which are rigidly connected to the end-effector on the right-hand side and to the manipulator's legs via a passive joint on the left-hand side. The stiffness model of this platform is described by the Eq. (4.67), which should be presented as

$$\left[\begin{array}{c|cccc} & \mathbf{K}_{11}^{(1e)} & 0 & 0 & \mathbf{K}_{12}^{(1e)} \\ -\mathbf{I}_{18\times18} & 0 & \mathbf{K}_{11}^{(2e)} & 0 & \mathbf{K}_{12}^{(2e)} \\ & 0 & 0 & \mathbf{K}_{11}^{(3e)} & \mathbf{K}_{12}^{(3e)} \\ \hline \mathbf{0}_{6\times18} & \mathbf{K}_{21}^{(1e)} & \mathbf{K}_{21}^{(2e)} & \mathbf{K}_{21}^{(3e)} & \sum_{i=1}^{3}\mathbf{K}_{22}^{(ie)} \end{array}\right]_{24\times42} \cdot \begin{bmatrix} \mathbf{W}_{agr} \\ \Delta\mathbf{t}_{agr} \end{bmatrix}_{42\times1} = \begin{bmatrix} \mathbf{0}_{18\times1} \\ \mathbf{W}_e \end{bmatrix}_{24\times1} \tag{4.138}$$

where the nodes connected to the legs are denoted as <1>, <2>, <3>, the end-effector node is denoted as <e>, and $\mathbf{W}_{agr} = col(\mathbf{W}_1, \mathbf{W}_2, \mathbf{W}_3)$, $\Delta\mathbf{t}_{agr} = col(\Delta\mathbf{t}_1, \Delta\mathbf{t}_2, \Delta\mathbf{t}_3, \Delta\mathbf{t}_e)$. Relevant boundary conditions take into account connections to the flexible legs described by the stiffness matrices $\mathbf{K}_C^{(1)}$, $\mathbf{K}_C^{(2)}$, $\mathbf{K}_C^{(3)}$ via the passive joints with the matrices specified in (4.125). Using adopted notation, they are written as

$$\begin{gathered} \mathbf{\Lambda}_*^r \mathbf{W}_i = \mathbf{\Lambda}_*^r \mathbf{K}_C^{(i)} \Delta\mathbf{t}_i; \qquad \mathbf{\Lambda}_*^p \cdot \mathbf{W}_i = 0 \\ i \in \{1, 2, 3\} \end{gathered} \tag{4.139}$$

that can be presented in the extended matrix form as

$$\left[\begin{array}{ccc|cccc} \mathbf{\Lambda}_*^p & 0 & 0 & & & & \\ 0 & \mathbf{\Lambda}_*^p & 0 & & \mathbf{0}_{3\times24} & & \\ 0 & 0 & \mathbf{\Lambda}_*^p & & & & \\ \hline -\mathbf{\Lambda}_*^r & 0 & 0 & \mathbf{\Lambda}_*^r\mathbf{K}_C^{(1)} & 0 & 0 & 0 \\ 0 & -\mathbf{\Lambda}_*^r & 0 & 0 & \mathbf{\Lambda}_*^r\mathbf{K}_C^{(2)} & 0 & 0 \\ 0 & 0 & -\mathbf{\Lambda}_*^r & 0 & 0 & \mathbf{\Lambda}_*^r\mathbf{K}_C^{(3)} & 0 \end{array}\right]_{18\times42} \cdot \begin{bmatrix} \mathbf{W}_{agr} \\ \Delta\mathbf{t}_{agr} \end{bmatrix}_{42\times1} = \mathbf{0}_{18\times1} \tag{4.140}$$

After assembling Eqs. (4.138) and (4.140), the aggregated stiffness model of the entire manipulator is presented as follows

$$\begin{bmatrix} -\mathbf{I}_{18\times18} & \mathbf{K}_{platf} \\ \mathbf{B}_{agr} & \mathbf{0}_{3\times24} \\ \mathbf{C}_{agr} & \mathbf{D}_{agr} \\ \mathbf{0}_{6\times18} & \mathbf{F}_{agr} \end{bmatrix}_{42\times42} \cdot \begin{bmatrix} \mathbf{W}_{agr} \\ \Delta\mathbf{t}_{agr} \end{bmatrix}_{42\times1} = \begin{bmatrix} \mathbf{0}_{36\times1} \\ \mathbf{W}_e \end{bmatrix}_{42\times1} \tag{4.141}$$

where

$$\mathbf{K}_{platf} = \begin{bmatrix} \mathbf{K}_{11}^{(1e)} & 0 & 0 & \mathbf{K}_{12}^{(1e)} \\ 0 & \mathbf{K}_{11}^{(2e)} & 0 & \mathbf{K}_{12}^{(2e)} \\ 0 & 0 & \mathbf{K}_{11}^{(3e)} & \mathbf{K}_{12}^{(3e)} \end{bmatrix}_{18\times 24} \qquad \mathbf{B}_{agr} = \begin{bmatrix} \mathbf{\Lambda}_*^p & 0 & 0 \\ 0 & \mathbf{\Lambda}_*^p & 0 \\ 0 & 0 & \mathbf{\Lambda}_*^p \end{bmatrix}_{3\times 18} \tag{4.142}$$

$$\mathbf{C}_{agr} = \begin{bmatrix} -\mathbf{\Lambda}_*^r & 0 & 0 \\ 0 & -\mathbf{\Lambda}_*^r & 0 \\ 0 & 0 & -\mathbf{\Lambda}_*^r \end{bmatrix}_{15\times 18} \qquad \mathbf{D}_{agr} = \begin{bmatrix} \mathbf{\Lambda}_*^r\mathbf{K}_C^{(1)} & 0 & 0 & 0 \\ 0 & \mathbf{\Lambda}_*^r\mathbf{K}_C^{(2)} & 0 & 0 \\ 0 & 0 & \mathbf{\Lambda}_*^r\mathbf{K}_C^{(3)} & 0 \end{bmatrix}_{15\times 24} \tag{4.143}$$

$$\mathbf{F}_{agr} = \begin{bmatrix} \mathbf{K}_{21}^{(1e)} & \mathbf{K}_{21}^{(2e)} & \mathbf{K}_{21}^{(3e)} & \sum_{i=1}^{3} \mathbf{K}_{22}^{(ie)} \end{bmatrix}_{6\times 24} \tag{4.144}$$

For computational convenience, this system can be also converted into the form (4.114), where

$$\mathbf{K}_{agr} = \begin{bmatrix} \mathbf{K}_{platf} \\ \mathbf{0}_{3\times 24} \\ \mathbf{D}_{agr} \end{bmatrix}_{36\times 24} \quad ; \quad \mathbf{S}_{agr} = \begin{bmatrix} -\mathbf{I}_{18\times 18} \\ \mathbf{B}_{agr} \\ \mathbf{C}_{agr} \end{bmatrix}_{36\times 18} \tag{4.145}$$

Further, after separating the node variables $\{\Delta\mathbf{t}_i\}$ in two groups corresponding internal nodes $\Delta\mathbf{t}_m$ and to the end effector $\Delta\mathbf{t}_e$ one can get the linear systems (4.115) and (4.116) with $\mathbf{b} = \mathbf{0}_{36\times 1}$, which allows us to compute the desired stiffness matrix for the entire manipulator.

Numerical results related to this example are presented in Tables 4.1, 4.2, 4.3, 4.4, 4.5 and 4.6, which contains the stiffness matrices of the manipulator and its legs corresponding to 3 different locations of the moving platform (see Figs. 4.13, 4.14 and 4.15). As follows from the obtained results, the proposed generalization of the MSA allows the user to obtain both full-rank and rank-deficient Cartesian stiffness matrices for complex architectures containing numerous passive, rigid and elastic joints. In particular, the stiffness matrices of the individual legs are rank-deficient for all considered manipulator configurations. In contrast, the stiffness matrices of the entire manipulator are non-singular in all examined cases. Moreover, these matrices are diagonal for configurations 1 and 2, which are in good agreement with the manipulator posture symmetry in these locations. Hence, the presented example confirms the advantages of the developed techniques, which enlarge the MSA application area from conventional truss structures to complex robotic mechanisms.

In order to estimate stiffness matrices for the second actuation mode it is required to make some changes in the leg model matrices $\mathbf{B}_{agr}$, $\mathbf{C}_{agr}$, $\mathbf{D}_{agr}$. They are presented below, where the actuation mode denoted by the superscripts '(2)'

Table 4.3 Stiffness matrices of NAVARO manipulator for platform location #1 (x = 0, y = 0, φ = 0°) [5]

Leg #1

$$\mathbf{K}_C^{(1)} = \begin{bmatrix} 1.87 \cdot 10^5 & -1.12 \cdot 10^5 & 0 & 0 & 0 & 2.48 \cdot 10^4 \\ -1.12 \cdot 10^5 & 6.70 \cdot 10^4 & 0 & 0 & 0 & -1.48 \cdot 10^4 \\ 0 & 0 & 2.98 \cdot 10^5 & -4.32 \cdot 10^4 & 5.25 \cdot 10^3 & 0 \\ 0 & 0 & -4.32 \cdot 10^4 & 8.57 \cdot 10^3 & -5.65 \cdot 10^2 & 0 \\ 0 & 0 & 5.25 \cdot 10^3 & -5.65 \cdot 10^2 & 8.64 \cdot 10^2 & 0 \\ 2.48 \cdot 10^4 & -1.48 \cdot 10^4 & 0 & 0 & 0 & 3.90 \cdot 10^3 \end{bmatrix}$$

$rank(\mathbf{K}_C^{(1)}) = 5$

Leg #2

$$\mathbf{K}_C^{(2)} = \begin{bmatrix} 1.94 \cdot 10^5 & 1.08 \cdot 10^5 & 0 & 0 & 0 & -2.52 \cdot 10^4 \\ 1.08 \cdot 10^5 & 6.00 \cdot 10^4 & 0 & 0 & 0 & -1.40 \cdot 10^4 \\ 0 & 0 & 2.98 \cdot 10^5 & 2.61 \cdot 10^4 & 3.47 \cdot 10^4 & 0 \\ 0 & 0 & 2.61 \cdot 10^4 & 3.28 \cdot 10^3 & 3.62 \cdot 10^3 & 0 \\ 0 & 0 & 3.47 \cdot 10^4 & 3.62 \cdot 10^3 & 6.16 \cdot 10^3 & 0 \\ -2.52 \cdot 10^4 & -1.40 \cdot 10^4 & 0 & 0 & 0 & 3.90 \cdot 10^3 \end{bmatrix}$$

$rank(\mathbf{K}_C^{(2)}) = 5$

(continued)

Table 4.3 (continued)

Leg #3

$$\mathbf{K}_C^{(3)} = \begin{bmatrix} 6.51\cdot10^1 & 4.06\cdot10^3 & 0 & 0 & 0 & 4.62\cdot10^2 \\ 4.06\cdot10^3 & 2.54\cdot10^5 & 0 & 0 & 0 & 2.89\cdot10^4 \\ 0 & 0 & 2.98\cdot10^5 & 1.70\cdot10^4 & -4.00\cdot10^4 & 0 \\ 0 & 0 & 1.70\cdot10^4 & 2.30\cdot10^3 & -3.06\cdot10^3 & 0 \\ 0 & 0 & -4.00\cdot10^4 & -3.06\cdot10^3 & 7.13\cdot10^3 & 0 \\ 4.62\cdot10^2 & 2.89\cdot10^4 & 0 & 0 & 0 & 3.89\cdot10^3 \end{bmatrix}$$

$rank(\mathbf{K}_C^{(3)}) = 5$

Entire manipulator

$$\mathbf{K}_C = \begin{bmatrix} 3.75\cdot10^5 & \sim0 & 0 & 0 & 0 & \sim0 \\ \sim0 & 3.75\cdot10^4 & 0 & 0 & 0 & \sim0 \\ 0 & 0 & 2.92\cdot10^5 & \sim0 & \sim0 & 0 \\ 0 & 0 & \sim0 & 4.46\cdot10^3 & \sim0 & 0 \\ 0 & 0 & \sim0 & \sim0 & 4.46\cdot10^3 & 0 \\ \sim0 & \sim0 & 0 & 0 & 0 & 8.00\cdot10^3 \end{bmatrix}$$

$rank(\mathbf{K}_C) = 6$

$$\mathbf{B}_{agr}^{(2)} = \left[\begin{array}{cc|cc|cc|cc|cc} 0 & \mathbf{\Lambda}_*^r & \mathbf{\Lambda}_*^r & 0 & 0 & 0 & 0 & 0 & 0 & 0 \\ 0 & \mathbf{\Lambda}_*^p & 0 & 0 & 0 & 0 & 0 & 0 & 0 & 0 \\ 0 & 0 & \mathbf{\Lambda}_*^p & 0 & 0 & 0 & 0 & 0 & 0 & 0 \\ \hline 0 & 0 & 0 & \mathbf{\Lambda}_*^r & \mathbf{\Lambda}_*^r & 0 & 0 & 0 & 0 & 0 \\ 0 & 0 & 0 & \mathbf{\Lambda}_*^p & 0 & 0 & 0 & 0 & 0 & 0 \\ 0 & 0 & 0 & 0 & \mathbf{\Lambda}_*^p & 0 & 0 & 0 & 0 & 0 \\ \hline 0 & 0 & 0 & 0 & 0 & \mathbf{\Lambda}_*^r & \mathbf{\Lambda}_*^r & 0 & \mathbf{\Lambda}_*^r & 0 \\ 0 & 0 & 0 & 0 & 0 & \mathbf{\Lambda}_*^p & 0 & 0 & \mathbf{\Lambda}_*^p & 0 \\ 0 & 0 & 0 & 0 & 0 & 0 & \mathbf{\Lambda}_*^p & 0 & 0 & 0 \\ \hline \mathbf{\Lambda}_*^p & 0 & 0 & 0 & 0 & 0 & 0 & 0 & 0 & 0 \end{array}\right]_{22\times60} \tag{4.146}$$

$$\mathbf{C}_{agr}^{(2)} = \left[\,0\;0\,|\,0\;0\,|\,0\;0\,|\,0\;-\mathbf{\Lambda}_*^e\,|\,0\;0\,\right]_{1\times60} \tag{4.147}$$

Table 4.4 Stiffness matrices of NAVARO manipulator for platform location #2 (x = 0, y = 0, φ = 60°) [5]

Leg #1

$$\mathbf{K}_C^{(1)} = \begin{bmatrix} 1.51\cdot10^5 & -9.90\cdot10^4 & 0 & 0 & 0 & 2.23\cdot10^4 \\ -9.90\cdot10^4 & 6.48\cdot10^4 & 0 & 0 & 0 & -1.46\cdot10^4 \\ 0 & 0 & 2.22\cdot10^5 & -1.88\cdot10^4 & 3.35\cdot10^4 & 0 \\ 0 & 0 & -1.88\cdot10^4 & 2.64\cdot10^3 & -3.34\cdot10^3 & 0 \\ 0 & 0 & 3.35\cdot10^4 & -3.34\cdot10^3 & 6.79\cdot10^3 & 0 \\ 2.23\cdot10^4 & -1.46\cdot10^4 & 0 & 0 & 0 & 3.90\cdot10^3 \end{bmatrix}$$

$rank(\mathbf{K}_C^{(1)}) = 5$

Leg #2

$$\mathbf{K}_C^{(2)} = \begin{bmatrix} 1.72\cdot10^5 & 8.69\cdot10^4 & 0 & 0 & 0 & -2.38\cdot10^4 \\ 8.69\cdot10^4 & 4.39\cdot10^4 & 0 & 0 & 0 & -1.20\cdot10^4 \\ 0 & 0 & 2.22\cdot10^5 & 3.84\cdot10^4 & -4.80\cdot10^2 & 0 \\ 0 & 0 & 3.84\cdot10^4 & 8.65\cdot10^3 & -1.26\cdot10^2 & 0 \\ 0 & 0 & -4.80\cdot10^2 & -1.26\cdot10^2 & 7.77\cdot10^2 & 0 \\ -2.38\cdot10^4 & -1.20\cdot10^4 & 0 & 0 & 0 & 3.90\cdot10^3 \end{bmatrix}$$

$rank(\mathbf{K}_C^{(2)}) = 5$

Leg #3

$$\mathbf{K}_C^{(3)} = \begin{bmatrix} 6.80\cdot10^2 & 1.21\cdot10^4 & 0 & 0 & 0 & 1.50\cdot10^3 \\ 1.21\cdot10^4 & 2.15\cdot10^5 & 0 & 0 & 0 & 2.66\cdot10^4 \\ 0 & 0 & 2.22\cdot10^5 & -1.96\cdot10^4 & -3.30\cdot10^4 & 0 \\ 0 & 0 & -1.96\cdot10^4 & 2.85\cdot10^3 & 3.47\cdot10^3 & 0 \\ 0 & 0 & -3.30\cdot10^4 & 3.47\cdot10^3 & 6.57\cdot10^3 & 0 \\ 1.50\cdot10^3 & 2.66\cdot10^4 & 0 & 0 & 0 & 3.90\cdot10^3 \end{bmatrix}$$

$rank(\mathbf{K}_C^{(3)}) = 5$

(continued)

$$\mathbf{D}_{agr}^{(2)} = \left[0\,0 \,\middle|\, 0\,0 \,\middle|\, 0\,0 \,\middle|\, 0\; K_e \boldsymbol{\Lambda}_*^e \,\middle|\, 0\,0\right]_{1\times60} \tag{4.148}$$

To evaluate the manipulator resistance with respect to the external wrench applied to the end-effector, let us estimate the maximum compliance errors corresponding to the force/torque of the limited magnitude. It allows us to use the singular value decomposition of relevant blocks of the Cartesian compliance matrix $\mathbf{K}_C^{-1}$.

Table 4.4 (continued)

Entire manipulator

$$\mathbf{K}_c = \begin{bmatrix} 3.25\cdot10^5 & \sim0 & 0 & 0 & 0 & \sim0 \\ \sim0 & 3.25\cdot10^4 & 0 & 0 & 0 & \sim0 \\ 0 & 0 & 2.02\cdot10^5 & \sim0 & \sim0 & 0 \\ 0 & 0 & \sim0 & 4.02\cdot10^3 & \sim0 & 0 \\ 0 & 0 & \sim0 & \sim0 & 4.02\cdot10^3 & 0 \\ \sim0 & \sim0 & 0 & 0 & 0 & 4.78\cdot10^3 \end{bmatrix}$$

$rank(\mathbf{K}_C) = 6$

Computational results for different actuation modes are presented in Tables 4.4, 4.5, 4.6 and 4.7, which shows the distribution of the maximum translational [mm] and rotational [deg] deflections within the manipulator workspace. It was assumed that the end-effector is subject to external loadings of the magnitude 100 N and

Table 4.5 Stiffness matrices of NAVARO manipulator for platform location #3 ($x = 150$, $y = 150$, $\varphi = 60°$) [5]

Leg #1

$$\mathbf{K}_C^{(1)} = \begin{bmatrix} 1.06\cdot10^6 & 5.39\cdot10^5 & 0 & 0 & 0 & 5.81\cdot10^4 \\ 5.39\cdot10^5 & 2.74\cdot10^5 & 0 & 0 & 0 & 2.96\cdot10^4 \\ 0 & 0 & 2.24\cdot10^5 & -2.92\cdot10^4 & 2.78\cdot10^4 & 0 \\ 0 & 0 & -2.92\cdot10^4 & 5.43\cdot10^3 & -4.69\cdot10^3 & 0 \\ 0 & 0 & 2.78\cdot10^4 & -4.69\cdot10^3 & 5.18\cdot10^3 & 0 \\ 5.81\cdot10^4 & 2.96\cdot10^4 & 0 & 0 & 0 & 3.80\cdot10^3 \end{bmatrix}$$

$rank(\mathbf{K}_C^{(1)}) = 5$

Leg #2

$$\mathbf{K}_C^{(2)} = \begin{bmatrix} 6.85\cdot10^2 & -1.59\cdot10^4 & 0 & 0 & 0 & 1.50\cdot10^3 \\ -1.59\cdot10^4 & 3.69\cdot10^5 & 0 & 0 & 0 & -3.48\cdot10^4 \\ 0 & 0 & 3.58\cdot10^5 & -3.07\cdot10^4 & 3.73\cdot10^4 & 0 \\ 0 & 0 & -3.07\cdot10^4 & 4.61\cdot10^3 & -4.14\cdot10^3 & 0 \\ 0 & 0 & 3.73\cdot10^4 & -4.14\cdot10^3 & 5.28\cdot10^3 & 0 \\ 1.50\cdot10^3 & -3.48\cdot10^4 & 0 & 0 & 0 & 3.89\cdot10^3 \end{bmatrix}$$

$rank(\mathbf{K}_C^{(2)}) = 5$

(continued)

Table 4.5 (continued)

Leg #3

$$\mathbf{K}_C^{(3)} = \begin{bmatrix} 1.05\cdot10^5 & 9.88\cdot10^4 & 0 & 0 & 0 & 1.86\cdot10^4 \\ 9.88\cdot10^4 & 9.28\cdot10^4 & 0 & 0 & 0 & 1.75\cdot10^4 \\ 0 & 0 & 2.59\cdot10^5 & -2.50\cdot10^4 & -3.20\cdot10^4 & 0 \\ 0 & 0 & -2.50\cdot10^4 & 2.45\cdot10^3 & 3.60\cdot10^3 & 0 \\ 0 & 0 & -3.20\cdot10^4 & 3.60\cdot10^3 & 5.77\cdot10^3 & 0 \\ 1.86\cdot10^4 & 1.75\cdot10^4 & 0 & 0 & 0 & 3.90\cdot10^3 \end{bmatrix}$$

$rank(\mathbf{K}_C^{(3)}) = 5$

Entire manipulator

$$\mathbf{K}_C = \begin{bmatrix} 1.17\cdot10^6 & 6.25\cdot10^5 & 0 & 0 & 0 & -3.99\cdot10^4 \\ 6.25\cdot10^5 & 7.33\cdot10^5 & 0 & 0 & 0 & -3.53\cdot10^4 \\ 0 & 0 & 2.95\cdot10^5 & -4.30\cdot10^3 & 2.47\cdot10^3 & 0 \\ 0 & 0 & -4.30\cdot10^3 & 3.53\cdot10^3 & -4.45\cdot10^2 & 0 \\ 0 & 0 & 2.47\cdot10^3 & -4.45\cdot10^2 & 3.53\cdot10^3 & 0 \\ -3.99\cdot10^4 & -3.53\cdot10^4 & 0 & 0 & 0 & 2.13\cdot10^3 \end{bmatrix}$$

$rank(\mathbf{K}_C) = 6$

100 Nm respectively. As follows from the obtained results, the first actuation mode ensures better stiffness properties for almost all configurations (excluding those close to the singular ones). For instance, for the second actuation mode, the translational compliance error within the circle of 200mm diameter can reach 0.94 mm while for the first actuation mode, it is less than 0.71 mm.

Table 4.6 Translational compliance errors for different actuation of NaVaRo manipulator

<table>
<tr><th>Actuation mode</th><th>Translational errors
mm/100N</th></tr>
<tr><td>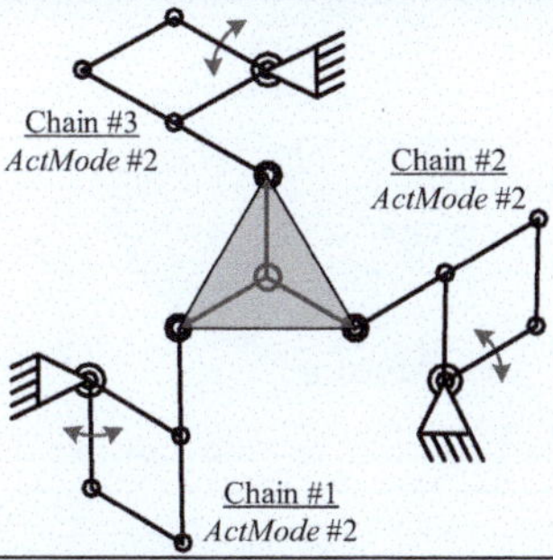</td><td>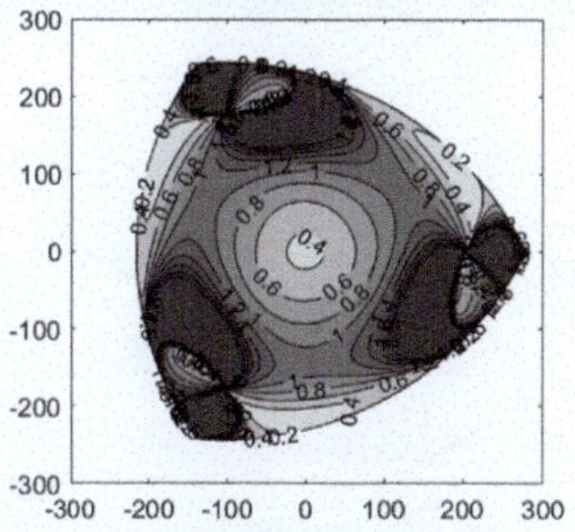</td></tr>
<tr><td>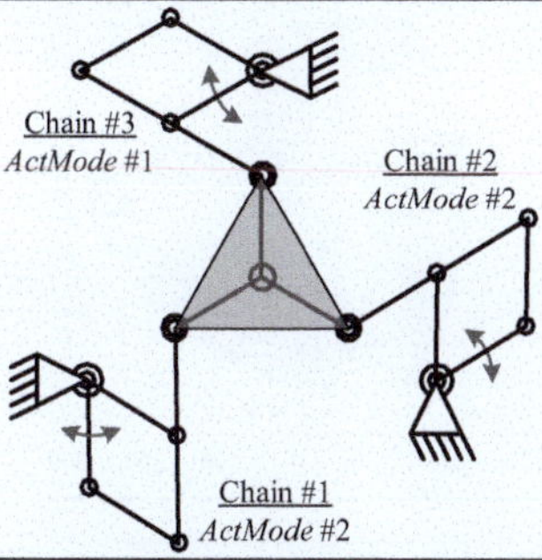</td><td>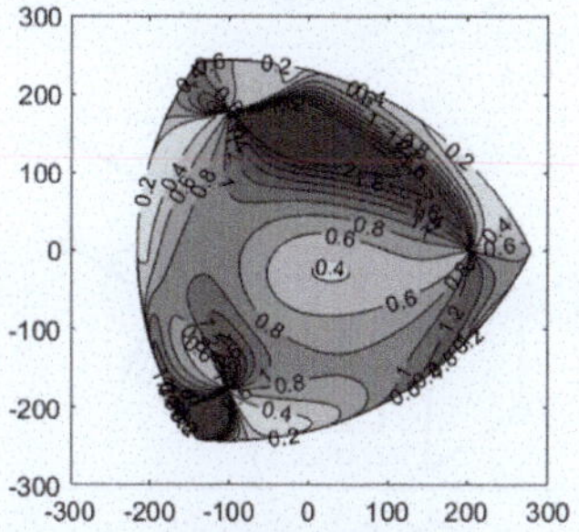</td></tr>
<tr><td>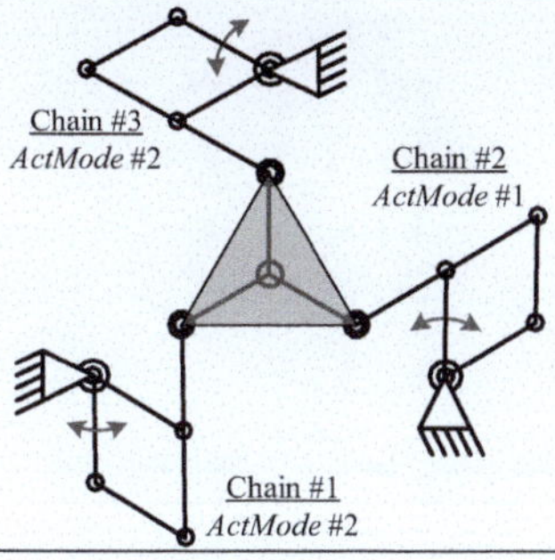</td><td>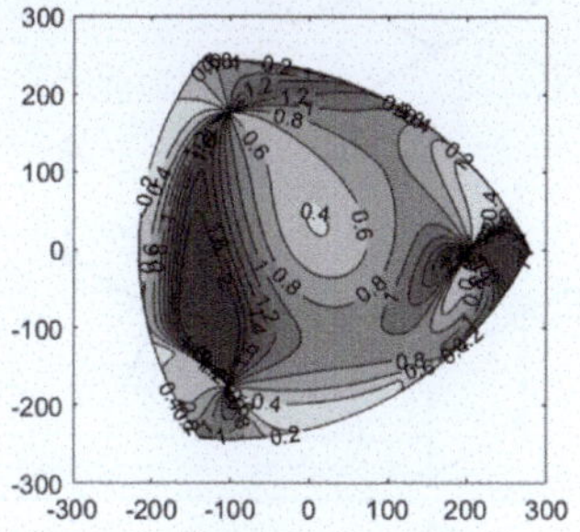</td></tr>
<tr><td>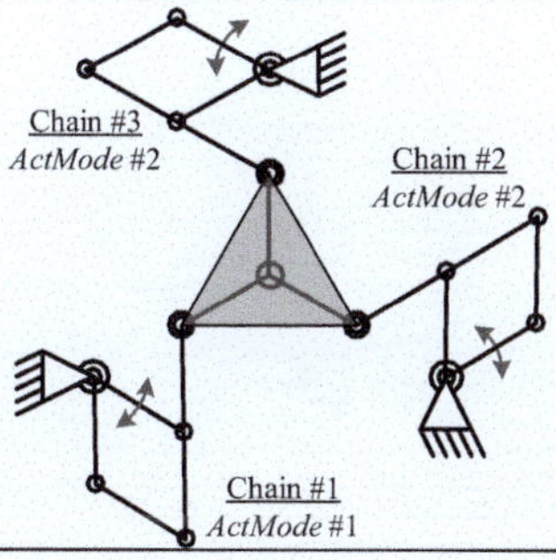</td><td>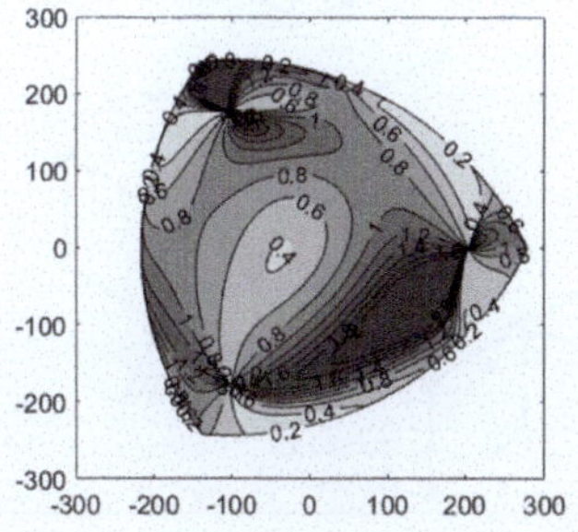</td></tr>
</table>

(continued)

Table 4.6 (continued)

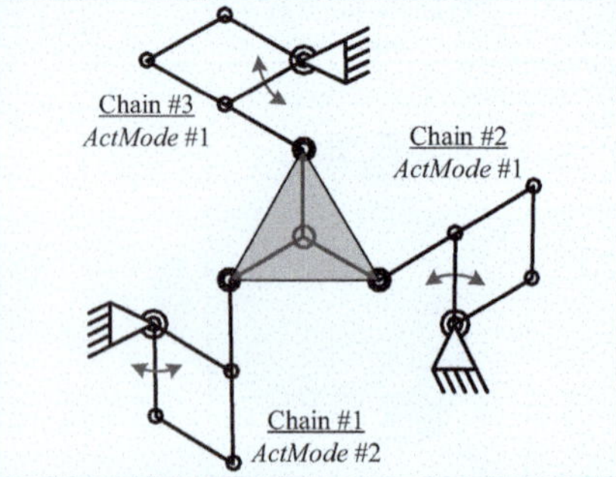

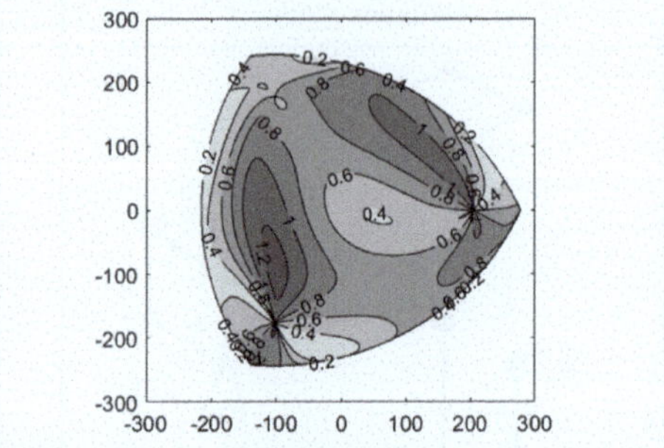

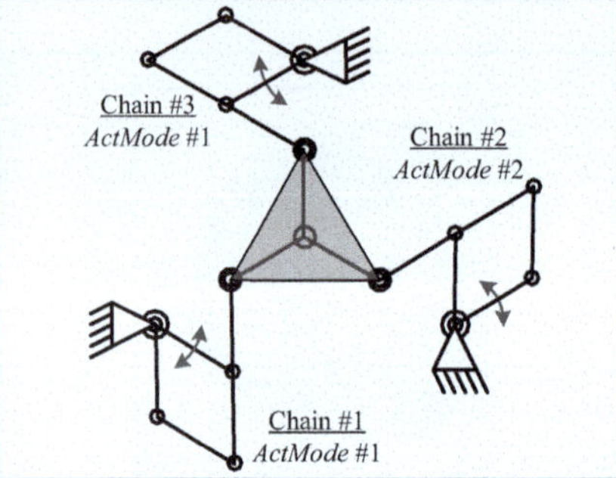

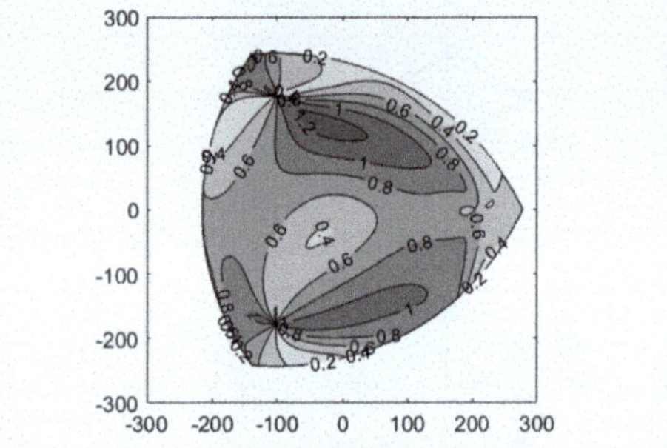

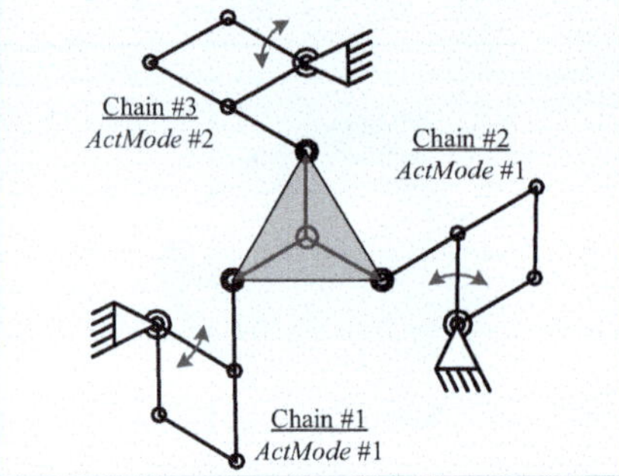

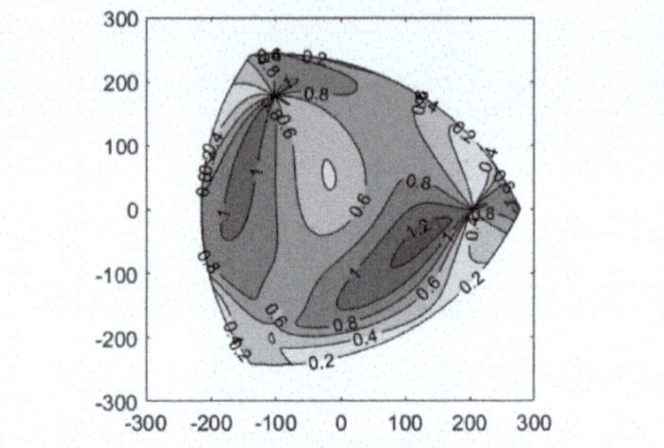

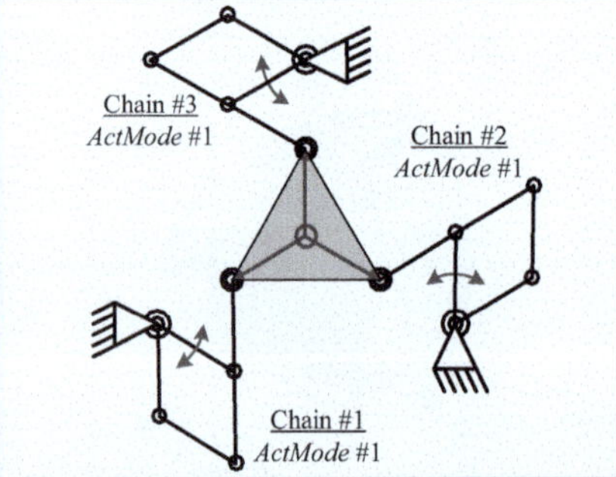

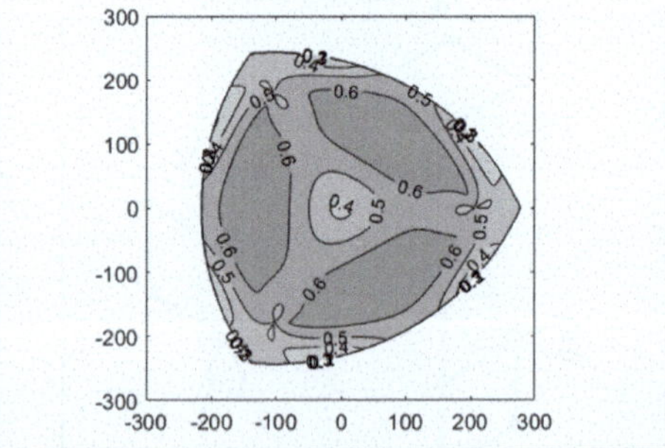

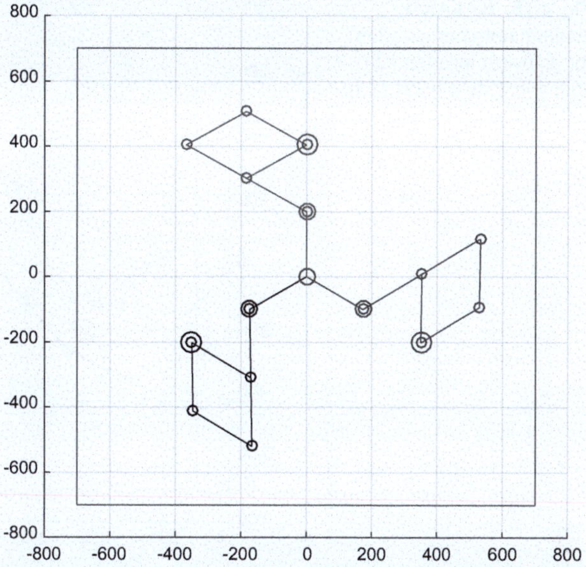

Fig. 4.13 NAVARO manipulator configuration for platform location #1 (x = 0, y = 0, φ = 0°) [5]

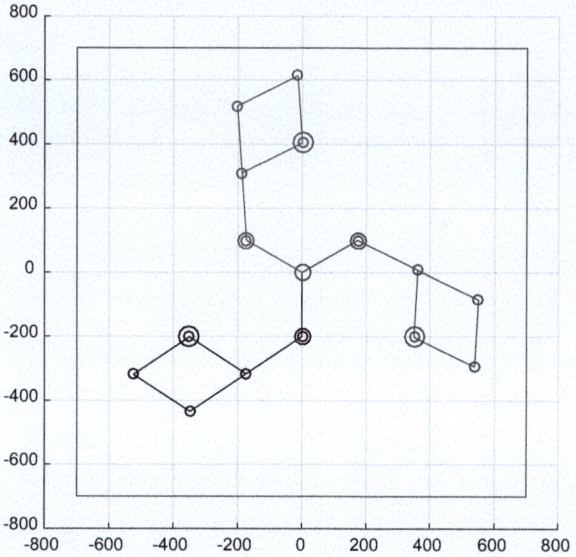

Fig. 4.14 NAVARO manipulator configuration for platform location #2 (x = 0, y = 0, φ = 60°) [5]

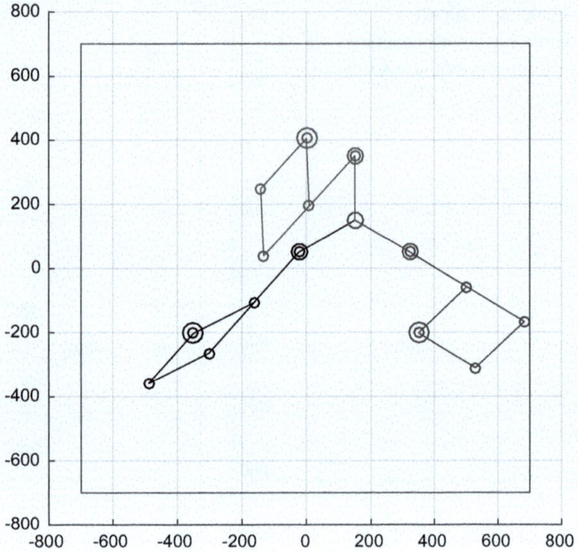

Fig. 4.15 NAVARO manipulator configuration for platform location #2 (x = 150, y = 150, φ = 60°) [5]

Table 4.7 Rotational compliance errors for different actuation of NaVaRo manipulator

Actuation mode	**Rotational errors deg/100Nm**
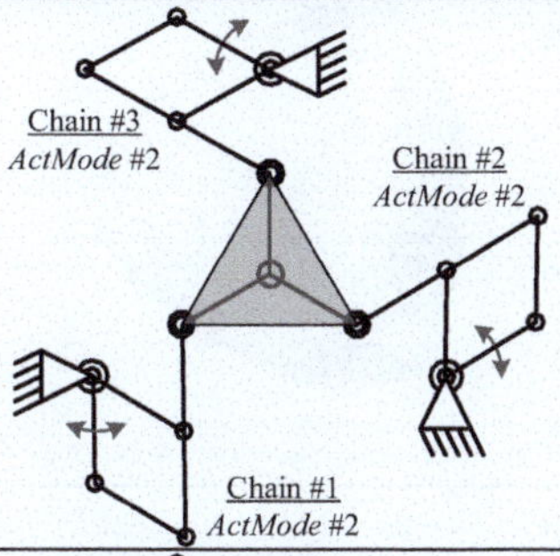	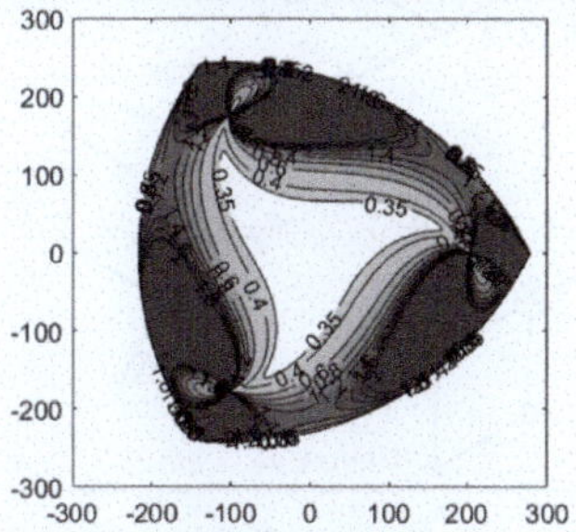
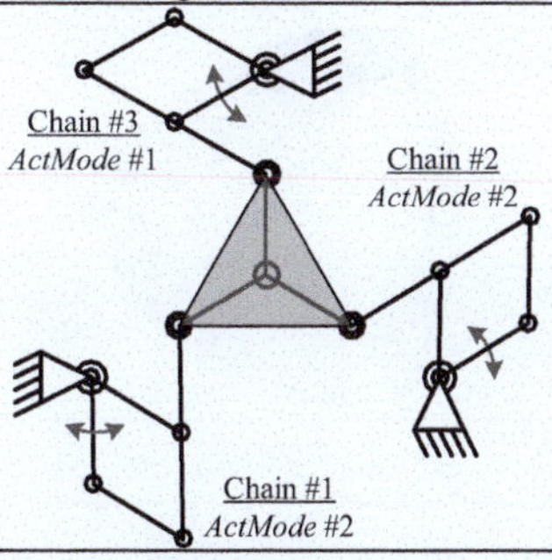	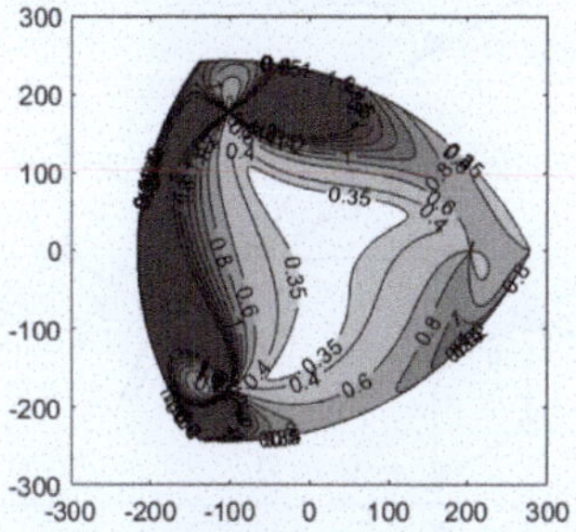
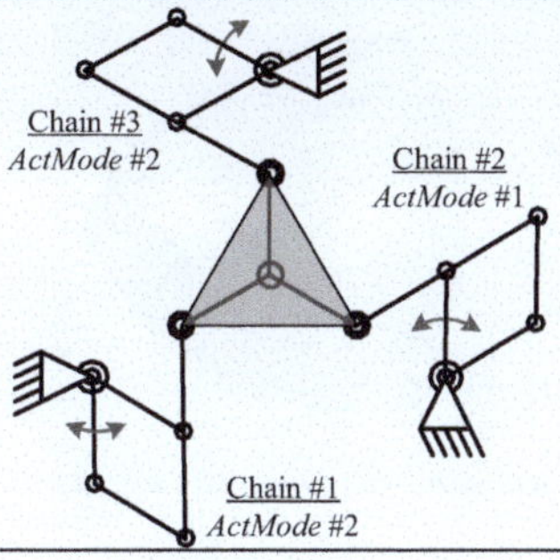	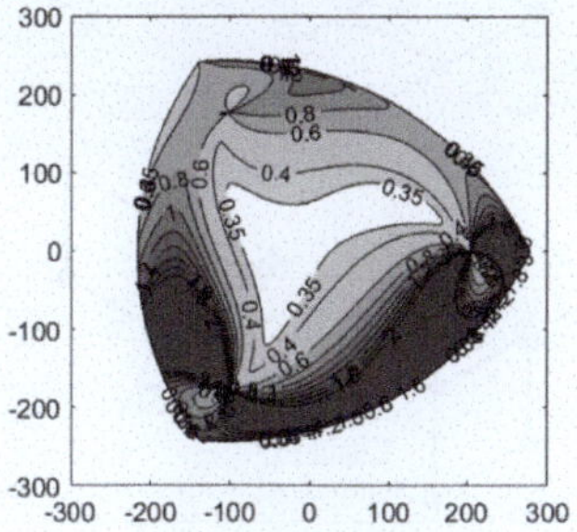
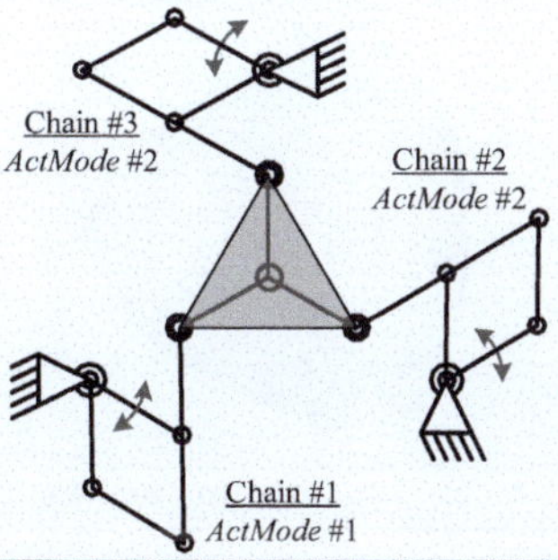	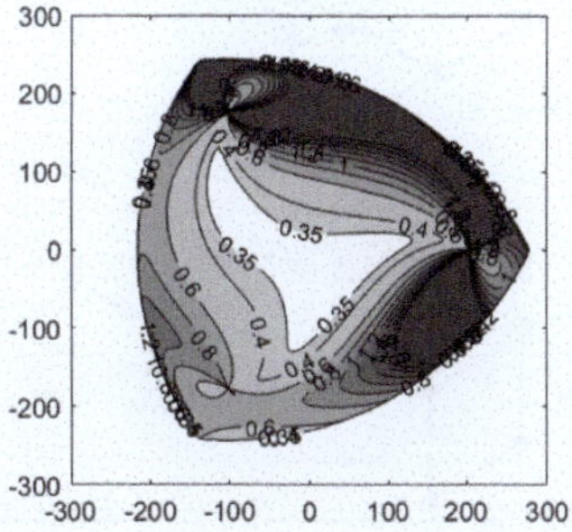

(continued)

Table 4.7 (continued)

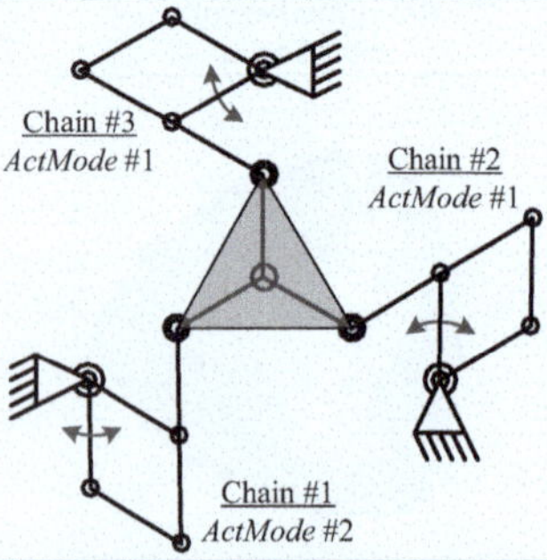

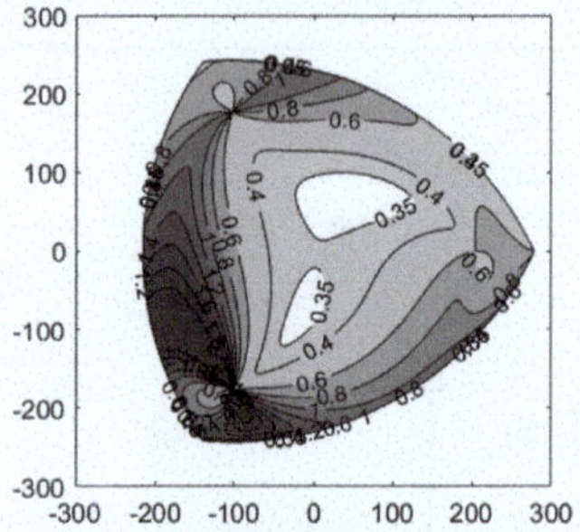

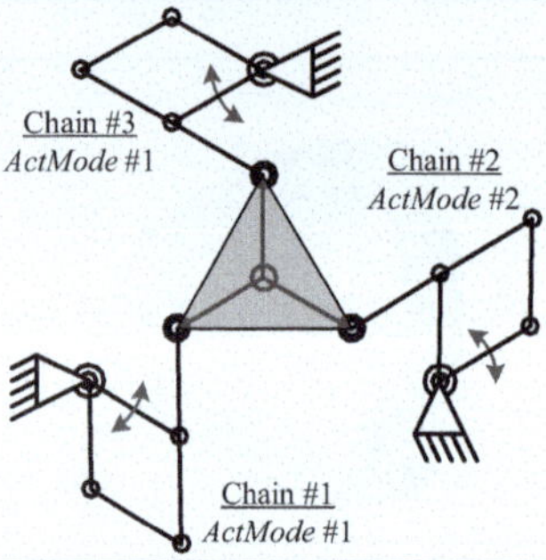

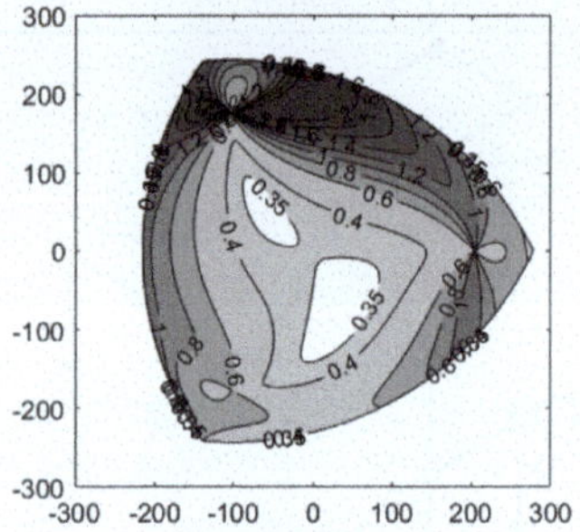

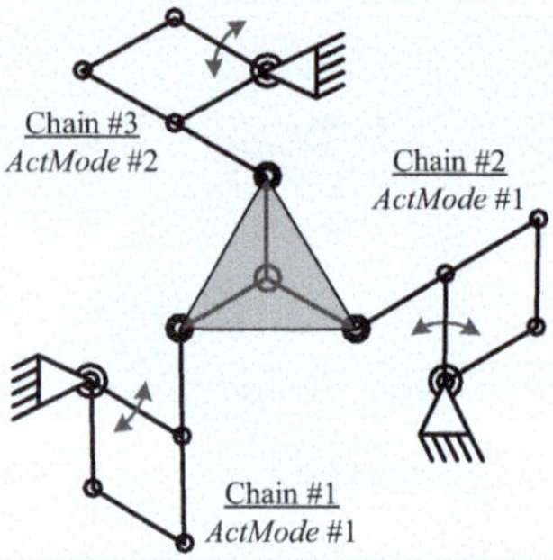

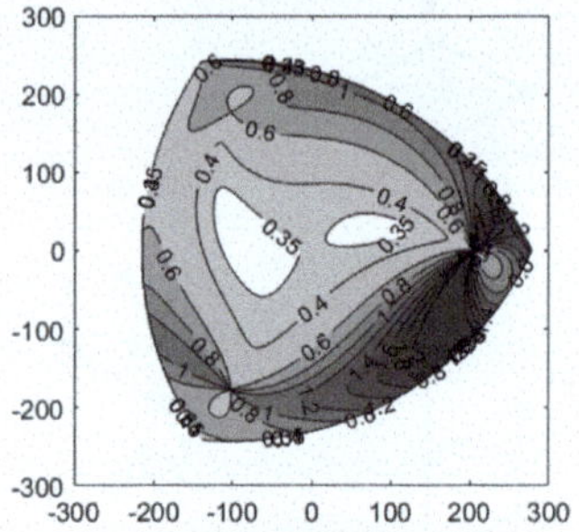

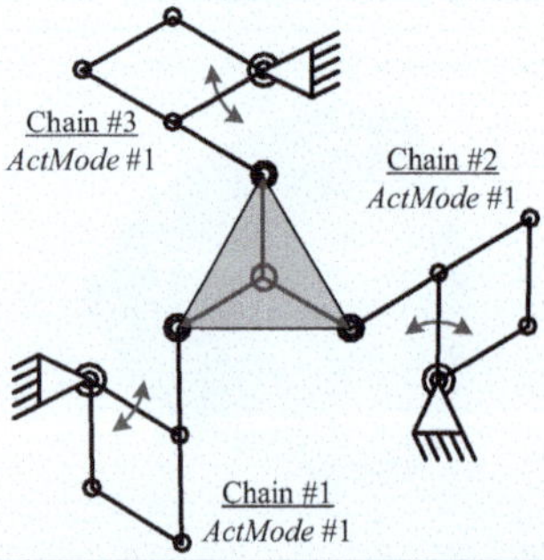

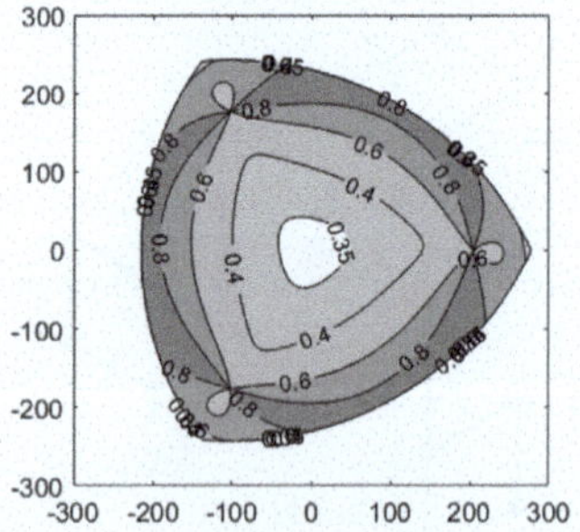

References

1. Marie S, Courteille E, Maurine P (2013) Elasto-geometrical modeling and calibration of robot manipulators: application to machining and forming applications. Mech Mach Theory 69:13–43
2. Nagai K, Liu Z A systematic approach to stiffness analysis of parallel mechanisms. In: Robotics and automation, 2008. ICRA 2008. IEEE International Conference on. IEEE, pp 1543–1548
3. Deblaise D, Hernot X, Maurine P A systematic analytical method for PKM stiffness matrix calculation. In: IEEE international conference on robotics and automation (ICRA 2006). IEEE, pp 4213–4219
4. Timoshenko S, Goodier JN (1970) Theory of elasticity. McGraw-Hill, Singapore
5. Klimchik A, Pashkevich A, Chablat D (2019) Fundamentals of manipulator stiffness modeling using matrix structural analysis. Mech Mach Theory 133:365–394
6. Deblaise D (2006) Contribution à la modélisation et à l'étalonnage élasto-géométriques des manipulateurs à structure parallèle. INSA de Rennes
7. Klimchik A, Pashkevich A, Chablat D (2013) CAD-based approach for identification of elasto-static parameters of robotic manipulators. Finite Elem Anal Design 75:19–30
8. Klimchik A, Chablat D, Pashkevich A (2014) Stiffness modeling for perfect and non-perfect parallel manipulators under internal and external loadings. Mech Mach Theory 79:1–28

Chapter 5
Virtual Joint Modeling: A Trade-Off Between the Complexity and Accuracy

5.1 VJM Method Background and Stiffness Model Parameters

5.1.1 VJM Method Background

In the frame of the VJM method, all types of compliance existing in a real manipulator (both distributed and lumped) are replaced by localized virtual springs located in its joints (Table 5.1). An example of VJM modelling for parallel manipulators is given in Fig. 5.1, where both the kinematic model of the Stewart Platform and its VJM model are presented. Then, for this mechanism consisting of rigid links and compliance joints, the static equilibrium equations are derived and linearized in order to obtain the Cartesian stiffness matrix, which in a general case depends on the manipulator configuration. Usually, it is assumed that the elastic deflections in virtual springs are relatively small and linearization is performed in the neighbourhood of the equilibrium configuration corresponding to zero forces and torques (unloaded mode).

To present basic ideas of the VJM technique, let us consider a general serial kinematic chain, which consists of a fixed "Base", a number of flexible actuated joints "Ac", a serial chain of flexible "Links", a number of passive joints "Ps" and a moving "Platform" at the end of the chain (Fig. 5.2). It is assumed that all links are separated by joints (actuated or passive, rotational or translational) and the joint type order is arbitrary. Besides, it is admitted that some links may be separated by actuated and passive joints simultaneously. Such sub-structures can be found in most of the parallel manipulators (Fig. 5.3) where several similar kinematic chains are connected to the same base and the platform in a different way (with the rotation of 90° or 120°, for instance), in order to eliminate the redundancy caused by the passive joints. It is obvious that such kinematic chains are statically *under-constrained* and

A. Klimchik et al., *Stiffness Modeling of Parallel Robots*, Parallel Robots: Theory and Applications, https://doi.org/10.1007/978-3-031-76320-5_5

Table 5.1 VJM-based modeling of manipulator components [1]

Manipulator components		VJM-model elements	
Mechanical component	Graphical presentation	VJM-model components	Graphical presentation
Actuated joint		Actuated joint and 6 dof. virtual spring	
Passive joint (non-actuated)		Passive joint	
Rigid link		Rigid link	
Elastic link		Rigid link and 6 dof. virtual spring	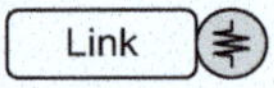
Rigid base	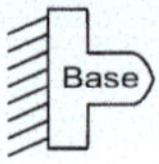	Rigid base	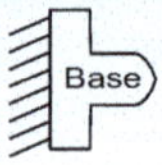
Elastic base	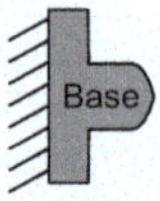	Rigid base and 6 dof. virtual spring	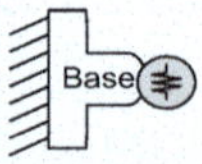
Rigid platform	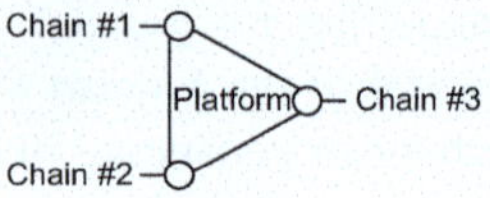	Rigid platform	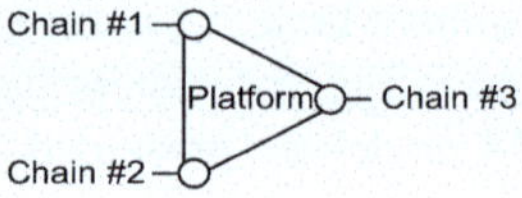
Elastic platform	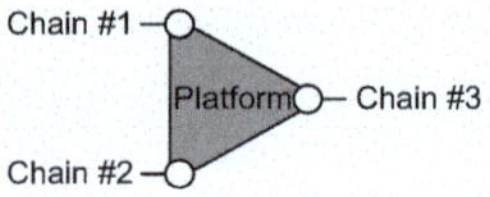	Rigid platform and three 6 dof virtual springs	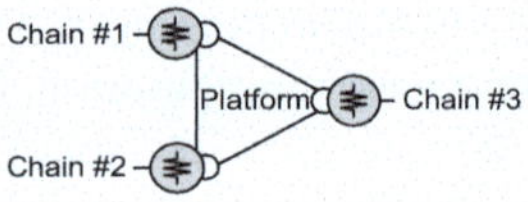

their stiffness analysis cannot be performed by the direct application of the standard methods.

In order to evaluate the stiffness of the considered serial chain, let us apply a modification of the virtual joint method (VJM), which is based on the lump modelling approach [4]. According to this approach, the original rigid model should be extended by adding virtual joints (localized springs), which describe elastic deformations of the links. Besides, virtual springs are included in the actuating joints, to take into account the stiffness of the control loop. Under such assumptions, the kinematic chain can be described by the following serial structure:

(a) a rigid link between the manipulator base and the first actuating joint described by the constant homogenous transformation matrix $\boldsymbol{T}_{Base}$;

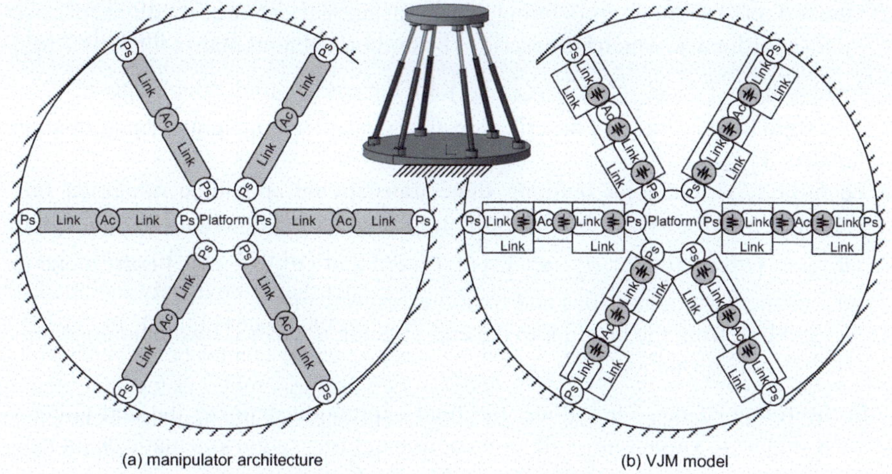

Fig. 5.1 Kinematic model of Stewart Platform and its VJM model [1]

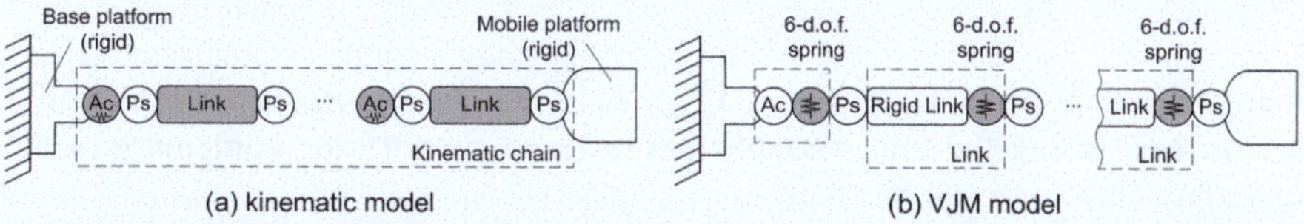

Fig. 5.2 General serial kinematic chain and its VJM model (Ac—actuated joint, Ps—passive joint) [2]

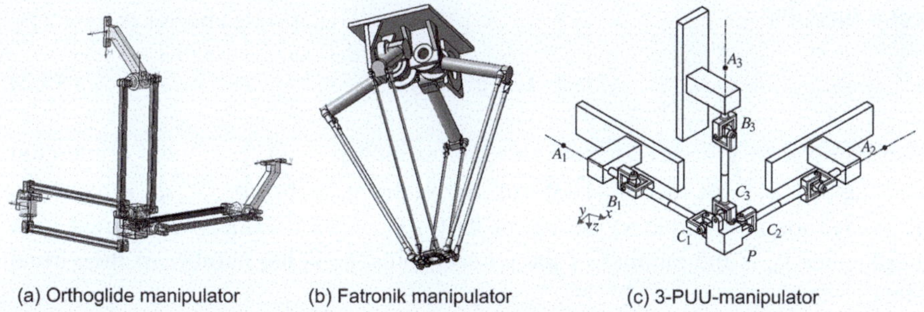

Fig. 5.3 Architecture of typical parallel manipulators and their kinematics chains [3]

(b) the 6-d.o.f. actuating joints defining three translational and three rotational actuator coordinates, which are described by the homogenous matrix function $\mathbf{T}\left(\boldsymbol{\theta}_{\mathbf{a}}^{\mathbf{i}}\right)$ where $\boldsymbol{\theta}_{\mathbf{a}}^{\mathbf{i}} = \left(\theta_x^{ai}, \theta_y^{ai}, \theta_z^{ai}, \theta_{\varphi x}^{ai}, \theta_{\varphi y}^{ai}, \theta_{\varphi z}^{ai}\right)$ are the virtual spring coordinates;

(c) the 6-d.o.f. passive joints defining three translational and three rotational passive joint coordinates, which are described by the homogenous matrix function $\boldsymbol{T}(\boldsymbol{q}_p^i)$ where $\mathbf{q}_p^i = \left(q_x^i, q_y^i, q_z^i, q_{\varphi x}^i, q_{\varphi y}^i, q_{\varphi z}^i\right)$ are the passive joint coordinates;
(d) the rigid links, which are described by the constant homogenous transformation matrix $\mathbf{T}_{\text{Link}}^i$;
(e) a 6-d.o.f. virtual joint defining three translational and three rotational link springs, which are described by the homogenous matrix function $\mathbf{T}(\boldsymbol{\theta}_{Link}^i)$, where $\boldsymbol{\theta}_{Link}^i = \left(\theta_x^i, \theta_y^i, \theta_z^i, \theta_{\varphi x}^i, \theta_{\varphi y}^i, \theta_{\varphi z}^i\right)$, $\left(\theta_x^i, \theta_y^i, \theta_z^i\right)$ and $\left(\theta_{\varphi x}^i, \theta_{\varphi y}^i, \theta_{\varphi z}^i\right)$ correspond to the elementary translations and rotations respectively;
(f) a rigid link from the last link to the end-effector, described by the homogenous matrix transformation $\mathbf{T}_{Tool}$.

In the frame of these notations, the final expression defining the end-effector location subject to variations of all joint coordinates of a single kinematic chain may be written as the product of the following homogenous matrices

$$\mathbf{T} = \mathbf{T}_{\text{Base}} \cdot \prod_i \left(\mathbf{T}(\boldsymbol{\theta}_{\text{a}}^i) \cdot \mathbf{T}\left(\mathbf{q}_{\text{p}}^{2i-1}\right) \cdot \mathbf{T}_{\text{Link}}^i \cdot \mathbf{T}(\boldsymbol{\theta}_{\text{Link}}^i) \cdot \mathbf{T}\left(\mathbf{q}_{\text{p}}^{2i}\right)\right) \cdot \mathbf{T}_{\text{Tool}} \tag{5.1}$$

where the components $\mathbf{T}_{\text{Base}}$, $\mathbf{T}(\ldots)$, $\mathbf{T}_{\text{Link}}^i$, $\mathbf{T}_{\text{Tool}}$ may be factorized with respect to the terms including the joint variables, in order to simplify the computing of the derivatives (Jacobian and Hessian).

This expression includes both traditional geometric variables (passive and active joint coordinates) and stiffness variables (virtual joint coordinates). The explicit position and orientation of the end-effector can be extracted from the matrix $\mathbf{T}$ in a standard way [5], so finally the extended kinematic model can be rewritten as the vector function

$$\mathbf{t} = \mathbf{g}(\mathbf{q}, \theta) \tag{5.2}$$

where the vector $\mathbf{t} = (\boldsymbol{p}, \boldsymbol{\varphi})^{\text{T}}$ includes the position $\mathbf{p} = (x, y, z)^{\text{T}}$ and orientation $\boldsymbol{\varphi} = \left(\varphi_x, \varphi_y, \varphi_z\right)^{\text{T}}$ of the end-platform, the vector $\mathbf{q} = \left(q_1, q_2, \ldots, q_{nq}\right)^T$ contains all passive joint coordinates, the vector $\boldsymbol{\theta}=(\theta_1,\theta_2,\ldots,\theta_{n\theta})^{\text{T}}$ collects all virtual joint coordinates, n_{q} is the number of the passive joins, n_θ is the number of the virtual joints.

5.1.2 VJM Model Parameters

In the first works, it was explicitly assumed that the main sources of elasticity are concentrated in actuated joints. Correspondingly, the links were assumed to be rigid and the VJM model included one-dimensional springs only. In other recent works, compliance with the links has been taken into account by introducing additional virtual joints describing their longitudinal elasticity [6] or stiffness properties in several directions [7]. Recent developments in this area use 6-dimensional virtual joints to describe the elasticity of each link [8, 9].

In the beginning, the *stiffness parameters* of the virtual joints describing the link elasticity (and incorporated in the matrix $\mathbf{K}_\theta$) were evaluated rather roughly, using a simplified representation of the link shape by regular beams. Besides, it was assumed that all linear and angular deflections (compression/tension, bending, torsion) are decoupled and are presented by independent one-dimensional springs that produce a diagonal stiffness matrix of size 6×6 for each link. Afterwards, this elasticity model was enhanced by using a complete 6×6 non-diagonal stiffness matrix of the cantilever beam [10, 11]. This allowed taking into account all types of translational/rotational compliance and relevant coupling between different deflections. Other enhancements include the link approximation by several beams, but it gives a rather modest improvement in accuracy.

Further advances in this direction (applicable to the links of complicated shapes) led to the *FEA-based identification technique* that involves virtual loading experiments in a CAD environment and stiffness matrix estimation using dedicated numerical routines [8]. The latter essentially increased the accuracy of the VJM modelling while preserving its high computational efficiency. It is worth mentioning that the usual high computational expenses of the FEA are not a critical issue here, because it is applied only once for each link (in contrast to the straightforward the FEA-modelling for the entire manipulator, which requires complete re-computing for each manipulator posture). As a result, this approach allowed the authors to integrate the accuracy of the FEA modelling into the VJM-modeling technique that provides high computational efficiency. A general methodology of this hybrid approach is presented in Fig. 5.4 and is described in detail in Sect. 7.1.

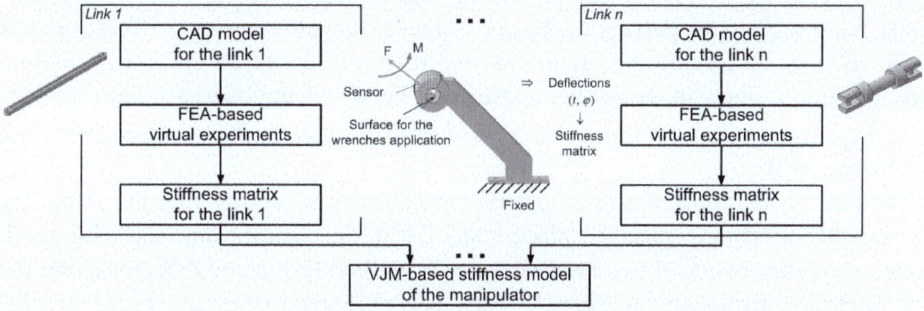

Fig. 5.4 Integration of VJM modeling approach with FEA-based identification technique [1]

5.2 VJM in Robotics: Modelling of Open and Closed Loop Architectures

5.2.1 Stiffness Modelling of Serial Chains with Passive Joints

Cartesian Stiffness Matrix of a Serial Chain

Using the notation introduced in the previous section, the unloaded configuration of the serial chain can be found from the extended kinematic equation $\mathbf{t}_0 = f(\mathbf{q}_0, \boldsymbol{\theta}_0)$, where $\mathbf{q}_0$ is computed via conventional inverse kinematics and $\boldsymbol{\theta}_0$ is assumed to be equal to zero (since there are no preloads in the springs). Further, let us assume that the external force $\mathbf{F}$ relocates the manipulator to the position $\mathbf{t} = f(\mathbf{q}_0 + \Delta\mathbf{q}, \boldsymbol{\theta}_0 + \Delta\boldsymbol{\theta})$, which for small displacements may be expressed as

$$\mathbf{t} = \mathbf{t}_0 + \mathbf{J}_\theta \cdot \Delta\boldsymbol{\theta} + \mathbf{J}_q \cdot \Delta\mathbf{q} \tag{5.3}$$

where $\mathbf{J}_\theta = \partial f(\mathbf{q}, \boldsymbol{\theta})/\partial\boldsymbol{\theta}$ and $\mathbf{J}_q = \partial f(\mathbf{q}, \boldsymbol{\theta})/\partial\mathbf{q}$ are the kinematic Jacobians with respect to the coordinates $\boldsymbol{\theta}$, $\mathbf{q}$, which may be computed from (5.1) analytically or semi-analytically, using the factorization technique proposed in [12].

For the kinetostatic model, which describes the force-and-motion relation, it is necessary to introduce additional equations that define the virtual joint reactions to the corresponding spring deformations. For analytical convenience, all relevant expressions may be collected in a single matrix equation

$$\boldsymbol{\tau}_\theta = \mathbf{K}_\theta \cdot \boldsymbol{\theta} \tag{5.4}$$

where $\boldsymbol{\tau}_\theta = \left(\tau_{\theta,1}, \tau_{\theta,2}, \ldots, \tau_{\theta,n\theta}\right)^{\mathrm{T}}$ is the aggregated vector of the virtual joint reactions, $\mathbf{K}_\theta = diag\left(\mathbf{K}_{\theta,1}, \mathbf{K}_{\theta,2}, \ldots, \mathbf{K}_{\theta,n\theta}\right)$ is the aggregated spring stiffness matrix of the size $n_\theta \times n_\theta$, and $\mathbf{K}_{\theta,i}$ is the spring stiffness matrix of the corresponding link. Similarly, one can define the aggregated vector of the passive joint reactions $\boldsymbol{\tau}_q = \left(\tau_{q,1}, \tau_{q,2}, \ldots, \tau_{q,nq}\right)^{\mathrm{T}}$ but, at the equilibrium, all its components must be equal to zero ($\boldsymbol{\tau}_q = \mathbf{0}$).

Further, let us apply the principle of virtual work assuming that the joints are given small, arbitrary virtual displacements $\Delta\boldsymbol{\theta}$ in the equilibrium neighbourhood. Then, the virtual work of the external force $\mathbf{F}$ applied to the end-effector along the corresponding displacement $\Delta\mathbf{t} = \mathbf{J}_\theta \cdot \Delta\boldsymbol{\theta} + \mathbf{J}_q \Delta\mathbf{q}$ is equal to the sum $\left(\mathbf{F}^{\mathrm{T}} \cdot \mathbf{J}_\theta\right) \cdot \Delta\boldsymbol{\theta} + \left(\mathbf{F}^{\mathrm{T}} \cdot \mathbf{J}_q\right) \cdot \Delta\mathbf{q}$. For the internal forces, the virtual work includes only one component $-\boldsymbol{\tau}_\theta^{\mathrm{T}} \cdot \Delta\boldsymbol{\theta}$, since the passive joints do not produce the force/torque reactions (the minus sign takes into account the adopted directions for the virtual spring forces/torques). Therefore, since in the static equilibrium, the total virtual work is equal to zero for any virtual displacement, the equilibrium conditions may be written as

$$\mathbf{J}_\theta^{\mathrm{T}} \cdot \mathbf{F} = \boldsymbol{\tau}_\theta; \mathbf{J}_q^{\mathrm{T}} \cdot \mathbf{F} = \boldsymbol{\tau}_q \tag{5.5}$$

This gives additional expressions describing the force/torque propagation from the joints to the end-effector.

Hence, the complete kinetostatic model consists of four matrix Eqs. (5.3–5.5) where either $\mathbf{F}$ or $\Delta\mathbf{t}$ are treated as known, and the remaining variables are considered as unknown. Thus, relevant equations of statics may be written as

$$\mathbf{J}_\theta \cdot \boldsymbol{\theta} + \mathbf{J}_\mathrm{q} \cdot \Delta\mathbf{q} = \Delta\mathbf{t}; \mathbf{F} \cdot \mathbf{J}_\theta^\mathrm{T} - \mathbf{K}_\theta \cdot \theta = 0; \mathbf{F} \cdot \mathbf{J}_\mathrm{q}^\mathrm{T} = 0 \tag{5.6}$$

here $\Delta\mathbf{t} = \mathbf{t} - \mathbf{t}_0$, and the general relation between the increments $\mathbf{F}$, $\Delta\mathbf{t}$, $\boldsymbol{\theta}$, $\mathbf{q}$ can be presented as

$$\begin{bmatrix} \mathbf{0} & \mathbf{J}_\theta & \mathbf{J}_\mathrm{q} \\ \mathbf{J}_\theta^\mathrm{T} & -\mathbf{K}_\theta & \mathbf{0} \\ \mathbf{J}_\mathrm{q}^\mathrm{T} & \mathbf{0} & \mathbf{0} \end{bmatrix} \cdot \begin{bmatrix} \mathbf{F} \\ \boldsymbol{\theta} \\ \Delta\mathbf{q} \end{bmatrix} = \begin{bmatrix} \Delta\mathbf{t} \\ \mathbf{0} \\ \mathbf{0} \end{bmatrix} \tag{5.7}$$

The latter gives a straightforward numerical technique for computing the desired stiffness matrix: direct inversion of the matrix on the left-hand side of (5.7) and extracting from it the upper-left sub-matrix of size 6×6

$$\begin{bmatrix} \mathbf{0} & \mathbf{J}_\theta & \mathbf{J}_\mathrm{q} \\ \mathbf{J}_\theta^\mathrm{T} & -\mathbf{K}_\theta & \mathbf{0} \\ \mathbf{J}_\mathrm{q}^\mathrm{T} & \mathbf{0} & \mathbf{0} \end{bmatrix}^{-1} = \left[\begin{array}{c|c|c} \mathbf{K}_\mathrm{C} & * & * \\ \hline * & * & * \\ \hline * & * & * \end{array}\right] \tag{5.8}$$

Since the matrix $\mathbf{K}_\theta$ is non-singular (it describes the stiffness of the virtual springs), the variables $\boldsymbol{\theta}$ can be expressed via $\mathbf{F}$ using the Eqs. (5.6) $\boldsymbol{\theta} = \mathbf{K}_\theta^{-1} \cdot \mathbf{J}_\theta^\mathrm{T} \cdot \mathbf{F}$. This leads to a reduced system of matrix equations with unknown $\mathbf{F}$ and $\Delta\mathbf{q}$

$$\begin{bmatrix} \mathbf{J}_\theta \cdot \mathbf{K}_\theta^{-1} \cdot \mathbf{J}_\theta^\mathrm{T} & \mathbf{J}_\mathrm{q} \\ \mathbf{J}_\mathrm{q}^\mathrm{T} & \mathbf{0} \end{bmatrix} \cdot \begin{bmatrix} \mathbf{F} \\ \Delta\mathbf{q} \end{bmatrix} = \begin{bmatrix} \Delta\mathbf{t} \\ \mathbf{0} \end{bmatrix} \tag{5.9}$$

that may be treated in a similar way, i.e. the desired stiffness matrix is also obtained by the direct inversion of the matrix on the left-hand side of (5.9) and extracting from it the upper-left sub-matrix of size 6×6:

$$\begin{bmatrix} \mathbf{J}_\theta \cdot \mathbf{K}_\theta^{-1} \cdot \mathbf{J}_\theta^\mathrm{T} & \mathbf{J}_\mathrm{q} \\ \mathbf{J}_\mathrm{q}^\mathrm{T} & \mathbf{0} \end{bmatrix}^{-1} = \left[\begin{array}{c|c} \mathbf{K}_\mathrm{C} & * \\ \hline * & * \end{array}\right] \tag{5.10}$$

where the sub-matrix $\mathbf{J}_\theta \cdot \mathbf{K}_\theta^{-1} \cdot \mathbf{J}_\theta^\mathrm{T}$ describes the spring compliance relative to the end-effector, and the sub-matrix $\mathbf{J}_\mathrm{q}$ takes into account the passive joint influence on the end-effector motions. Therefore, for a separate kinematic chain, the desired stiffness matrix $\mathbf{K}_\mathrm{C}$ can be computed by the direct inversion of the $(6 + n_q) \times (6 + n_q)$ the matrix on the left-hand side of (5.10) and extracting from it the 6×6 sub-matrix.

Deflections in a Serial Chain with Passive Joints

Cartesian stiffness matrix $\mathbf{K}_C$ computed in the previous sub-section, allows us to obtain the linear relation between the end-effector displacement $\Delta\mathbf{t}$ and the external loading $\mathbf{F}$

$$\mathbf{F} = \mathbf{K}_C \cdot \Delta\mathbf{t}, \tag{5.11}$$

that defines the kinematic chain elastostatic properties in a traditional way. However, due to the specificity of the robotic mechanism, it is useful to propose additional performance measures that evaluate variations in passive $\Delta\mathbf{q}$ and virtual joint coordinates $\boldsymbol{\theta}$ because of the end-point displacement. These values provide the designer a supplementary information on the manipulator properties, such as the weakest and strongest links with respect to the compliance errors, the most significant displacements in the passive joints caused by the shifting of the end-point location etc. It is worth mentioning that these values cannot be directly computed from kinematic equations due to the redundant number of the joint variables (in fact, it is necessary to take into account additional relations derived from static equilibrium equations).

The desired performance measures can be introduced similarly, via the stiffness matrices $\mathbf{K}_{C\theta}$ and $\mathbf{K}_{Cq}$ defining linear mappings of the end-point displacement to the internal coordinates deflections $\boldsymbol{\theta}$ and $\Delta\mathbf{q}$:

$$\boldsymbol{\theta} = \mathbf{K}_{C\theta}\Delta\mathbf{t};\ \Delta\mathbf{q} = \mathbf{K}_{Cq}\Delta\mathbf{t} \tag{5.12}$$

These matrices $\mathbf{K}_{C\theta}$ and $\mathbf{K}_{Cq}$ can be computed from the full-scale force–deflection relation (5.7) by the direct inversion of $(6 + n_q + n_\theta) \times (6 + n_q + n_\theta)$ the matrix on the left-hand side of (5.8) and extracting from it $n_\theta \times 6$ and $n_q \times 6$ sub-matrices, which correspond to $\mathbf{K}_{C\theta}$ and $\mathbf{K}_{Cq}$ respectively (see (5.8) for details). Moreover, since the virtual joint coordinates θ and external loading $\mathbf{F}$ are defined, the internal loadings $\boldsymbol{\tau}_\theta$ in virtual joints can be computed using the expressions (5.4) and (5.5).

To avoid high-dimensional matrix inversion, some transformations can be performed analytically assuming that the matrix K_C is known. In particular, using expressions $\boldsymbol{\theta} = \mathbf{K}_\theta^{-1} \cdot \mathbf{J}_\theta^T \cdot \mathbf{F}$ and $\mathbf{F} = \mathbf{K}_C \cdot \Delta\mathbf{t}$ that follow from (5.4) and (5.5), the desired matrix $\mathbf{K}_{C\theta}$ can be expressed as

$$\mathbf{K}_{C\theta} = \mathbf{K}_\theta^{-1} \cdot \mathbf{J}_\theta^T \cdot \mathbf{K}_C \tag{5.13}$$

and relevant torques in the virtual joints can be computed as

$$\boldsymbol{\tau}_\theta = \mathbf{J}_\theta^T \cdot \mathbf{K}_C \cdot \Delta\mathbf{t} \tag{5.14}$$

Further, to compute the matrix $\mathbf{K}_{Cq}$, let us consider the first equation of the system (5.6) and substitute $\boldsymbol{\theta} = \mathbf{K}_{C\theta}\Delta\mathbf{t}$. This yields the redundant system of linear equations

$$\mathbf{J}_\theta\mathbf{K}_{C\theta} \cdot \Delta\mathbf{t} + \mathbf{J}_q \cdot \Delta\mathbf{q} = \Delta\mathbf{t}, \tag{5.15}$$

which can be solved using the common pseudo-inversion technique that gives the following expression for the second matrix

$$\mathbf{K}_{\mathrm{Cq}} = \left(\mathbf{J}_{\mathrm{q}}^{\mathrm{T}} \cdot \mathbf{J}_{\mathrm{q}}\right)^{-1} \cdot \mathbf{J}_{\mathrm{q}}^{\mathrm{T}} \cdot (\mathbf{I} - \mathbf{J}_{\theta}\mathbf{K}_{\mathrm{C}\theta}) \tag{5.16}$$

that includes inversion of dimension $n_{\mathrm{q}} \times n_{\mathrm{q}}$. Hence, using expressions (5.13) and (5.16), the designer can obtain additional information allowing us to evaluate the resistance of the internal joint variables with respect to the deflection of the end-platform and to make the relevant modifications to the manipulator parameters.

It should be stressed that the above expressions allow us to compute the Cartesian stiffness matrix $\mathbf{K}_{\mathrm{C}}$ for serial kinematic chains with the passive joints in the numerical form, by the inversion of $(6 + n_q + n_\theta) \times (6 + n_q + n_\theta)$ matrix (5.8) and extracting the 6×6 sub-matrix from it. Another more efficient solution requires matrix inversion of dimension $(6 + n_q) \times (6 + n_q)$ in accordance with (5.10). This Section gives the analytical solution to the problem, which requires several matrix inversions of dimension 6×6 only. Besides, it allows us to explicitly separate the influence of virtual elastic joints and passive joints. For illustration purposes, it also includes several case studies that show how the stiffness matrix is modified due to the presence of the passive joints.

Summarizing this Section, it should be stressed that the described technique can be applied to a serial kinematic chain with the arbitrary location of passive joints and allows us to obtain even rank-deficient stiffness matrices caused by the presence of the passive joints or singular configuration of the chain. Also, it evaluates internal deflections (and corresponding forces/torques) that are useful for mechanical design. This technique is computationally efficient and includes low-order matrix inversion only. However, in some cases, it is useful to obtain an analytical solution.

5.2.2 *Analytical Computing of Cartesian Stiffness Matrix for Serial Chains*

Stiffness Mapping Equation for a Serial Chain with Passive Joints

In some works, stiffness modelling is associated with slightly different terminology that operates with the so-called "Stiffness mapping equation" [4, 13], which defines the relation between the stiffness properties in joint and Cartesian spaces. For the manipulators without the passive joints, such an equation is expressed as $\mathbf{K}_{\mathrm{C}} = \mathbf{J}_{\theta}^{-1} \cdot \mathbf{K}_{\theta} \cdot \mathbf{J}_{\theta}^{-\mathrm{T}}$ or $\mathbf{K}_{\theta} = \mathbf{J}_{\theta} \cdot \mathbf{K}_{\mathrm{C}} \cdot \mathbf{J}_{\theta}^{\mathrm{T}}$. It is worth mentioning that in most of the works, it was implicitly assumed that the Jacobian $\mathbf{J}_{\theta}$ is a 6×6 invertible non-singular matrix, while in a general case, the considered kinematic chains may produce non-square Jacobians of size $6 \times n_{\theta}$, but $rank(\mathbf{J}_{\theta}) = 6$ due to the special arrangement of virtual springs. Let us generalize the stiffness mapping equation for the case of the manipulators with passive joints and a redundant number of virtual joints $n_{\theta} > 6$.

In order to derive the desired equation, it is reasonable to consider the matrix expression (5.10) and apply the blockwise inversion technique of Frobenius [14]

$$\begin{bmatrix} \mathbf{A} & \mathbf{B} \\ \mathbf{C} & \mathbf{D} \end{bmatrix}^{-1} = \left[\begin{array}{c|c} \mathbf{A}^{-1} + \mathbf{A}^{-1}\mathbf{B}\mathbf{H}^{-1}\mathbf{C}\mathbf{A}^{-1} & -\mathbf{A}^{-1}\mathbf{B}\mathbf{H}^{-1} \\ \hline -\mathbf{H}^{-1}\mathbf{C}\mathbf{A}^{-1} & \mathbf{H}^{-1} \end{array}\right]; \qquad \mathbf{H} = \mathbf{D} - \mathbf{C}\mathbf{A}^{-1}\mathbf{B} \tag{5.17}$$

where the sub-matrices **A**,…,**H** are expressed as

$$\begin{aligned} &\mathbf{A} = \mathbf{J}_\theta \cdot \mathbf{K}_\theta^{-1} \cdot \mathbf{J}_\theta^{\mathrm{T}};\ \mathbf{B} = \mathbf{J}_\mathrm{q};\ \mathbf{C} = \mathbf{J}_\mathrm{q}^{\mathrm{T}};\ \mathbf{D} = \mathbf{0}; \\ &\mathbf{H} = -\mathbf{J}_\mathrm{q}^{\mathrm{T}} \cdot \left(\mathbf{J}_\theta \cdot \mathbf{K}_\theta^{-1} \cdot \mathbf{J}_\theta^{\mathrm{T}}\right)^{-1} \cdot \mathbf{J}_\mathrm{q} \end{aligned} \tag{5.18}$$

This yields the following presentation of the left-hand side of (5.10)

$$\begin{bmatrix} \mathbf{J}_\theta \cdot \mathbf{K}_\theta^{-1} \cdot \mathbf{J}_\theta^{T} & \mathbf{J}_\mathrm{q} \\ \mathbf{J}_\mathrm{q}^{\mathrm{T}} & \mathbf{0} \end{bmatrix}^{-1} = \begin{bmatrix} \mathbf{A}' & \mathbf{B}' \\ \mathbf{C}' & \mathbf{D}' \end{bmatrix} \tag{5.19}$$

where

$$\begin{aligned} \mathbf{A}' &= \left(\mathbf{J}_\theta \cdot \mathbf{K}_\theta^{-1} \cdot \mathbf{J}_\theta^{\mathrm{T}}\right)^{-1} \\ &\quad - \left(\mathbf{J}_\theta \cdot \mathbf{K}_\theta^{-1} \cdot \mathbf{J}_\theta^{\mathrm{T}}\right)^{-1} \cdot \mathbf{J}_\mathbf{q} \cdot \left(\mathbf{J}_\mathbf{q}^{\mathbf{T}} \cdot \left(\mathbf{J}_\theta \cdot \mathbf{K}_\theta^{-1} \cdot \mathbf{J}_\theta^{\mathrm{T}}\right)^{-1} \cdot \mathbf{J}_\mathbf{q}\right)^{-1} \cdot \mathbf{J}_\mathbf{q}^{\mathbf{T}} \cdot \left(\mathbf{J}_\theta \cdot \mathbf{K}_\theta^{-1} \cdot \mathbf{J}_\theta^{\mathrm{T}}\right)^{-1} \\ \mathbf{B}' &= \left(\mathbf{J}_\theta \cdot \mathbf{K}_\theta^{-1} \cdot \mathbf{J}_\theta^{\mathrm{T}}\right)^{-1} \cdot \mathbf{J}_\mathbf{q} \cdot \left(\mathbf{J}_\mathbf{q}^{\mathbf{T}} \cdot \left(\mathbf{J}_\theta \cdot \mathbf{K}_\theta^{-1} \cdot \mathbf{J}_\theta^{\mathrm{T}}\right)^{-1} \cdot \mathbf{J}_\mathbf{q}\right)^{-1}; \\ \mathbf{C}' &= \left(\mathbf{J}_\mathbf{q}^{\mathbf{T}} \cdot \left(\mathbf{J}_\theta \cdot \mathbf{K}_\theta^{-1} \cdot \mathbf{J}_\theta^{\mathrm{T}}\right)^{-1} \cdot \mathbf{J}_\mathbf{q}\right)^{-1} \cdot \mathbf{J}_\mathbf{q}^{\mathbf{T}} \cdot \left(\mathbf{J}_\theta \cdot \mathbf{K}_\theta^{-1} \cdot \mathbf{J}_\theta^{\mathrm{T}}\right)^{-1} \\ \mathbf{D}' &= -\left(\mathbf{J}_\mathbf{q}^{\mathbf{T}} \cdot (\mathbf{J}_\theta \cdot \mathbf{K}_\theta^{-1} \cdot \mathbf{J}_\theta^{\mathrm{T}})^{-1} \cdot \mathbf{J}_\mathbf{q}\right)^{-1} \end{aligned} \tag{5.20}$$

Hence, the desired Cartesian stiffness matrix $\mathbf{K}_{\mathrm{Cq}}$ can be presented analytically as

$$\begin{aligned} \mathbf{K}_\mathrm{C} &= \left(\mathbf{J}_\theta \cdot \mathbf{K}_\theta^{-1} \cdot \mathbf{J}_\theta^{T}\right)^{-1} \\ &\quad - \left(\mathbf{J}_\theta \cdot \mathbf{K}_\theta^{-1} \cdot \mathbf{J}_\theta^{T}\right)^{-1} \cdot \mathbf{J}_\mathrm{q} \cdot \left(\mathbf{J}_\mathrm{q}^{\mathrm{T}} \cdot \left(\mathbf{J}_\theta \cdot \mathbf{K}_\theta^{-1} \cdot \mathbf{J}_\theta^{\mathrm{T}}\right)^{-1} \cdot \mathbf{J}_\mathrm{q}\right)^{-1} \cdot \mathbf{J}_\mathrm{q}^{\mathrm{T}} \cdot \left(\mathbf{J}_\theta \cdot \mathbf{K}_\theta^{-1} \cdot \mathbf{J}_\theta^{\mathrm{T}}\right)^{-1} \end{aligned} \tag{5.21}$$

Similarly, the matrix $\mathbf{K}_{\mathrm{Cq}}$ defining mapping from the Cartesian space to the passive joint space can also be expressed analytically as

$$\mathbf{K}_\mathrm{Cq} = \left(\mathbf{J}_\mathrm{q}^{\mathrm{T}}\left(\mathbf{J}_\theta \mathbf{K}_\theta^{-1} \mathbf{J}_\theta^{\mathrm{T}}\right)^{-1} \mathbf{J}_\mathrm{q}\right)^{-1} \mathbf{J}_\mathrm{q}^{\mathrm{T}}\left(\mathbf{J}_\theta \mathbf{K}_\theta^{-1} \mathbf{J}_\theta^{\mathrm{T}}\right)^{-1} \tag{5.22}$$

To compare these results with the classical stiffness mapping equation $\mathbf{K}_C = \mathbf{J}_\theta^{-1} \cdot \mathbf{K}_\theta \cdot \mathbf{J}_\theta^{-T}$, the expression (5.21) can be rewritten as the difference of two 6×6 matrices

$$\mathbf{K}_C = \mathbf{K}_C^0 - \mathbf{K}_C^0 \mathbf{J}_q \mathbf{K}_{Cq} \tag{5.23}$$

where the first term $\mathbf{K}_C^0$ is the stiffness matrix of the corresponding serial chain without passive joints

$$\mathbf{K}_C^0 = \left(\mathbf{J}_\theta \mathbf{K}_\theta^{-1} \mathbf{J}_\theta^T\right)^{-1} \tag{5.24}$$

and the second term $\mathbf{K}_C^0 \cdot \mathbf{J}_q \cdot \mathbf{K}_{Cq}$ defines the influence of the passive joints on the Cartesian stiffness matrix $\mathbf{K}_C$. Hence, the obtained expression (5.23) extends the classical stiffness mapping technique for the case of manipulators with passive joints by adding an additional component to it. The latter allows us to estimate the impact of passive joints on the kinematic chain stiffness in an explicit form and to obtain some useful analytical results presented in the following sub-section.

It is worth mentioning that in contrast to previous works, the proposed extended stiffness mapping Eq. (5.23) can be also applied to the kinematic chains with non-square Jacobians J_θ of full rank, since the matrix inversion is applied to the 6×6 matrix $\mathbf{J}_\theta \cdot \mathbf{K}_\theta^{-1} \cdot \mathbf{J}_\theta^T$, which is always invertible (even for singular postures of the kinematic chain). Besides, it is in good agreement with other relevant works [15–17] where K_C was presented as the difference between the two matrices but the second one was computed in a different way (to take into account the impact of geometrical constraints).

Singularities in Cartesian Stiffness Matrix Computing

Let us analyze in detail the expression (5.21) that allows computing the stiffness matrix of a serial chain with passive joints $\mathbf{K}_C$ from the corresponding matrix of the chain without passive joints $\mathbf{K}_C^0$. Assuming that $rank\left(\mathbf{K}_C^0\right) = 6$, this matrix can be factorised into the product of two non-singular square matrices $\mathbf{K}_C^0 = \mathbf{S}^T\mathbf{S}$. This yields a compact presentation of the desired matrix in the form

$$\mathbf{K}_C = \mathbf{S}^T \cdot \mathbf{M} \cdot \mathbf{S} \tag{5.25}$$

where

$$\mathbf{M} = \mathbf{I} - (\mathbf{S} \cdot \mathbf{J}_q) \cdot \left(\mathbf{J}_q^T \cdot \mathbf{S}^T \cdot \mathbf{S} \cdot \mathbf{J}_q\right)^{-1} \cdot \left(\mathbf{J}_q^T \cdot \mathbf{S}^T\right) \tag{5.26}$$

and the inverse $\left(\mathbf{J}_q^T \cdot \mathbf{S}^T \cdot \mathbf{S} \cdot \mathbf{J}_q\right)^{-1}$ exists due to the assumption $rank(\mathbf{J}_q) = n_q$. Further, the product $\mathbf{S} \cdot \mathbf{J}_q$ can be also factorized using the SVD-technique [18] as

$$\mathbf{SJ}_{\mathrm{q}} = \mathbf{U}\boldsymbol{\Sigma}\mathbf{V}^{\mathrm{T}} = \left[u_{ij}\right]_{6\times 6}\left[\frac{\cdot\boldsymbol{\Sigma}_{\mathrm{q}}}{\mathbf{0}}\right]_{6\times \mathrm{n_q}} \cdot \left[v_{ij}\right]_{\mathrm{n_q}\times \mathrm{n_q}} \tag{5.27}$$

where $\mathbf{U}$, $\mathbf{V}$ are the orthogonal matrices and the matrix $\boldsymbol{\Sigma}_{\mathrm{q}} = diag(\sigma_1, \sigma_2, ...)$ is composed of n_{q} non-zero singular values σ_i (provided that $rank\left(\mathbf{J}_{\mathrm{q}}\right) = n_{\mathrm{q}}$). The latter gives the following presentation of $\mathbf{M}$

$$\mathbf{M} = \mathbf{U} \cdot \left(\mathbf{I} - \boldsymbol{\Sigma} \cdot \left(\boldsymbol{\Sigma}^{\mathrm{T}} \cdot \boldsymbol{\Sigma}\right)^{-1} \cdot \boldsymbol{\Sigma}^{\mathrm{T}}\right) \cdot \mathbf{U}^{\mathrm{T}} \tag{5.28}$$

where the product $\boldsymbol{\Sigma} \cdot \left(\boldsymbol{\Sigma}^{T} \cdot \boldsymbol{\Sigma}\right)^{-1} \cdot \boldsymbol{\Sigma}^{\mathrm{T}}$ maybe computed straightforwardly:

$$\begin{aligned}\boldsymbol{\Sigma} \cdot \left(\boldsymbol{\Sigma}^{\mathrm{T}} \cdot \boldsymbol{\Sigma}\right)^{-1} \cdot \boldsymbol{\Sigma}^{\mathrm{T}} &= \begin{bmatrix}\boldsymbol{\Sigma}_{\mathrm{q}} \\ \mathbf{0}\end{bmatrix} \cdot \left(\left[\boldsymbol{\Sigma}_{\mathrm{q}}\, \mathbf{0}\right] \cdot \begin{bmatrix}\boldsymbol{\Sigma}_{\mathrm{q}} \\ \mathbf{0}\end{bmatrix}\right)^{-1} \cdot \left[\boldsymbol{\Sigma}_{\mathrm{q}}\, \mathbf{0}\right] \\ &= \begin{bmatrix}\mathbf{I}_{\mathrm{n_q}} & \mathbf{0} \\ \mathbf{0} & \mathbf{0}\end{bmatrix}_{6\times 6}\end{aligned} \tag{5.29}$$

So, the inner part of $\mathbf{M}$ may be presented as

$$\mathbf{M}' = \mathbf{I} - \boldsymbol{\Sigma} \cdot \left(\boldsymbol{\Sigma}^{T} \cdot \boldsymbol{\Sigma}\right)^{-1} \cdot \boldsymbol{\Sigma}^{\mathrm{T}} = \left[\begin{array}{c|c}\mathbf{0}_{n_q} & \mathbf{0} \\ \hline \mathbf{0} & \mathbf{I}_{6\,-\,\mathrm{n_q}}\end{array}\right]_{6\times 6} \tag{5.30}$$

This leads to the final expression

$$\mathbf{K}_C = \mathbf{S}^{T} \cdot \mathbf{U} \cdot \mathbf{M}' \cdot \mathbf{U}^{T} \cdot \mathbf{S} \tag{5.31}$$

that allows to evaluate the rank of the stiffness matrix $\mathbf{K}_{\mathrm{C}}$

$$rank(\mathbf{K}_{\mathrm{C}}) = rank\left(\mathbf{M}'\right) = 6 - n_{\mathrm{q}} \tag{5.32}$$

For computational convenience, the orthogonal matrix $\mathbf{U}$ maybe split into six vector columns $[\mathbf{u}_1, ..., \mathbf{u}_6]$ and the matrix product $\mathbf{U} \cdot \mathbf{M}' \cdot \mathbf{U}^{\mathbf{T}}$ is expressed as a sub-sum of $\mathbf{u}_i\mathbf{u}_i^T$ corresponding to the non-zero elements of $\mathbf{M}'$. This gives another presentation of the desired Cartesian stiffness matrix

$$\mathbf{K}_{\mathrm{C}} = \mathbf{S}^{\mathrm{T}} \cdot \left(\sum_{i=nq+1}^{6} \mathbf{u}_i\mathbf{u}_i^{\mathrm{T}}\right) \cdot \mathbf{S} \tag{5.33}$$

where the middle term includes only those unit vectors $\mathbf{u}_i$ that are not "compensated" by the passive joints (for this notation, the directions $\mathbf{u}_1, ...\mathbf{u}_{nq}$ correspond to the end-effector motions, which do not produce elastostatic reactions because of the passive joints). It should be noted that in the case of $n_{\mathrm{q}} = 0$, i.e. without passive joints, the total sum of $\mathbf{u}_i\mathbf{u}_i^{\mathrm{T}}$ produces a unit matrix $\mathbf{I}_6$ and the latter expression is reduced to $\mathbf{K}_{\mathrm{C}}^{0} = \mathbf{S}^{\mathrm{T}} \cdot \mathbf{S}$.

All this leads to the following conclusion concerning computational singularities:

Statement 1. The first term of the expression (5.21) is non-singular if and only if $rank(\mathbf{J}_\theta) = 6$, i.e. if the VJM model of the chain includes at least 6 independent virtual springs.

Statement 2. The second term of the expression (5.21) is non-singular if and only if $rank\left(\mathbf{J}_{\mathrm{q}}\right) = n_{\mathrm{q}}$, i.e. if the VJM model of the chain does not include redundant passive joints.

Statement 3. If both terms of (5.21) are non-singular, their difference produces a symmetrical stiffness matrix, which is always singular and $rank(\mathbf{K}_C) = 6 - n_q$.

Statement 4. If the matrix $\mathbf{K}_C^0$ of the chain without passive joints is symmetrical and positive-definite, the stiffness matrix of the chain with passive joints $\mathbf{K}_C$ is also symmetrical but positive-semidefinite.

Hence, in practice, the expression (5.21) does not cause any computational difficulties and always produces a singular stiffness matrix of rank $6 - n_{\mathrm{q}}$. In analytical computations, it can be also useful in modifying the original stiffness matrix K_C^0 sequentially, which will be the focus of the next sub-section.

Recursive Computation of the Stiffness Matrix

To simplify the Cartesian **stiffness** matrix computation technique and to specify the impact of each passive joint on the resulting matrix $\mathbf{K}_{\mathrm{C}}$, it is useful to obtain a recursive procedure which allows us sequentially modify the original stiffness matrix $\mathbf{K}_{\mathrm{C}}^{0}$. To obtain a relevant recursive procedure, let us assume that it is possible to take into account passive joints sequentially, in two steps. Let us group these joints into two vectors $\mathbf{q}_1$ and $\mathbf{q}_2$ and decompose the Jacobian $\mathbf{J}_{\mathrm{q}}$ of size $6 \times n_{\mathrm{q}}$ into two sub-matrices $\left[J_{q1}, J_{q2}\right]$ of sizes $6 \times n_{\mathrm{q1}}$ and $6 \times n_{\mathrm{q2}}$ correspondingly. Then, recursive expressions corresponding to (5.21) can be written as

$$\begin{aligned} \mathbf{K}_{\mathrm{C}}^{1} &= \mathbf{K}_{\mathrm{C}}^{0} - \mathbf{K}_{\mathrm{C}}^{0}\mathbf{J}_{\mathrm{q1}}\left(\mathbf{J}_{\mathrm{q1}}^{\mathrm{T}}\mathbf{K}_{\mathrm{C}}^{0}\mathbf{J}_{\mathrm{q1}}\right)^{-1}\mathbf{J}_{\mathrm{q1}}^{T}\mathbf{K}_{\mathrm{C}}^{0} \\ \mathbf{K}_{\mathrm{C}}^{2} &= \mathbf{K}_{\mathrm{C}}^{1} - \mathbf{K}_{\mathrm{C}}^{1}\mathbf{J}_{\mathrm{q2}}\left(\mathbf{J}_{\mathrm{q2}}^{\mathrm{T}}\mathbf{K}_{\mathrm{C}}^{1}\mathbf{J}_{\mathrm{q2}}\right)^{-1}\mathbf{J}_{\mathrm{q2}}^{\mathrm{T}}\mathbf{K}_{\mathrm{C}}^{1} \end{aligned} \tag{5.34}$$

In order to evaluate the obtained matrix $\mathbf{K}_{\mathrm{C}}^{2}$, let us substitute the first expression in the second one and perform some equivalent transformations using notations

$$\begin{aligned} &\mathbf{A} = \mathbf{J}_{\mathrm{q1}}^{\mathrm{T}}\mathbf{K}_{\mathrm{C}}^{0}\mathbf{J}_{\mathrm{q1}};\, \mathbf{B} = \mathbf{J}_{\mathrm{q1}}^{\mathrm{T}}\mathbf{K}_{\mathrm{C}}^{0}\mathbf{J}_{\mathrm{q2}};\, \mathbf{C} = \mathbf{J}_{\mathrm{q2}}^{\mathrm{T}}\mathbf{K}_{\mathrm{C}}^{0}\mathbf{J}_{\mathrm{q1}};\, \mathbf{D} = \mathbf{J}_{\mathrm{q2}}^{\mathrm{T}}\mathbf{K}_{\mathrm{C}}^{0}\mathbf{J}_{\mathrm{q2}} \\ &\mathbf{H} = \mathbf{D} - \mathbf{C}\mathbf{A}^{-1}\mathbf{B} = \mathbf{J}_{\mathrm{q2}}^{\mathrm{T}}\mathbf{K}_{\mathrm{C}}^{0}\mathbf{J}_{\mathrm{q2}} - \mathbf{J}_{\mathrm{q2}}^{\mathrm{T}}\mathbf{K}_{\mathrm{C}}^{0}\mathbf{J}_{\mathrm{q1}} \cdot \left(\mathbf{J}_{\mathrm{q1}}^{\mathrm{T}}\mathbf{K}_{\mathrm{C}}^{0}\mathbf{J}_{\mathrm{q1}}\right)^{-1} \cdot \mathbf{J}_{\mathrm{q1}}^{\mathrm{T}}\mathbf{K}_{\mathrm{C}}^{0}\mathbf{J}_{\mathrm{q2}} \end{aligned} \tag{5.35}$$

This allows converting the original bulky expression

$$
\begin{aligned}
\mathbf{K}_{\mathrm{C}}^{2} = & \left\{\mathbf{K}_{\mathrm{C}}^{0} - \mathbf{K}_{\mathrm{C}}^{0} \cdot \mathbf{J}_{\mathrm{q1}} \cdot \left(\mathbf{J}_{\mathrm{q1}}^{\mathrm{T}} \cdot \mathbf{K}_{\mathrm{C}}^{0} \cdot \mathbf{J}_{\mathrm{q1}}\right)^{-1} \cdot \mathbf{J}_{\mathrm{q1}}^{\mathrm{T}} \cdot \mathbf{K}_{\mathrm{C}}^{0}\right\} \\
& - \left\{\mathbf{K}_{\mathrm{C}}^{0} - \mathbf{K}_{\mathrm{C}}^{0} \cdot \mathbf{J}_{\mathrm{q1}} \cdot \left(\mathbf{J}_{\mathrm{q1}}^{\mathrm{T}} \cdot \mathbf{K}_{\mathrm{C}}^{0} \cdot \mathbf{J}_{\mathrm{q1}}\right)^{-1} \cdot \mathbf{J}_{\mathrm{q1}}^{T} \cdot \mathbf{K}_{\mathrm{C}}^{0}\right\} \\
& \times \mathbf{J}_{\mathrm{q2}} \cdot \left(\mathbf{J}_{\mathrm{q2}}^{\mathrm{T}} \cdot \left\{\mathbf{K}_{\mathrm{C}}^{0} - \mathbf{K}_{\mathrm{C}}^{0} \cdot \mathbf{J}_{\mathrm{q1}} \cdot \left(\mathbf{J}_{\mathrm{q1}}^{\mathrm{T}} \cdot \mathbf{K}_{\mathrm{C}}^{0} \cdot \mathbf{J}_{\mathrm{q1}}\right)^{-1} \cdot \mathbf{J}_{\mathrm{q1}}^{\mathrm{T}} \cdot \mathbf{K}_{\mathrm{C}}^{0}\right\} \cdot \mathbf{J}_{\mathrm{q2}}\right)^{-1} \\
& \times \mathbf{J}_{\mathrm{q2}}^{\mathrm{T}} \cdot \left\{\mathbf{K}_{\mathrm{C}}^{0} - \mathbf{K}_{\mathrm{C}}^{0} \cdot \mathbf{J}_{\mathrm{q1}} \left(\mathbf{J}_{\mathrm{q1}}^{\mathrm{T}} \cdot \mathbf{K}_{\mathrm{C}}^{0} \cdot \mathbf{J}_{\mathrm{q1}}\right)^{-1} \cdot \mathbf{J}_{\mathrm{q1}}^{\mathrm{T}} \cdot \mathbf{K}_{\mathrm{C}}^{0}\right\}
\end{aligned}
\tag{5.36}
$$

into a more compact form

$$
\mathbf{K}_{\mathrm{C}}^{2} = \mathbf{K}_{\mathrm{C}}^{0} - \mathbf{K}_{\mathrm{C}}^{0} \cdot \begin{pmatrix} \mathbf{J}_{\mathrm{q1}}\mathbf{A}^{-1}\mathbf{J}_{\mathrm{q1}}^{\mathrm{T}} + \mathbf{J}_{\mathrm{q2}}\mathbf{H}^{-1}\mathbf{J}_{\mathrm{q2}}^{\mathrm{T}} + \mathbf{J}_{\mathrm{q2}}\mathbf{H}^{-1}\mathbf{C}\mathbf{A}^{-1}\mathbf{J}_{\mathrm{q1}}^{\mathrm{T}} \\ -\mathbf{J}_{\mathrm{q1}}\mathbf{A}^{-1}\mathbf{B}\,\mathbf{H}^{-1}\mathbf{J}_{\mathrm{q2}}^{\mathrm{T}} + \mathbf{J}_{\mathrm{q1}}\mathbf{A}^{-1}\mathbf{B}\,\mathbf{H}^{-1}\mathbf{C}\mathbf{A}^{-1}\mathbf{J}_{\mathrm{q1}}^{\mathrm{T}} \end{pmatrix} \cdot \mathbf{K}_{\mathrm{C}}^{0} \tag{5.37}
$$

or, using block-matrix presentation, it can be written as

$$
\mathbf{K}_{\mathrm{C}}^{2} = \mathbf{K}_{\mathrm{C}}^{0} - \mathbf{K}_{\mathrm{C}}^{0} \cdot \left[\mathbf{J}_{q1}\ \mathbf{J}_{q2}\right] \cdot \left[\begin{array}{c|c} \mathbf{A}^{-1} + \mathbf{A}^{-1}\mathbf{B}\mathbf{H}^{-1}\mathbf{C}\mathbf{A}^{-1} & -\mathbf{A}^{-1}\mathbf{B}\mathbf{H}^{-1} \\ \hline -\mathbf{H}^{-1}\mathbf{C}\mathbf{A}^{-1} & \mathbf{H}^{-1} \end{array}\right] \cdot \begin{bmatrix} \mathbf{J}_{q1}^{T} \\ \mathbf{J}_{q2}^{T} \end{bmatrix} \cdot \mathbf{K}_{\mathrm{C}}^{0} \tag{5.38}
$$

Furthermore, using the Frobenius formula for the blockwise matrix inverse (5.17), the derived expression can be presented in the form

$$
\mathbf{K}_{\mathrm{C}}^{2} = \mathbf{K}_{\mathrm{C}}^{0} - \mathbf{K}_{\mathrm{C}}^{0} \cdot \left[\mathbf{J}_{\mathrm{q1}}\mathbf{J}_{\mathrm{q2}}\right] \left[\begin{array}{c|c} \mathbf{J}_{\mathrm{q1}}^{\mathrm{T}}\mathbf{K}_{\mathrm{C}}^{0}\mathbf{J}_{\mathrm{q1}} & \mathbf{J}_{\mathrm{q1}}^{\mathrm{T}}\mathbf{K}_{\mathrm{C}}^{0}\mathbf{J}_{\mathrm{q2}} \\ \hline \mathbf{J}_{\mathrm{q2}}^{\mathrm{T}}\mathbf{K}_{\mathrm{C}}^{0}\mathbf{J}_{\mathrm{q1}} & \mathbf{J}_{\mathrm{q2}}^{\mathrm{T}}\mathbf{K}_{\mathrm{C}}^{0}\mathbf{J}_{\mathrm{q2}} \end{array}\right]^{-1} \begin{bmatrix} \mathbf{J}_{\mathrm{q1}}^{\mathrm{T}} \\ \mathbf{J}_{\mathrm{q2}}^{\mathrm{T}} \end{bmatrix} \mathbf{K}_{\mathrm{C}}^{0} \tag{5.39}
$$

or

$$
\mathbf{K}_{\mathrm{C}}^{2} = \mathbf{K}_{\mathrm{C}}^{0} - \mathbf{K}_{\mathrm{C}}^{0} \cdot \left[\mathbf{J}_{\mathrm{q1}}\mathbf{J}_{\mathrm{q2}}\right] \cdot \left(\begin{bmatrix} \mathbf{J}_{\mathrm{q1}}^{\mathrm{T}} \\ \mathbf{J}_{\mathrm{q2}}^{\mathrm{T}} \end{bmatrix} \cdot \mathbf{K}_{\mathrm{C}}^{0} \cdot \left[\mathbf{J}_{\mathrm{q1}}\ \mathbf{J}_{\mathrm{q2}}\right]\right)^{-1} \cdot \begin{bmatrix} \mathbf{J}_{\mathrm{q1}}^{\mathrm{T}} \\ \mathbf{J}_{\mathrm{q2}}^{\mathrm{T}} \end{bmatrix} \cdot \mathbf{K}_{\mathrm{C}}^{0} \tag{5.40}
$$

that exactly coincides with the expression for the stiffness matrix $\mathbf{K}_{\mathrm{C}}$ corresponding to the aggregated Jacobian $\mathbf{J}_q = \left[\mathbf{J}_{q1}, \mathbf{J}_{q2}\right]$. Hence, it proves that the desired stiffness matrix of the kinematic chain with passive joints can be computed recursively, using arbitrary partitioning of the Jacobian $\mathbf{J}_{\mathrm{q}}$. Obviously, it is more convenient to apply column-wise splitting that allows sequential modification of the matrix $\mathbf{K}_{\mathrm{C}}^{0}$ taking into account the geometry of each passive joint separately (and reducing sequentially by one the rank of the Cartesian stiffness matrix).

Statement. If the chain does not include redundant passive joints, the expression (5.21) allows recursive presentation

$$\mathbf{K}_{\mathrm{C}}^{i+1} = \mathbf{K}_{\mathrm{C}}^{i} - \mathbf{K}_{\mathrm{C}}^{i}\mathbf{J}_{\mathrm{q}}^{i}\left(\mathbf{J}_{\mathrm{q}}^{i\mathrm{T}}\mathbf{K}_{\mathrm{C}}^{i}\mathbf{J}_{\mathrm{q}}^{i}\right)^{-1}\mathbf{J}_{\mathrm{q}}^{i\mathrm{T}}\mathbf{K}_{\mathrm{C}}^{i} \tag{5.41}$$

in which the sub-Jacobians $\mathbf{J}_{\mathrm{q}}^{i} \subset \mathbf{J}_{\mathrm{q}}$ are extracted from $\mathbf{J}_{\mathrm{q}} = \left[\mathbf{J}_{\mathrm{q}}^{1}, \mathbf{J}_{\mathrm{q}}^{2}, \ldots\right]$ in arbitrary order (column-by-column, or by groups of columns).

Corollary *The desired stiffness matrix* $\mathbf{K}_{\mathrm{C}}$ *can be computed in* n_{q} *steps, by sequential application of expression* (5.41) *for each column of the Jacobian* $\mathbf{J}_{\mathrm{q}}$ *(i.e. for each passive joint separately).*

The latter motivates an additional analysis that is presented below and gives very simple and practically convenient expressions for the modifications of the original stiffness matrix $\mathbf{K}_{\mathrm{C}}^{0}$.

Single-Joint Decomposition in Stiffness Matrix Computing

Recursive Eq. (5.41) allows us to essentially simplify the computational procedure using the sequential modification of the original stiffness matrix $\mathbf{K}_{\mathrm{C}}^{0}$ for each passive joint independently, using separate columns of $\mathbf{J}_{\mathrm{q}} = [\mathbf{J}_{\mathrm{q}1}, \mathbf{J}_{\mathrm{q}2}, \ldots]$. Moreover, for some typical cases, relevant computations may be easily performed analytically. Here, some useful techniques related to this approach are presented.

Let us assume that a current recursion deals with a single passive joint corresponding to the *i*-th column of the Jacobian $\mathbf{J}_{\mathrm{q}}$, which is denoted as $\mathbf{J}_{\mathrm{q}}^{i}$ and has size 6×1. In this case, the matrix expression $\left(\mathbf{J}_{\mathrm{q}}^{i\mathrm{T}}\mathbf{K}_{\mathrm{C}}^{i}\mathbf{J}_{\mathrm{q}}^{i}\right)^{-1}$ is reduced to the size of 1×1 and the matrix inversion is replaced by a simple scalar division. Besides, the term $\mathbf{K}_{\mathrm{C}}^{i}\mathbf{J}_{\mathrm{q}}^{i}$ has size 6×1, so the recursion (5.41) is simplified to

$$\mathbf{K}_{\mathrm{C}}^{i+1} = \mathbf{K}_{\mathrm{C}}^{i} - \frac{1}{\mu}\mathbf{u}_{i} \cdot \mathbf{u}_{i}^{\mathrm{T}} or \left[K_{jk}^{(i+1)}\right] = \left[K_{jk}^{(i)}\right] - \frac{1}{\mu}\left[u_{j}^{(i)} u_{k}^{(i)}\right] \tag{5.42}$$

where $\mathbf{u}_{i} = \mathbf{K}_{C}^{i}\mathbf{J}_{q}^{i}$ is a 6×1 vector and $\mu = \mathbf{J}_{\mathrm{q}}^{i\mathrm{T}}\mathbf{K}_{\mathrm{C}}^{i}\mathbf{J}_{\mathrm{q}}^{i}$ is a scalar. Using the methodology presented in Section (a), it can be also proved that each recursion reduces the rank of the stiffness matrix by 1

$$rank\left(\mathbf{K}_{\mathrm{C}}^{i+1}\right) = rank\left(\mathbf{K}_{\mathrm{C}}^{i}\right) - 1 \tag{5.43}$$

provided that the current Jacobian $\mathbf{J}_{\mathrm{q}}^{i}$ is independent of the previous ones $\mathbf{J}_{\mathrm{q}}^{1}, \mathbf{J}_{\mathrm{q}}^{2}\ldots$ (i.e. the *i*-th passive joint is not redundant relative to joints $1, \ldots, i-1$).

Since in practice, any combination of passive joints can be decomposed into elementary translational and rotational ones, it is enough to consider only two types of Jacobian columns $\mathbf{J}_{\mathrm{q}i}$:

$$\mathbf{J}_{\mathrm{tran}} = \left[e_1 \; e_2 \; e_3 \; 0 \; 0 \; 0\right]^{\mathrm{T}}; \mathbf{J}_{\mathrm{rot}} = \left[d_1 \; d_2 \; d_3 \; e_1 \; e_2 \; e_3\right]^{\mathrm{T}} \tag{5.44}$$

where the unit vector $\mathbf{e} = \left[e_1 \; e_2 \; e_3 \right]$, $\mathbf{e}^T\mathbf{e} = 1$ defines the orientation of the passive joint axis (both for translational and rotational ones) and the vector $\mathbf{d} = \left[d_1 \; d_2 \; d_3 \right]$ defines the influence of the rotational passive joints on the linear velocity at the reference point, i.e. $\mathbf{d} = \mathbf{e} \times \mathbf{r}$ where $\mathbf{r}$ is a vector from the joint centre point to the reference point. After relevant substitutions, one can obtain the following expression for the translational joint

$$\mu = \sum_{j=1}^{3}\sum_{k=1}^{3} K_{jk}^{(i)} e_j e_k; \; u_j = \sum_{k=1}^{3} K_{jk}^{(i)} e_k \tag{5.45}$$

For the rotational joint

$$\begin{aligned} \mu &= \sum_{j=1}^{3}\sum_{k=1}^{3} K_{jk}^{(i)} d_j d_k + 2\sum_{j=1}^{3}\sum_{k=4}^{6} K_{jk}^{(i)} d_j e_{k-3} + \sum_{j=4}^{6}\sum_{k=4}^{6} K_{jk}^{(i)} e_{j-3} e_{k-3}; \\ u_j &= \sum_{k=1}^{3} K_{jk}^{(i)} d_k + \sum_{k=4}^{6} K_{jk}^{(i)} e_{k-3} \end{aligned} \tag{5.46}$$

Hence, the single-joint decomposition allows to replace of the matrix computations (5.41) by simple scalar calculations (5.42) that deal with the modification of the stiffness matrix elements, in accordance with the passive joint location defined by the vectors **d** and **e**. It is worth mentioning that in general cases, the transformation (5.42) affects all elements of the matrix $\mathbf{K}_C^i$, which differs from simple setting to zero the corresponding row and/or column that has been proposed in some publications [19]. Let us consider now several specific (but rather typical) cases where the transformation rules are simpler and more elegant.

Analytical Computations for Serial Chains with Trivial Passive Joints

For practical purposes, let us consider a special type of kinematic chain where the passive joint axes are collinear to the axes x, *y* or *z* of the Cartesian coordinate system. Such passive joints will be further referred to as "trivial" or "quasi-trivial" ones. For such cases, the vector-columns of the Jacobian $\mathbf{J}_\mathrm{q}$ include a number of zero elements, so the expressions (5.42) can be essentially simplified.

Let us start from a set of *trivial* cases where the Jacobians $\mathbf{J}_\mathrm{q}^i$ are created from the columns of the identity matrix $\mathbf{I}_{6\times6}$:

$$\mathbf{J}_{\mathrm{q}}^{(1)} = \begin{bmatrix} 1 \\ 0 \\ 0 \\ 0 \\ 0 \\ 0 \end{bmatrix}, \mathbf{J}_{\mathrm{q}}^{(2)} = \begin{bmatrix} 0 \\ 1 \\ 0 \\ 0 \\ 0 \\ 0 \end{bmatrix}, \ldots, \mathbf{J}_{\mathrm{q}}^{(6)} = \begin{bmatrix} 0 \\ 0 \\ 0 \\ 0 \\ 0 \\ 1 \end{bmatrix} \tag{5.47}$$

It can be easily proved that they cover the following types of the joint geometry:

- *translational passive joints* with arbitrary spatial position (but with the joint axis directed along x, y or z);
- *rotational passive joints* positioned at the reference point (and with the joint axis directed along x, y or z) or rotational passive joint $\mathrm{R_x}$ located at any point at the x-axis.

For these trivial cases, assuming that $\mathbf{J}_{\mathrm{q}}^{(p)}$ denotes the vector-column with a single non-zero element in the p-th position, a straightforward substitution into (5.42) yields $u_j = K_{jp}^{(i)}$; $\mu = K_{pp}^{(i)}$. So, the recursive expression for the Cartesian stiffness matrix is simplified to

$$\left[K_{jk}^{(i+1)}\right] = \left[K_{jk}^{(i)}\right] - \left[\frac{K_{jk}^{(i)} K_{kj}^{(i)}}{K_{pp}^{(i)}}\right] \tag{5.48}$$

and the p–th row and column of the matrix $\mathbf{K}_{\mathrm{C}}^{i+1}$ become equal to zero

$$K_{pk}^{(i+1)} = 0;\ K_{kp}^{(i+1)} = 0;\ \forall p = 1, \ldots 6\cdot \tag{5.49}$$

So, the recursive computations are easily performed analytically.

For the trivial passive joints, the results are summarized in Table 5.2 where the original stiffness matrix $[K_{ij}]$ is assumed to be sparse in accordance with the structure corresponding to a symmetric beam-type link. As it follows from them, here in all cases the stiffness matrix modifications include setting to zero of an appropriate row and column, but in some cases, some elements of other rows and columns are also modified (by simple division by 4).

Further, let us consider a set of *quasi-trivial* cases where the Jacobians $\mathbf{J}_{\mathrm{q}}^{i}$ are of a more complicated structure compared to $\mathbf{J}_{\mathrm{q}}^{(4)}, \mathbf{J}_{\mathrm{q}}^{(5)}, \mathbf{J}_{\mathrm{q}}^{(6)}$

$$\mathbf{J}_{q}^{(4+)} = \begin{bmatrix} 0 \\ d_y \\ d_z \\ 1 \\ 0 \\ 0 \end{bmatrix}, \mathbf{J}_{q}^{(5+)} = \begin{bmatrix} d_x \\ 0 \\ d_z \\ 0 \\ 1 \\ 0 \end{bmatrix}, \mathbf{J}_{q}^{(6+)} = \begin{bmatrix} d_x \\ d_y \\ 0 \\ 0 \\ 0 \\ 1 \end{bmatrix} \tag{5.50}$$

Table 5.2 Stiffness matrix transformations caused by trivial passive joints

Type of joint	Jacobian	Modified Stiffness matrix
Translational T_x 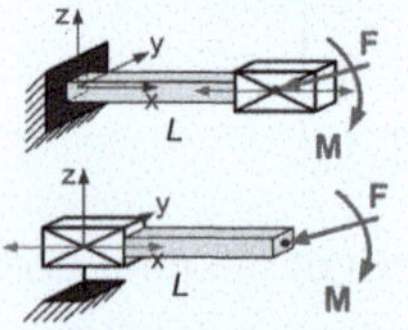	$\mathbf{J}_q^{(1)} = \begin{bmatrix}1\\0\\0\\0\\0\\0\end{bmatrix}$	$\begin{bmatrix}0&0&0&0&0&0\\0&K_{22}&0&0&0&K_{26}\\0&0&K_{33}&0&K_{35}&0\\0&0&0&K_{44}&0&0\\0&0&K_{53}&0&K_{55}&0\\0&K_{62}&0&0&0&K_{66}\end{bmatrix}$
Translational T_y 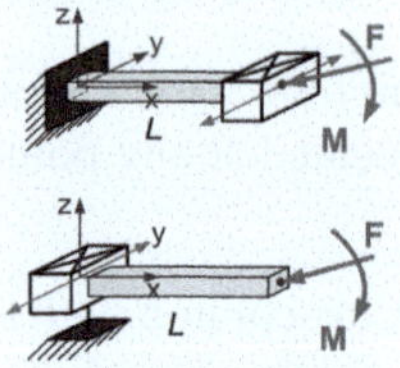	$\mathbf{J}_q^{(2)} = \begin{bmatrix}0\\1\\0\\0\\0\\0\end{bmatrix}$	$\begin{bmatrix}K_{11}&0&0&0&0&0\\0&0&0&0&0&0\\0&0&K_{33}&0&K_{35}&0\\0&0&0&K_{44}&0&0\\0&0&K_{53}&0&K_{55}&0\\0&0&0&0&0&\frac{K_{66}}{4}\end{bmatrix}$
Translational T_z 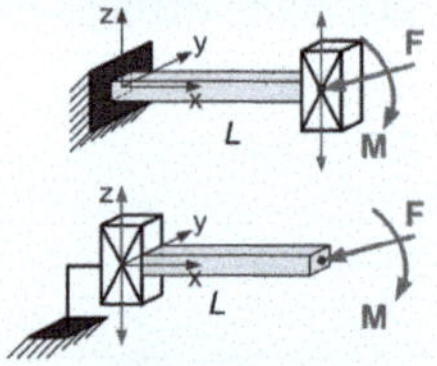	$\mathbf{J}_q^{(3)} = \begin{bmatrix}0\\0\\1\\0\\0\\0\end{bmatrix}$	$\begin{bmatrix}K_{11}&0&0&0&0&0\\0&K_{22}&0&0&0&K_{26}\\0&0&0&0&0&0\\0&0&0&K_{44}&0&0\\0&0&0&0&\frac{K_{55}}{4}&0\\0&K_{62}&0&0&0&K_{66}\end{bmatrix}$
Rotational R_x 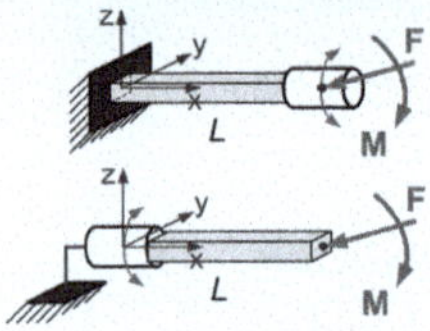	$\mathbf{J}_q^{(4+)} = \begin{bmatrix}0\\0\\0\\1\\0\\0\end{bmatrix}$	$\begin{bmatrix}K_{11}&0&0&0&0&0\\0&K_{22}&0&0&0&K_{26}\\0&0&K_{33}&0&K_{35}&0\\0&0&0&0&0&0\\0&0&K_{53}&0&K_{55}&0\\0&K_{62}&0&0&0&K_{66}\end{bmatrix}$
Rotational R_y 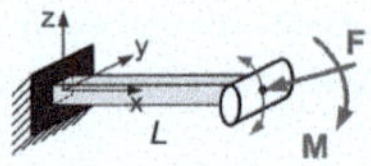	$\mathbf{J}_q^{(5+)} = \begin{bmatrix}0\\0\\0\\0\\1\\0\end{bmatrix}$	$\begin{bmatrix}K_{11}&0&0&0&0&0\\0&K_{22}&0&0&0&K_{26}\\0&0&\frac{K_{33}}{4}&0&0&0\\0&0&0&K_{44}&0&0\\0&0&0&0&0&0\\0&K_{62}&0&0&0&K_{66}\end{bmatrix}$
Rotational R_z 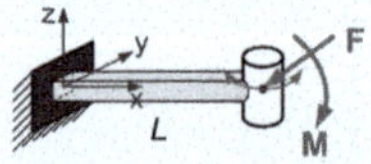	$\mathbf{J}_q^{(6+)} = \begin{bmatrix}0\\0\\0\\0\\0\\1\end{bmatrix}$	$\begin{bmatrix}K_{11}&0&0&0&0&0\\0&\frac{K_{22}}{4}&0&0&0&0\\0&0&K_{33}&0&K_{35}&0\\0&0&0&K_{44}&0&0\\0&0&K_{53}&0&K_{55}&0\\0&0&0&0&0&0\end{bmatrix}$

Table 5.3 Examples of stiffness matrix transformations for quasi-trivial passive joints

Type of joint	Jacobian	Modified Stiffness matrix
Rotational R_y 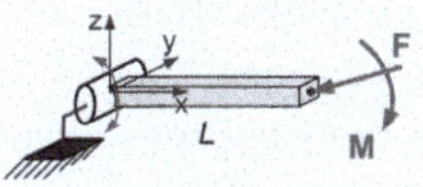	$\mathbf{J}_q^{(5+)} = \begin{bmatrix} 0 \\ 0 \\ -L \\ 0 \\ 1 \\ 0 \end{bmatrix}$	$\begin{bmatrix} K_{11} & 0 & 0 & 0 & 0 & 0 \\ 0 & K_{22} & 0 & 0 & 0 & K_{26} \\ 0 & 0 & \frac{K_{33}}{4} & 0 & \frac{K_{35}}{2} & 0 \\ 0 & 0 & 0 & K_{44} & 0 & 0 \\ 0 & 0 & \frac{K_{53}}{2} & 0 & \frac{3 \cdot K_{55}}{4} & 0 \\ 0 & K_{62} & 0 & 0 & 0 & K_{66} \end{bmatrix}$
Rotational R_z 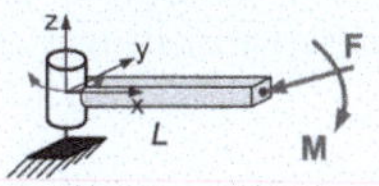	$\mathbf{J}_q^{(6+)} = \begin{bmatrix} 0 \\ L \\ 0 \\ 0 \\ 0 \\ 1 \end{bmatrix}$	$\begin{bmatrix} K_{11} & 0 & 0 & 0 & 0 & 0 \\ 0 & \frac{K_{22}}{4} & 0 & 0 & 0 & \frac{K_{26}}{2} \\ 0 & 0 & K_{33} & 0 & K_{35} & 0 \\ 0 & 0 & 0 & K_{44} & 0 & 0 \\ 0 & 0 & K_{53} & 0 & K_{55} & 0 \\ 0 & \frac{K_{62}}{2} & 0 & 0 & 0 & \frac{3 \cdot K_{66}}{4} \end{bmatrix}$

where $d_x,\ d_y,\ d_z$ denote the elements of the vector $\mathbf{d}$ defining the passive joint location. It can be proved that such types of Jacobians describe *rotational passive joints shifted* by the distance L with respect to the reference point in the direction either x, y or z (and with the joint axis directed along x, y or z), and the values of $d_x,\ d_y,\ d_z$ are equal to either $\pm L$ or 0.

For these case studies, relevant results are presented in Table 5.3 where the original stiffness matrix $[K_{ij}]$ is assumed to be the same as above. As it follows from them, here the resulting stiffness matrix is singular but does not include purely zero rows and columns. Hence, the existing technique [19] based on simple zeroing of the rows and columns cannot be applied for such types of passive joints.

Using results obtained for the trivial and quasi-trivial passive joints, in many practically significant cases we can easily modify analytically the original stiffness matrix by sequentially taking into account each passive joint in accordance with the developed recursive technique. Relevant examples will be presented below.

5.2.3 Stiffness Modelling of Parallel Manipulators: Serial Chains Aggregation Technique

Stiffness Model Aggregation for Perfect Serial Chains

The VJM-based stiffness modelling technique presented in previous sections allows us to compute the stiffness matrix for a serial chain with an arbitrary location of passive joints and to evaluate its properties. It also provides an analytical solution to the problem. However, this technique is limited by the stiffness modelling of serial kinematic chains or serial manipulators. Hence, it is reasonable to extend this technique for parallel architectures and to propose the method of aggregation of the elastostatic models of separate kinematic chains to the stiffness model of the parallel manipulator. Let us consider this problem in detail.

Let us assume that a parallel manipulator may be presented as a strictly parallel system of the actuated serial legs connecting the base and the end-platform (Fig. 5.3) [20]. Using the methodology described in the previous sections and applying it to each leg, a set of m Cartesian stiffness matrices $\mathbf{K}_{\mathrm{C}}^{(i)}$ expressed with respect to the same coordinate system but corresponding to different platform points can be computed. If initially the chain stiffness matrices are computed in local coordinate systems, their transformation is performed in a standard way [5], as

$$\mathbf{K}_{\mathrm{C}}^{\mathrm{glob}} = \begin{bmatrix} \mathbf{R} & \mathbf{0} \\ \mathbf{0} & \mathbf{R} \end{bmatrix} \mathbf{K}_{\mathrm{C}}^{\mathrm{loc}} \begin{bmatrix} \mathbf{R}^{\mathrm{T}} & \mathbf{0} \\ \mathbf{0} & \mathbf{R}^{\mathrm{T}} \end{bmatrix} \tag{5.51}$$

where $\mathbf{R}$ is a 3×3 rotation matrix describing the orientation of the local coordinate system with respect to the global one.

To aggregate these matrices $\mathbf{K}_{\mathrm{C}}^{(i)}$, they must be also re-computed with respect to the same reference point of the platform. Assuming that the platform is rigid enough (as compared to the compliance of the legs), this conversion can be performed by extending the legs by a virtual rigid link connecting the end-point of the leg and the reference point of the platform (see Fig. 5.5 where these extensions are defined by the vectors $\mathbf{v}_i$).

After such extension, an equivalent stiffness matrix of the leg may be expressed using the relevant expression for a usual serial chain, i.e. as $\mathbf{J}_{\mathrm{v}}^{(i)-\mathrm{T}}\mathbf{K}_{\mathrm{C}}^{(i)}\mathbf{J}_{\mathrm{v}}^{(i)-1}$, where the Jacobian $\mathbf{J}_{\mathrm{v}}^{(i)}$ defines the differential relation between the coordinates of the i-th virtual spring and the reference frame of the end-platform. Hence, the final expression for the stiffness matrix of the considered parallel manipulator can be written as

$$\mathbf{K}_{\mathrm{C}}^{(m)} = \sum_{i=1}^{m} \left(\mathbf{J}_{\mathrm{v}}^{(i)^{-\mathrm{T}}} \cdot \mathbf{K}_{\mathrm{C}}^{(i)} \cdot \mathbf{J}_{\mathrm{v}}^{(i)^{-1}} \right) \tag{5.52}$$

where m is the number of serial kinematic chains in the manipulator architecture.

As a result, the expression (5.52) allows us to compute the Cartesian stiffness matrix for the parallel manipulator based on stiffness matrices for serial chains and transformation Jacobians $\mathbf{J}_{\mathrm{v}}^{(i)}$, which defines a geometrical mapping between end-points of serial chains and reference point frames (end-effector) [21]. Moreover, it is implicitly assumed here that all stiffness matrices (both for the serial chains and for the whole manipulator) are expressed in the same global coordinate system, otherwise, they should be recomputed to requisite coordinates. Hence, the axes of all virtual springs are parallel to the axes x, y, and z of this system. This allows us to evaluate Jacobians $\mathbf{J}_{\mathrm{v}}^{(i)}$ and their inverses from the geometry of the end-platform analytically

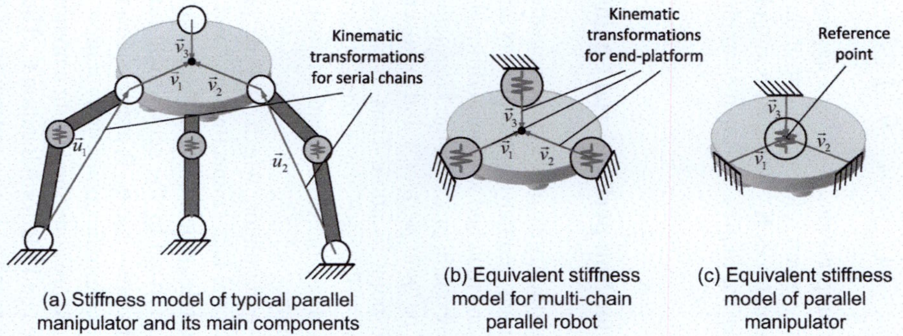

Fig. 5.5 Typical parallel manipulator (**a**) and transformation of its VJM models (**b, c**)

$$\mathbf{J}_{\mathrm{V}}^{(i)} = \begin{bmatrix} \mathbf{I}_3 & (\mathbf{v}_i\times) \\ \mathbf{0} & \mathbf{I}_3 \end{bmatrix}_{6\times 6}, \mathbf{J}_{\mathrm{V}}^{(i)-1} = \begin{bmatrix} \mathbf{I}_3 & -(\mathbf{v}_i\times) \\ \mathbf{0} & \mathbf{I}_3 \end{bmatrix}_{6\times 6} \tag{5.53}$$

where $\mathbf{I}_3$ is 3×3 identity matrix, and $(\mathbf{v}\times)$ is a skew-symmetric matrix corresponding to the vector $\mathbf{v}$:

$$(\mathbf{v}\times) = \begin{bmatrix} 0 & -v_{\mathrm{z}} & v_{\mathrm{y}} \\ v_{\mathrm{z}} & 0 & -v_{\mathrm{x}} \\ -v_{\mathrm{y}} & v_{\mathrm{x}} & 0 \end{bmatrix} \tag{5.54}$$

Therefore, expression (5.52) represents an explicit aggregation of the leg stiffness matrices with respect to any given reference point of the platform. It is worth mentioning that in practice, the matrices $\mathbf{K}_{\mathrm{C}}^{(i)}$ are always singular while the aggregation usually produces a non-singular matrix. Relevant examples are presented in the following sub-section.

Stiffness Model Aggregation for Non-perfect Serial Chains

In the previous Section, it was implicitly assumed that kinematic chains of the parallel manipulator are "*perfect*", i.e. their geometrical parameters are strictly nominal and their end-frames can be aligned and matched without additional efforts and/or internal stresses while assembling. However, in practice, the kinematic chain geometry may differ from the nominal one (i.e. be "*non-perfect*"), and the assembling causes internal stresses and shifting of the end-point location. Thus, it is reasonable to generalize the above aggregation procedure to the case of non-perfect serial chains and develop a technique, which is able to evaluate deflections and internal forces/torques caused by geometrical errors in the chains.

Let us consider a parallel manipulator, which may be presented as a strictly parallel system of the actuated serial legs connecting the fixed base and mobile end-platform [20]. Using the methodology described in the previous sections, each i-th leg (serial chain) may be characterized by the geometrical function $\mathbf{t}_i = \mathbf{g}_i(\mathbf{q}_i, \boldsymbol{\theta}_i)$, where $\mathbf{t}_i$ defines the end-frame location and $\mathbf{q}_i$, $\boldsymbol{\theta}_i$ are passive and virtual joint coordinates respectively, which define kinematic chain configuration. For perfect kinematic chains, the joint coordinates $\mathbf{q}_i$, $\boldsymbol{\theta}_i$ are computed using a nominal geometrical model of the leg $\mathbf{g}_i^0(\mathbf{q}_i, \boldsymbol{\theta}_i)$, for the given end-platform location $\mathbf{t}^0$. These notations are illustrated by Fig. 5.6a where A_i correspond to the end-points of the perfect chains and O is the platform reference point. Similar to Sect. 3.4, the vectors $\overrightarrow{A_iO}$ are denoted as $\mathbf{v}_i$. Using the above-presented technique, it is possible to compute the Cartesian stiffness matrices of the chains and express them with respect to the same reference point O. Let us denote this matrix as $K_C^{(i)}$.

For the non-perfect chain before assembling, similar configurations $\mathbf{q}_i$, $\boldsymbol{\theta}_i$ produce slightly different end-point locations of the chains $\mathbf{t}_i = \mathbf{g}_i^0(\mathbf{q}_i, \boldsymbol{\theta}_i) + \boldsymbol{\varepsilon}_i$, which are denoted as A_i' in Fig. 5.6a. Correspondently, assuming that the platform is rigid

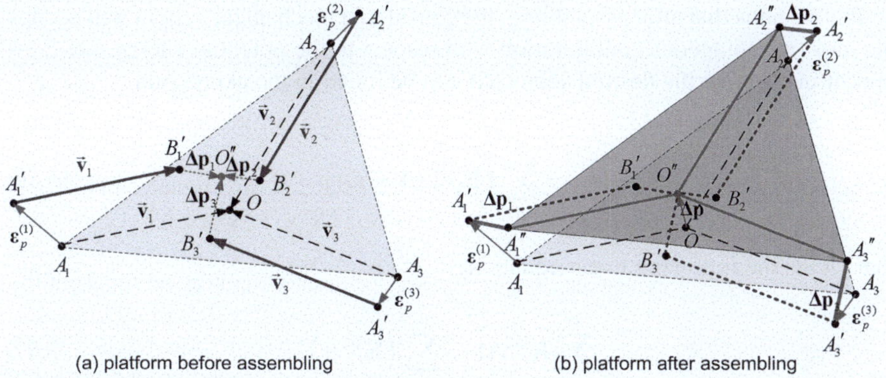

Fig. 5.6 Transformation of characteristic points of serial chains in assembling the non-perfect parallel manipulator; (A_i, A_i'—end-point locations of the serial chain before assembling for perfect and non-perfect manipulators respectively, A_i''—end-point location of the serial chain after assembling for non-perfect manipulator) [1]

enough, the points A_i' can be mapped to B_i' that differ from O by $\boldsymbol{\varepsilon}_i$. Hence, geometrical errors do not allow the assembling of parallel manipulators at the original reference point.

To assemble all chains in the same reference point, it is required to apply additional efforts. Geometrically, it leads to the relocation of the frames corresponding to the points B_i' to a new position B_i'', and relocations of the point A_i' to A_i'' that is defined by the vector $\Delta\mathbf{t}_i$. In Fig. 5.6b, the relevant reference point of the platform is denoted as O''. Hence, the stiffness model aggregation problem is reduced to computing a new end-platform location $t^0 + \Delta t$ for which end-frames of all kinematic chains are aligned and matched. Consequently, a manipulator will occupy the most advantageous configuration with respect to the potential energy of the elastic elements.

To compute the end-platform deflection $\Delta\mathbf{t}$, let us assume that the geometrical errors are small enough and corresponding shifting of the chain end-points and end-effector do not change Cartesian stiffness matrices $K_C^{(i)}$ or their influences are negligible. So, the stiffness matrices of the serial chains $K_C^{(i)}$ are the same at the points O, B_i' and O'' and computed for the nominal configurations q_i, θ_i. Let us also assume that the point O'' is shifted from O by Δp and the mobile platform orientations for the parallel manipulator with perfect and non-perfect serial chains differ by $\Delta\boldsymbol{\varphi}$. This allows us to express the potential energy of the parallel manipulator with geometrical errors in kinematic chains as

$$E = \frac{1}{2}\sum_{i=1}^{m}\left((\boldsymbol{\varepsilon}_i - \Delta\mathbf{t})^{\mathrm{T}}\mathbf{K}_{\mathrm{C}}^{(i)}(\boldsymbol{\varepsilon}_i - \Delta\mathbf{t})\right) \tag{5.55}$$

where $\Delta\mathbf{t} = (\Delta\mathbf{p}, \Delta\boldsymbol{\varphi})$ is displacement (position and orientation) of the reference point.

It is obvious that after assembling, the considered mechanical system will occupy the most advantageous configuration with respect to the potential energy, i.e. $E \underset{\Delta\mathbf{t}}{\rightarrow} \min$. It means that the desired vector $\Delta\mathbf{t}$ can be found in the expression

$$\frac{\partial E}{\partial \Delta\mathbf{t}} = \sum_{i=1}^{m} \left(\mathbf{K}_{\mathrm{C}}^{(i)} (\Delta\mathbf{t} - \boldsymbol{\varepsilon}_i) \right) = 0 \tag{5.56}$$

that yields the following linear equation

$$\sum_{i=1}^{m} \mathbf{K}_{\mathrm{C}}^{(i)} \Delta\mathbf{t} = \sum_{i=1}^{m} \left(\mathbf{K}_{\mathrm{C}}^{(i)} \boldsymbol{\varepsilon}_i \right) \tag{5.57}$$

which allows us to evaluate the end-platform deflection

$$\Delta\mathbf{t} = \left(\sum_{i=1}^{m} \mathbf{K}_{\mathrm{C}}^{(i)} \right)^{-1} \sum_{i=1}^{m} \left(\mathbf{K}_{\mathrm{C}}^{(i)} \boldsymbol{\varepsilon}_i \right) \tag{5.58}$$

and the end-platform location after assembling

$$\mathbf{t}' = \mathbf{t}^0 + \Delta\mathbf{t} \tag{5.59}$$

For each separate kinematic chain, the end-frame deflections due to assembling can be expressed as

$$\Delta\mathbf{t}_i = \Delta\mathbf{t} - \boldsymbol{\varepsilon}_i = \left(\sum_{i=1}^{m} \mathbf{K}_{\mathrm{C}}^{(i)} \right)^{-1} \sum_{i=1}^{m} \left(\mathbf{K}_{\mathrm{C}}^{(i)} \boldsymbol{\varepsilon}_i \right) - \boldsymbol{\varepsilon}_i \tag{5.60}$$

This allows us to compute the loading for each kinematic chain applied to the end-point (due to interaction with other non-perfect chains)

$$\mathbf{F}_i = \mathbf{K}_C^{(i)} \cdot \Delta\mathbf{t}_i \tag{5.61}$$

and corresponding loadings in the virtual joints $\boldsymbol{\tau}_{\theta}^{(i)}$

$$\tau_{\theta}^{(i)} = \mathbf{F}_{\theta}^{(i)T} \mathbf{F}_i = \mathbf{J}_{\theta}^{(i)T} \mathbf{K}_C^{(i)} \Delta t_i \tag{5.62}$$

It is worth mentioning that here $\sum_{i=1}^{m} \mathbf{F}_i = \mathbf{0}$, since there is no external loading applied to the platform reference point after the assembly. Besides, it is possible to compute relevant deflections of the virtual and passive joint coordinates of the chains

$$\boldsymbol{\theta}_i = \mathbf{K}_{\mathrm{C}\theta}^{(i)} \Delta\mathbf{t}_i;\ \Delta\mathbf{q}_i = \mathbf{K}_{\mathrm{Cq}}^{(i)} \Delta\mathbf{t}_i \tag{5.63}$$

caused by the assembling.

Thus, the above expressions allow us to evaluate the end-platform deflection and internal forces/torques caused by the assembling of kinematic chains with geometrical errors. However, the total manipulator Cartesian stiffness matrix K_C is assumed to be the same as in Sect. 5.2.3(a), since the geometrical errors are assumed to be small enough.

5.3 Illustrative Example: VJM-Based Stiffness Modelling of the Gougth Platform

Let us apply now the developed technique to computing the stiffness matrix for two versions of a general Stewart–Gough platform presented in Fig. 5.7 [17, 22–25]. It is assumed that in both cases the manipulator base and the moving plate (platform) are connected by six similar extensible legs (Fig. 5.8) but their spatial arrangements are different:

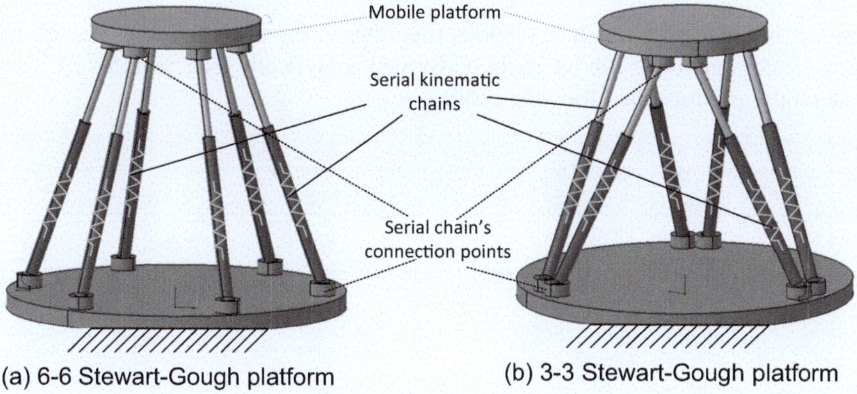

Fig. 5.7 Geometry of the Stewart–Gough platforms under study

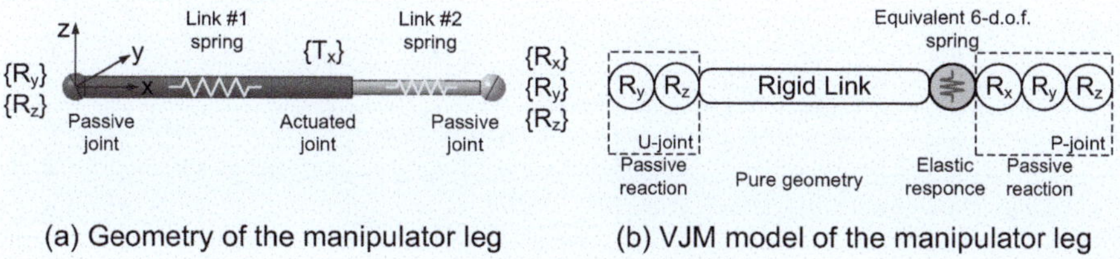

Fig. 5.8 Geometry of the manipulator's leg and its VJM model

- Case A: the legs are regularly connected to the base and platform, with the same angular distance of 60° (it is obviously a degenerated design, where the stiffness matrix should be singular)
- Case B: the legs are connected to the base and platform in three pairs, with an angular distance of 120° between the mounting points (it is a classical design of Stewart-Gough where the stiffness matrix should be non-singular).

For both designs, the original leg stiffness (i.e. without the passive joints) can be described by the sparse matrix corresponding to the symmetric beam. Further, to take into account the influence of the passive joints, it should be applied recursively to the procedure (5.41) with the elementary Jacobians

$$\mathbf{J}_{\mathrm{Rx}}^{(1)}=\begin{bmatrix}0\\0\\0\\1\\0\\0\end{bmatrix};\mathbf{J}_{\mathrm{Ry}}^{(2)}=\begin{bmatrix}0\\0\\0\\0\\1\\0\end{bmatrix};\mathbf{J}_{\mathrm{Rz}}^{(3)}=\begin{bmatrix}0\\0\\0\\0\\0\\1\end{bmatrix};\mathbf{J}_{\mathrm{Ry}+}^{(4)}=\begin{bmatrix}0\\0\\-L\\0\\1\\0\end{bmatrix};\mathbf{J}_{\mathrm{Rz}+}^{(5)}=\begin{bmatrix}0\\L\\0\\0\\0\\1\end{bmatrix} \tag{5.64}$$

where L is the leg length. It is obvious that, due to the trivial structure of J_q^i, the recursive computations can be easily performed analytically (see Sect. 5.2.2). They sequentially produce the following results:

$$\mathbf{K}_C^0=\left[\begin{array}{ccc|ccc}K_{11}&0&0&0&0&0\\0&K_{22}&0&0&0&K_{26}\\0&0&K_{33}&0&K_{35}&0\\\hline 0&0&0&K_{44}&0&0\\0&0&K_{53}&0&K_{55}&0\\0&K_{62}&0&0&0&K_{66}\end{array}\right]\qquad \mathbf{K}_C^1=\left[\begin{array}{ccc|ccc}K_{11}&0&0&0&0&0\\0&K_{22}&0&0&0&K_{26}\\0&0&K_{33}&0&K_{35}&0\\\hline 0&0&0&0&0&0\\0&0&K_{53}&0&K_{55}&0\\0&K_{62}&0&0&0&K_{66}\end{array}\right]$$

$$\mathbf{K}_C^2=\left[\begin{array}{ccc|ccc}K_{11}&0&0&0&0&0\\0&K_{22}&0&0&0&K_{26}\\0&0&K_{33}/4&0&0&0\\\hline 0&0&0&0&0&0\\0&0&0&0&0&0\\0&K_{62}&0&0&0&K_{66}\end{array}\right]\qquad \mathbf{K}_C^3=\left[\begin{array}{ccc|ccc}K_{11}&0&0&0&0&0\\0&K_{22}/4&0&0&0&0\\0&0&K_{33}/4&0&0&0\\\hline 0&0&0&0&0&0\\0&0&0&0&0&0\\0&0&0&0&0&0\end{array}\right]$$

$$\mathbf{K}_C^4=\left[\begin{array}{ccc|ccc}K_{11}&0&0&0&0&0\\0&K_{22}/4&0&0&0&0\\0&0&0&0&0&0\\\hline 0&0&0&0&0&0\\0&0&0&0&0&0\\0&0&0&0&0&0\end{array}\right]\qquad \mathbf{K}_C^5=\left[\begin{array}{ccc|ccc}K_{11}&0&0&0&0&0\\0&0&0&0&0&0\\0&0&0&0&0&0\\\hline 0&0&0&0&0&0\\0&0&0&0&0&0\\0&0&0&0&0&0\end{array}\right]$$

Hence, in the final form, the derived matrix includes only the traction/compression terms (and not bending, torsion, etc.) that agree perfectly with other results on Stewart–Gough platforms.

Further, in order to be applied to each leg, the obtained matrix must be transformed from the local to the global coordinate system. In this specific case, due to the special structure of $\mathbf{K}_{\mathrm{C}}^{5}$, relevant transformation [5, 26]

$$\mathbf{K}_{\mathrm{C}_i} = \begin{bmatrix} \mathbf{R}_i & \mathbf{0} \\ \mathbf{0} & \mathbf{R}_i \end{bmatrix} \cdot K_{11} \left[\begin{array}{ccc|c} 1 & & & \\ & 0 & & \mathbf{0}_{3\times3} \\ & & 0 & \\ \hline & \mathbf{0}_{3\times3} & & \mathbf{0}_{3\times3} \end{array} \right] cdot \begin{bmatrix} \mathbf{R}_i^{\mathrm{T}} & \mathbf{0} \\ \mathbf{0} & \mathbf{R}_i^{\mathrm{T}} \end{bmatrix} \tag{5.65}$$

expressed via the orthogonal rotation matrix $\mathbf{R}_i$ describing the orientation of the i-th local coordinate system with respect to the global one is easily reduced to

$$\mathbf{K}_{\mathrm{C}_i} = K_{11} \cdot \left[\begin{array}{c|c} \mathbf{u}_i^0 \cdot \mathbf{u}_i^{0\mathrm{T}} & \mathbf{0}_{3\times3} \\ \hline \mathbf{0}_{3\times3} & \mathbf{0}_{3\times3} \end{array} \right] \tag{5.66}$$

where $\mathbf{u}_i^0$ is the unit vector directed along the leg axis $\mathbf{u}_i$ (see Fig. 5.5). Besides, before aggregating, the stiffness matrices of the separate legs $\mathbf{K}_{\mathrm{C}_i}$ must be re-computed with respect to the same reference point in accordance with expressions (5.51) (5.52) which yields

$$\mathbf{K}_{\mathrm{C}} = K_{11} \sum_{i=1}^{6} \left[\begin{array}{c|c} \mathbf{I}_3 & \mathbf{0} \\ \hline (\mathbf{v}_i\times) & \mathbf{I}_3 \end{array} \right] \left[\begin{array}{c|c} \mathbf{u}_i^0\mathbf{u}_i^{0\mathrm{T}} & \mathbf{0}_{3\times3} \\ \hline \mathbf{0}_{3\times3} & \mathbf{0}_{3\times3} \end{array} \right] \left[\begin{array}{c|c} \mathbf{I}_3 & (\mathbf{v}_i\times)^{\mathrm{T}} \\ \hline \mathbf{0} & \mathbf{I}_3 \end{array} \right] \tag{5.67}$$

where $\mathbf{v}_i$ is the vector from the leg end-point to the platform reference point (see Fig. 5.5). So, after relevant transformations, one can get the final expression of the manipulator stiffness matrix

$$\mathbf{K}_{\mathrm{C}} = K_{11} \sum_{i=1}^{6} \left[\begin{array}{c|c} (\mathbf{u}_i^0\mathbf{u}_i^{0\mathrm{T}}) & (\mathbf{u}_i^0\mathbf{u}_i^{0\mathrm{T}}) \cdot (\mathbf{v}_i\times)^{\mathrm{T}} \\ \hline (\mathbf{v}_i\times)(\mathbf{u}_i^0\mathbf{u}_i^{0\mathrm{T}}) & (\mathbf{v}_i\times)(\mathbf{u}_i^0\mathbf{u}_i^{0\mathrm{T}})(\mathbf{v}_i\times)^{\mathrm{T}} \end{array} \right] \tag{5.68}$$

or

$$\mathbf{K}_{\mathrm{C}} = K_{11} \sum_{i=1}^{6} \left[\begin{array}{c} \mathbf{u}_i^0 \\ \hline (\mathbf{v}_i \times \mathbf{u}_i^0) \end{array} \right] \left[\begin{array}{c|c} \mathbf{u}_i^{0\mathrm{T}} & (\mathbf{v}_i \times \mathbf{u}_i^0)^{\mathrm{T}} \end{array} \right] \tag{5.69}$$

where the vector $\mathbf{u}_i$, $\mathbf{v}_i$ describing spatial locations of the legs and computed via the direct kinematics, and $\mathbf{v}_i \times \mathbf{u}_i^0$ denotes the vector product (in contrast to the above notation $(\mathbf{v}_i\times)$ which is referred to as the corresponding skew-symmetric matrix).

The derived equation has been applied to both case studies, assuming that the manipulators are in their "*home*" configurations when the platform is parallel to the base and it is symmetrical with respect to the vertical axis. Corresponding expressions for the leg vectors are

$$\mathbf{u}_i = \begin{bmatrix} r\cos(\psi_i) - R\cos(\phi_i) \\ r\sin(\psi_i) - R\sin(\phi_i) \\ h \end{bmatrix}, \mathbf{v}_i = \begin{bmatrix} -r\cos(\psi_i) \\ -r\sin(\psi_i) \\ 0 \end{bmatrix} \tag{5.70}$$

where $\phi_i = \psi_i \in \{0,\ 60°,\ 120°,\ 180°,\ 240°,\ 300°\}$ for the case A, and $\phi_i \in \{0,\ 120°,\ 120°,\ 240°,\ 240°,\ 360°\}$; $\psi_i \in \{60, 60, 180°, 180°, 300°, 300°\}$ for the case B. Substitution of these vectors into the expression (5.68) leads to the following stiffness matrices

$$\mathbf{K}_{\mathrm{C}}^{(A)} = \frac{3K_{11}}{L^2}\left[\begin{array}{ccc|ccc} d_a^2 & 0 & 0 & 0 & rhd_a & 0 \\ 0 & d_a^2 & 0 & -rhd_a & 0 & 0 \\ 0 & 0 & 2h^2 & 0 & 0 & 0 \\ \hline 0 & -rhd_a & 0 & r^2h^2 & 0 & 0 \\ rhd_a & 0 & 0 & 0 & r^2h^2 & 0 \\ 0 & 0 & 0 & 0 & 0 & 0 \end{array}\right] \tag{5.71}$$

and

$$\mathbf{K}_{\mathrm{C}}^{(B)} = \frac{3K_{11}}{L^2}\left[\begin{array}{ccc|ccc} d_{\mathrm{a}}^2 + Rr & 0 & 0 & 0 & rhd_{\mathrm{b}} & 0 \\ 0 & d_{\mathrm{a}}^2 + Rr & 0 & -rhd_{\mathrm{b}} & 0 & 0 \\ 0 & 0 & 2h^2 & 0 & 0 & 0 \\ \hline 0 & -rhd_{\mathrm{b}} & 0 & r^2h^2 & 0 & 0 \\ rhd_{\mathrm{b}} & 0 & 0 & 0 & r^2h^2 & 0 \\ 0 & 0 & 0 & 0 & 0 & 1.5r^2R^2 \end{array}\right] \tag{5.72}$$

where R and r denote the circle radius which comprises the leg connection point at the base and moving platform respectively, $d_{\mathrm{a}} = R - r$; $d_{\mathrm{b}} = R/2 - r$; L is the leg length, h is the vertical distance between the base and the platform, and the superscripts 'A' and 'B' define the relevant case study. As it follows from these expressions, in the 'zero' location the matrix $\mathbf{K}_{\mathrm{C}}^{(A)}$ is singular and allows "free" rotation of the end-platform around the vertical axis. In contrast, for the same location the matrix $\mathbf{K}_{\mathrm{C}}^{(B)}$ is non-singular and the manipulator resists all external forces/torques applied to the platform. These results are in good agreement with previous research on the Stewart–Gough platforms and confirm the efficiency of the developed computational technique for manipulator stiffness modelling [25, 27].

Hence, the proposed stiffness model assembling technique allows us to aggregate elastostatic models of separate kinematic chains in the stiffness model of the parallel manipulator. It is also able to take into account the geometry of the end-platform and its connection with kinematic chains in an explicit form.

References

1. Klimchik A, Chablat D, Pashkevich A (2014) Stiffness modeling for perfect and non-perfect parallel manipulators under internal and external loadings. Mech Mach Theory 79:1–28
2. Pashkevich A, Klimchik A, Chablat D (2011) Enhanced stiffness modeling of manipulators with passive joints. Mech Mach Theory 46:662–679
3. Anatol P, Alexandr K, Damien C (2010) Enhanced stiffness modeling of serial manipulators with passive joints. In: Ernest H (ed) Advances in robot manipulators. IntechOpen, Rijeka, pp Ch. 17
4. Gosselin C (1990) Stiffness mapping for parallel manipulators. Robot Autom IEEE Trans 6:377–382
5. Angeles J (2007) Fundamentals of robotic mechanical systems: theory, methods, and algorithms. Springer
6. Gosselin C, Zhang D (2002) Stiffness analysis of parallel mechanisms using a lumped model. Int J Robot Autom 17:17–27
7. Mamedov S, Popov D, Mikhel S, Klimchik A, Pashkevich A Stiffness modeling of planar robotic manipulators: model reduction and identifiability of parameters. In: ROMANSY 23—robot design, dynamics and control. Springer International Publishing, pp 476–485
8. Klimchik A, Pashkevich A, Chablat D (2013) CAD-based approach for identification of elastostatic parameters of robotic manipulators. Finite Elem Anal Des 75:19–30
9. Pashkevich A, Klimchik A, Chablat D, Wenger P Accuracy improvement for stiffness modeling of parallel manipulators. In: 42nd CIRP conference on manufacturing systems, pp 8
10. http://en.wikipedia.org/wiki/Stewart_platform
11. Carricato M, Duffy J, Parenti-Castelli V (2002) Catastrophe analysis of a planar system with flexural pivots. Mech Mach Theory 37:693–716
12. Pashkevich A, Chablat D, Wenger P (2009) Stiffness analysis of overconstrained parallel manipulators. Mech Mach Theory 44:966–982
13. Chen S, Kao I The conservative congrue nce transformation of stiffness control in robotic grasping and manipulation. In: Proceedings of the 9th international symposium of robotics research, pp 7–14
14. Gantmakher FR (1959) The theory of matrices. 1. Chelsea Publishing Company
15. Quennouelle C, Gosselin Cá (1959) Stiffness matrix of compliant parallel mechanisms. In: Advances in robot kinematics: analysis and design. Springer, pp 331–341
16. Quennouelle C, Gosselin CM (2008) Instantaneous kinemato-static model of planar compliant parallel mechanisms. In: ASME 2008 international design engineering technical conferences and computers and information in engineering conference. American Society of Mechanical Engineers, pp 163–173
17. Chen J, Lan F (2008) Instantaneous stiffness analysis and simulation for hexapod machines. Simul Modell Pract Theory 16:419–428
18. Strang G, Strang G, Strang G, Strang G (1993) Introduction to linear algebra. Wellesley-Cambridge Press Wellesley, MA
19. Taghvaeipour A, Angeles J, Lessard L (2012) On the elastostatic analysis of mechanical systems. Mech Mach Theory 58:202–216
20. Merlet JP (2006) Parallel robots. Springer
21. Klimchik A, Pashkevich A, Caro S, Chablat D (2012) Stiffness matrix of manipulators with passive joints: computational aspects. Robot IEEE Trans 28:955–958

22. Arumugam HK, Voyles RM, Bapat S (2012) Stiffness analysis of a class of parallel mechanisms for micro-positioning applications. In: Intelligent robots and systems, 2004.(IROS 2004). Proceedings. 2004 IEEE/RSJ international conference on. IEEE, pp 1826–1831
23. El-Khasawneh BS, Ferreira PM Computation of stiffness and stiffness bounds for parallel link manipulators. Int J Mach Tools Manufact 39:321–342
24. Shi H, Su H-J, Dagalakis N, Kramar JA (2013) Kinematic modeling and calibration of a flexure based hexapod nanopositioner. Precis Eng 37:117–128
25. Yu-Wen L, Jin-Song W, Li-Ping W Stiffness analysis of a Stewart platform-based parallel kinematic machine. In: Proceedings 2002 IEEE international conference on robotics and automation (Cat. No.02CH37292), pp 3672–3677, vol 3674
26. Deblaise D, Hernot X, Maurine P A systematic analytical method for PKM stiffness matrix calculation. In: IEEE international conference on robotics and automation (ICRA 2006). IEEE, pp 4213–4219
27. Svinin MM, Hosoe S, Uchiyama M On the stiffness and stability of Gough-Stewart platforms. In: Proceedings 2001 ICRA. IEEE international conference on robotics and automation (Cat. No.01CH37164), pp 3268–3273, vol 3264

Chapter 6
Non-linear Stiffness Models for Robotic Manipulators Under the Loading

6.1 Influence of External and Internal Loadings on the Cartesian Stiffness Matrix

In the previous Chapter, it was explicitly assumed that the loading causes rather *small changes* in all manipulator variables (and also in equilibrium configuration). It allowed us to obtain *linear relations* between all considered variables and to limit our study by computing *stiffness matrices*. This case was also referred to as the '*unloaded mode*'. However, in practice, the loading may cause essential deflections in manipulator elastic elements, so the stiffness model should be definitely non-linear. For this reason, in this Chapter, the 'small deflections' assumption is released and the study focuses on obtaining *non-linear force–deflection relations* and their linearization in the neighbourhood of the elastostatic equilibriums, which produces relevant stiffness matrices. Below, the 'large deflections' case will be referred to as the '*loaded mode*'.

In literature, the manipulator stiffness modelling for the loaded mode has not been studied in detail yet. The majority of the related works focus on computing of Cartesian stiffness matrix taking into account the loading value but implicitly assuming that the *equilibrium* is exactly the same as for the unloaded mode. To our knowledge, at present, there are no robotic papers where a non-linear force–deflection relation has been obtained strictly and for a wide range of external loading. In particular, in the works [1–3] one could obtain an *extended-expression* for the stiffness matrix of a robotic manipulator, where in addition to the classical term $\mathbf{J}_\theta^{-T} \cdot \mathbf{K}_\theta \cdot \mathbf{J}_\theta^{-1}$, the second term $\mathbf{J}_\theta^{-T} \cdot \mathbf{K}_\theta \cdot \mathbf{J}_\theta^{-1}$ that takes into account the configuration changes due to the loading has been introduced (here, $\mathbf{J}_\theta$ is the Jacobian, $\mathbf{K}_\theta$ is the joint stiffness matrix and $\mathbf{K}_F$ is computed in a special way). From a mathematical point of view, this result can be treated as the second-order approximation of the force–deflection relation in the neighbourhood of the unloaded equilibrium. It means that strictly speaking, this expression cannot be applied to a wide range of external loading. Another limitation is related to the influence of the passive joints that have

A. Klimchik et al., *Stiffness Modeling of Parallel Robots*, Parallel Robots: Theory and Applications, https://doi.org/10.1007/978-3-031-76320-5_6

not been considered within the frame in the above-mentioned papers. Other related works include [1, 3–7]. In our opinion, the most advanced results were obtained in [8, 9], but the issue of computing loaded equilibrium has been omitted.

In industrial robotics, it is usually assumed that the loading is caused by an interaction between the robot end-effector and a workpiece. However, there may exist other types of external or internal forces/torques that affect the manipulator stiffness properties. For instance, for heavy manipulators [10, 11], the gravity forces produce essential end-effector deflections. So, they are usually compensated by supplementary mechanisms that generate additional loadings applied to the manipulator elements. It is obvious that such a type of *external loading* (both gravity itself and gravity compensation) must be taken into account in the stiffness model. Further, these types of forces/torques are referred to as the '*auxiliary loading*'.

Furthermore, kinematic chains of robotic manipulators may include some additional elastic elements in the actuated or/and passive joints that are intended to increase the robot positioning accuracy or to improve the stiffness properties in certain workspace areas. For example, to eliminate the backlash, the gear trains may include spring-loaded scissor elements that generate the internal forces, which must be also integrated into the stiffness model [12]. Similar forces may also arise in the parallel manipulators with antagonistic actuating [13, 14]. Other examples include parallel manipulators with springs interposed in the passive joints in order to improve stiffness in the singularity neighbourhood. For this type of force/torque the term '*internal preloading*' will be used below.

As it follows from our analysis, non-linear stiffness modelling of robotic manipulators must include explicit computation of the coordinates corresponding to the *loaded equilibrium*. These coordinates should be used while calculating Jacobians (and Hessians) of the manipulator geometrical model that is obviously included in the desired stiffness model. One of the potential problems here is that, in general, the static equilibrium equation of the loaded manipulator cannot be solved unambiguously, i.e. for the same loading multiple equilibriums may exist, either stable or unstable. Hypothetically, this may create very specific manipulator behaviour while the loading is increasing or decreasing (sudden jump from one equilibrium configuration to another). Some aspects of this problem have been studied in [15, 16].

In the frame of the non-linear stiffness modelling technique, it is also worth revising the *stiffness models aggregation* technique. Here, in addition to the aggregated Cartesian stiffness matrix, it is required to obtain the aggregated force–deflection relation assuming that the kinematic chain may include geometrical errors. Moreover, in practice, it is often necessary to solve the inverse problem that is equivalent to obtaining the non-linear deflection-force relation. These problems have not been studied in literature yet. However, they are very important for *compliance error compensation* (via proper adjusting of the actuated coordinates). At present, the compliance error compensation problems have been solved in the following works [17–25].

As it follows from the above analysis, the manipulator stiffness modelling for the loaded mode needs further improvement. Thus, the goal of this chapter is to extend the results of Chap. 5 for the case of large deflections (loaded mode).

6.1.1 *Serial Kinematic Chain of Parallel Manipulator with Loading at the End-Point and Preloading in Joints*

Let us start the development of a non-linear stiffness modelling technique assuming that the most essential loading is applied to the end-point of the kinematic chain. Besides, let us assume that some joints may include special springs, which eliminate backlash and/or modify the manipulator stiffness properties. Under these assumptions, the results of linear stiffness modelling cannot be applied straightforwardly due to significant changes in the manipulator configuration under the loading. In particular, for loaded equilibrium, the values of virtual and passive joint coordinates $\boldsymbol{\theta}, \mathbf{q}$ cannot be computed purely geometrically. Moreover, here the stiffness properties should depend on the loading magnitude [2, 8, 26] and some non-linear effects in the stiffness behaviour can be discovered [27]. Let us focus on these issues.

Problem Statement

The serial chain under study includes flexible actuated joints and flexible links that may be separated by passive joints. Its geometry is described by the vector function

$$\mathbf{t} = \mathbf{g}(\mathbf{q}, \boldsymbol{\theta}) \tag{6.1}$$

where the vector $\mathbf{t} = (\mathbf{p}, \boldsymbol{\varphi})^{\mathrm{T}}$ includes the position $\mathbf{p} = (x, y, z)^{\mathrm{T}}$ and orientation $\boldsymbol{\varphi} = (\varphi_x, \varphi_y, \varphi_z)^{\mathrm{T}}$ of the end-effector in the Cartesian space; the vector $\mathbf{q} = (q_1, q_2, \ldots, q_{\mathrm{nq}})^{\mathrm{T}}$ contains "perfect passive joint" coordinates (i.e., without internal preloading); the vector $\boldsymbol{\theta} = (\theta_1, \theta_2, \ldots, \theta_{\mathrm{n}\theta})^{\mathrm{T}}$ collects coordinates of all virtual joints and "preloaded passive joints" (with auxiliary internal springs); n_{q}, n_{θ} are the sizes of $\mathbf{q}$ and $\boldsymbol{\theta}$ respectively.

Several types of passive joints with and without internal preloading are presented in Fig. 6.1. Such joints include internal springs, as such their statics are described by the following expression

$$\tau_{\theta i} = K_{\theta i} \cdot (\theta_i - \theta_{0i}) \tag{6.2}$$

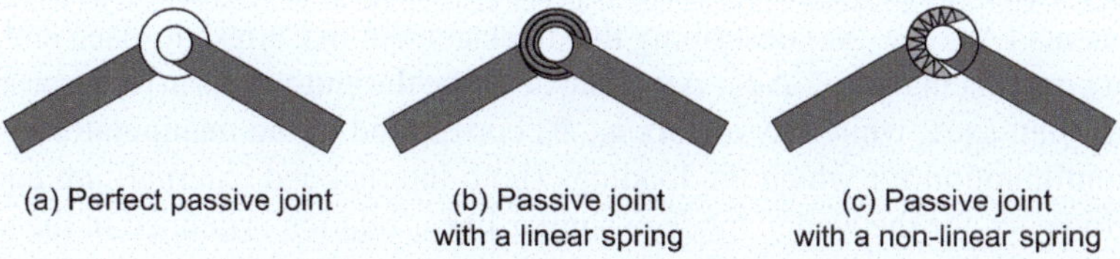

Fig. 6.1 Examples of auxiliary springs in preloaded passive joints [28]

where $\tau_{\theta i}$ is the torque/force caused by the deviation of the joint coordinate θ_i from its unloaded ("zero") value θ_{0i}, and coefficient $K_{\theta i}$ defines the spring stiffness. For the purpose of generality, let us introduce similar "zero" values θ_{0i} for the virtual springs that described the flexibility of the links (obviously they are equal to zero for this subset of $\boldsymbol{\theta}$). This allows us to define the vector $\boldsymbol{\theta}_0$ of the same size as $\boldsymbol{\theta}$ and to present the static equations corresponding to all variables (corresponding to perfect and preloaded passive joints, virtual springs of links and actuators) in general form

$$\boldsymbol{\tau}_\theta = \mathbf{K}_\theta \cdot (\boldsymbol{\theta} - \boldsymbol{\theta}_0),\ \boldsymbol{\tau}_q = \mathbf{0} \tag{6.3}$$

Here $\boldsymbol{\tau}_\theta$, $\boldsymbol{\tau}_q$ are the generalized torque/force in joints corresponding to the variables $\boldsymbol{\theta}$ and $\mathbf{q}$; the matrix $\mathbf{K}_\theta$ collects stiffness coefficients of all springs of the kinematic chain.

It is worth mentioning that in the case without internal preloading, the vector $\boldsymbol{\theta}$ describes only the flexibility of manipulator links/actuators that are presented by virtual springs, while vector $\mathbf{q}$ collects an entire set of passive joint coordinates. In contrast, here, the passive joint coordinates are divided into two subsets: (i) "perfect passive joints" included in $\mathbf{q}$, and (ii) "preloaded passive joints" included in $\boldsymbol{\theta}$ together with traditional virtual springs. Besides, if a passive joint includes a nonlinear spring (see Fig. 6.1), the corresponding joint variable may be included either in $\boldsymbol{\theta}$ or $\mathbf{q}$, depending on the current configuration of the manipulator. However, for each configuration, this assignment is strictly unique.

In contrast to the linear case described in Sect. 5.2, here the desired stiffness model is defined by a non-linear relation

$$\mathbf{F} = f(\Delta\mathbf{t}), \tag{6.4}$$

where $f(\ldots)$ is a so-called "force-deflections" function that associates a deflection $\Delta\mathbf{t}$ with an external force $\mathbf{F}$ that causes deformations. It is worth mentioning that the function $f(\ldots)$ can be determined even for the singular configurations (or redundant kinematics) while the inverse statement is not generally true. Hence, enhanced stiffness analysis must include the computation of this function and the detailed analysis of its singularities that may provoke various nonlinear phenomena (such as buckling). In the unloaded case, this function is usually defined through the "stiffness matrix" $\mathbf{K}$, which describes the linear relation $\mathbf{F} = \mathbf{K}(\mathbf{q}_0, \boldsymbol{\theta}_0) \cdot \Delta\mathbf{t}$ between small six-dimensional translational/rotational displacements $\Delta\mathbf{t}$, and the external forces/torques $\mathbf{F}$ causing this transition. Here, it is assumed that $\Delta\mathbf{t}$ includes three positional components $(\Delta x, \Delta y, \Delta z)$ describing the displacement in Cartesian space and three angular components $\left(\Delta\varphi_x, \Delta\varphi_y, \Delta\varphi_z\right)$ that describe the end-platform rotation around the Cartesian axes, while the vectors $\mathbf{q}_0$, $\boldsymbol{\theta}_0$ correspond to the manipulator equilibrium configuration for which the loadings (both internal and external) are equal to zero. However, for the loaded mode, a similar linear relation is defined in the neighbourhood of another static equilibrium, which corresponds to a different manipulator configuration $(\mathbf{q}, \boldsymbol{\theta})$, that is modified by external forces/torques $\mathbf{F}$. Respectively, in this case, the stiffness model describes the relation between the increments of the force $\delta\mathbf{F}$ and the position $\delta\mathbf{t}$

$$\delta\mathbf{F} = \mathbf{K}^F(\mathbf{q}, \boldsymbol{\theta}) \cdot \delta\mathbf{t} \tag{6.5}$$

where $\mathbf{q} = \mathbf{q}_0 + \boldsymbol{\Delta q}$ and $\boldsymbol{\theta} = \boldsymbol{\theta}_0 + \Delta\theta$ denote the new configuration of the manipulator, and $\Delta\mathbf{q}$, $\Delta\theta$ are the deviations in the coordinates $\mathbf{q}$, $\boldsymbol{\theta}$ respectively.

Hence, the considered problem of enhanced stiffness modelling may be separated into two sequential sub-problems: (a) computing *full-scale "force-deflections relation"* for the loaded manipulator; (b) the *linearization* of the relevant force–deflection relations in the neighbourhood of the equilibrium and computing corresponding stiffness matrix (that in a general case can be singular due to the presence of passive joints). Let us focus on these sub-problems.

Static Equilibrium Configuration for a Serial Chain in the Loaded Mode

To solve the first sub-problem, it is desired to obtain a relation between the external loading $\mathbf{F}$ and internal coordinates of the kinematic chain ($\mathbf{q}$, $\boldsymbol{\theta}$) corresponding to the static equilibrium. Based on this data, the desired value of the end-point displacement can be computed straightforwardly, using a geometric Eq. (6.1). In previous works, this issue was usually ignored and the linearization was performed in the neighborhood of the unloaded configuration assuming that the external load is small enough. It is obvious that the latter essentially limits relevant results and does not allow the detection of non-linear effects such as buckling. From a mathematical point of view, the problem is reduced to solving a system of non-linear static equilibrium equations that may produce unique or non-unique, stable or unstable solutions.

For computational reasons, let us consider the *dual problem* that deals with determining the external force $\mathbf{F}$ and the manipulator equilibrium configuration ($\mathbf{q}$, $\boldsymbol{\theta}$) that corresponds to the end-effector location $\mathbf{t}$ taking into account internal preloading in the joints. Let us assume that the joints are given small, arbitrary virtual displacements $\delta\mathbf{q}$, $\delta\boldsymbol{\theta}$ in the neighbourhood of ($\mathbf{q}$, $\boldsymbol{\theta}$). According to the principle of virtual displacements, the virtual work of the external force $\mathbf{F}$ applied to the end-effector along the corresponding displacement $\delta\mathbf{t} = \mathbf{J}_\theta \cdot \delta\boldsymbol{\theta} + \mathbf{J}_q \cdot \delta\mathbf{q}$ is equal to the sum $\left(\mathbf{F}^{\mathrm{T}} \cdot \mathbf{J}_\theta\right) \cdot \delta\boldsymbol{\theta} + \left(\mathbf{F}^{\mathrm{T}} \cdot \mathbf{J}_q\right) \cdot \delta\mathbf{q}$ (here $\mathbf{J}_\theta$ and $\mathbf{J}_q$ are the kinematic Jacobians with respect to the coordinates $\boldsymbol{\theta}$, $\mathbf{q}$). Since the passive joints do not produce any force/torque reactions, the virtual work corresponding to the generalized forces/torques $\boldsymbol{\tau}_\theta$, $\boldsymbol{\tau}_q$ includes only one component $-\boldsymbol{\tau}_\theta^{\mathrm{T}} \cdot \delta\boldsymbol{\theta}$. The minus sign takes into account the force–displacement directions for the virtual springs. In the static equilibrium, the total virtual work of all forces is equal to zero for any virtual displacement, therefore, the equilibrium conditions may be written as $\mathbf{J}_\theta^{\mathrm{T}} \cdot \mathbf{F} = \boldsymbol{\tau}_\theta$; $\mathbf{J}_q^{\mathrm{T}} \cdot \mathbf{F} = \mathbf{0}$, and taking into account assumptions and notations from the previous section, the static equilibrium conditions can be presented as

$$\mathbf{J}_\theta^{\mathrm{T}} \cdot \mathbf{F} = \mathbf{K}_\theta \cdot (\boldsymbol{\theta} - \boldsymbol{\theta}_0); \mathbf{J}_q^{\mathrm{T}} \cdot \mathbf{F} = \mathbf{0}; \mathbf{t} = \mathbf{g}(\mathbf{q}, \boldsymbol{\theta}) \tag{6.6}$$

where the vector $\boldsymbol{\theta}_0$ defines internal preloading in the joints, the matrix K_θ describes the stiffness of all springs in the adopted VJM model, while the external loading $\mathbf{F}$ and the configuration $(\mathbf{q}, \boldsymbol{\theta})$ are treated as the unknown for the given end-effector location $\mathbf{t} = \mathbf{g}(\mathbf{q}, \boldsymbol{\theta})$. Hence, the designed equilibrium configuration must satisfy the system of nonlinear Eq. (6.6).

Since there is no general method to determine an analytical solution for this system, a numerical technique is required. Let us linearize the kinematic equation in the neighborhood of the current position $(\mathbf{q}, \boldsymbol{\theta})$

$$\mathbf{t} = \mathbf{g}(\mathbf{q}_i, \boldsymbol{\theta}_i) + \mathbf{J}_\mathrm{q}(\mathbf{q}_i, \boldsymbol{\theta}_i) \cdot (\mathbf{q}_{i+1} - \mathbf{q}_i) + \mathbf{J}_\theta(\mathbf{q}_i, \boldsymbol{\theta}_i) \cdot (\boldsymbol{\theta}_{i+1} - \boldsymbol{\theta}_i) \tag{6.7}$$

and rewrite the static equilibrium equations as

$$\mathbf{J}_\theta^\mathrm{T}(\mathbf{q}_i, \boldsymbol{\theta}_i) \cdot \mathbf{F}_{i+1} = \mathbf{K}_\theta(\boldsymbol{\theta}_{i+1} - \boldsymbol{\theta}_0); \ \mathbf{J}_\mathrm{q}^\mathrm{T}(\mathbf{q}_i, \boldsymbol{\theta}_i) \cdot \mathbf{F}_{i+1} = \mathbf{0} \tag{6.8}$$

which leads to a system of linear algebraic equations with respect to $(\mathbf{q}_{i+1}, \boldsymbol{\theta}_{i+1}, \mathbf{F}_{i+1})$ that includes the Jacobians $\mathbf{J}_\theta(\mathbf{q}_i, \boldsymbol{\theta}_i)$, $\mathbf{J}_\theta(\mathbf{q}_i, \boldsymbol{\theta}_i)$ and the geometrical location function $\mathbf{g}(\mathbf{q}_i, \boldsymbol{\theta}_i)$ computed in the previous point $(\mathbf{q}_i, \boldsymbol{\theta}_i)$:

$$\begin{aligned}
&\mathbf{J}_\mathrm{q}(\mathbf{q}_i, \boldsymbol{\theta}_i) \cdot \mathbf{q}_{i+1} + \mathbf{J}_\theta(\mathbf{q}_i, \boldsymbol{\theta}_i) \cdot \boldsymbol{\theta}_{i+1} = \mathbf{t} - \mathbf{g}(\mathbf{q}_i, \boldsymbol{\theta}_i) + \mathbf{J}_\mathrm{q}(\mathbf{q}_i, \boldsymbol{\theta}_i) \cdot \mathbf{q}_i + \mathbf{J}_\theta(\mathbf{q}_i, \boldsymbol{\theta}_i) \cdot \boldsymbol{\theta}_i; \\
&\mathbf{J}_\mathrm{q}^\mathrm{T}(\mathbf{q}_i, \boldsymbol{\theta}_i) \cdot \mathbf{F}_{i+1} = \mathbf{0} \\
&\mathbf{J}_\theta^\mathrm{T}(\mathbf{q}_i, \boldsymbol{\theta}_i) \cdot \mathbf{F}_{i+1} - \mathbf{K}_\theta \cdot \boldsymbol{\theta}_{i+1} = -\mathbf{K}_\theta \cdot \boldsymbol{\theta}_0
\end{aligned} \tag{6.9}$$

This gives the following iterative scheme

$$\begin{aligned}
\begin{bmatrix} \mathbf{F}_{i+1} \\ \mathbf{q}_{i+1} \\ \boldsymbol{\theta}_{i+1} \end{bmatrix} &= \begin{bmatrix} \mathbf{0} & \mathbf{J}_q(\mathbf{q}_i, \boldsymbol{\theta}_i) & \mathbf{J}_\theta(\mathbf{q}_i, \boldsymbol{\theta}_i) \\ \mathbf{J}_\mathrm{q}^\mathrm{T}(\mathbf{q}_i, \boldsymbol{\theta}_i) & \mathbf{0} & \mathbf{0} \\ \mathbf{J}_\theta^\mathrm{T}(\mathbf{q}_i, \boldsymbol{\theta}_i) & \mathbf{0} & -\mathbf{K}_\theta \end{bmatrix}^{-1} \cdot \\
&\cdot \begin{bmatrix} \mathbf{t} - \mathbf{g}(\mathbf{q}_i, \boldsymbol{\theta}_i) + \mathbf{J}_\mathrm{q}(\mathbf{q}_i, \boldsymbol{\theta}_i) \cdot \mathbf{q}_i + \mathbf{J}_\theta(\mathbf{q}_i, \boldsymbol{\theta}_i) \cdot \boldsymbol{\theta}_i \\ \mathbf{0} \\ -\mathbf{K}_\theta \cdot \boldsymbol{\theta}_0 \end{bmatrix}
\end{aligned} \tag{6.10}$$

that may be reduced down to

$$\begin{aligned}
\begin{bmatrix} \mathbf{F}_{i+1} \\ \mathbf{q}_{i+1} \end{bmatrix} &= \begin{bmatrix} \mathbf{J}_\theta(\mathbf{q}_i, \theta_i) \cdot \mathbf{K}_\theta^{-1} \cdot \mathbf{J}_\theta^T(\mathbf{q}_i, \theta_i) & \mathbf{J}_q(\mathbf{q}_i, \theta_i) \\ \mathbf{J}_q^T(\mathbf{q}_i, \theta_i) & 0 \end{bmatrix}^{-1} \\
&\cdot \begin{bmatrix} \mathbf{t} - \mathbf{g}(\mathbf{q}_i, \theta_i) + \mathbf{J}_q(\mathbf{q}_i, \theta_i) \cdot \mathbf{q}_i + \mathbf{J}_\theta(\mathbf{q}_i, \theta_i) \cdot \theta_i \\ 0 \end{bmatrix} \\
\theta_{i+1} &= \mathbf{K}_\theta^{-1} \cdot \mathbf{J}_\theta^T(\mathbf{q}_i, \theta_i) \cdot \mathbf{F}_{i+1} + \theta_0
\end{aligned} \tag{6.11}$$

The latter is more convenient computationally since it requires the inversion of a lower-dimension matrix $(n + 6) \times (n + 6)$ instead of $(n + m + 6) \times (n + m + 6)$, where n, m are the sizes of the vectors $\mathbf{q}$ and $\boldsymbol{\theta}$ respectively. For instance, for the kinematic chains of the Orthoglide manipulator (see example in Chap. 5), the expression (6.10) requires the inversion of 34×34 matrix, while the iterative scheme (6.11) needs the inversion of 10×10 matrix only. It should be mentioned that K_θ^{-1} is computed only once, outside of the iterative loop.

Similar to the other iterative schemes, the convergence of this algorithm highly depends on the starting point. However, due to the physical nature of the considered problem, it is possible to start iterations from the non-loaded configuration $(\mathbf{q}_0, \ \boldsymbol{\theta}_0)$. Besides, it is useful to modify the target point for each iteration in accordance with the expression $\mathbf{t}_i = \alpha_i \cdot \mathbf{t} + (1 - \alpha_i) \cdot \mathbf{t}_0$ using scalar variable α_i that is monotonically increasing from 0 up to 1. Another approach can be used for computing the force–deflection curve. Here, the starting point can be taken from a previously computed loaded configuration corresponding to another value of deflection that is very close to the target one. For typical values of deformations, the proposed iterative procedure converges in 3–5 iterations if the configuration is stable. In the simulation studies, the convergence has been evaluated as the weighted sum of residual norms corresponding to Eq. (6.6) and the algorithm stops when this criterion achieves the prescribed value.

However, some computational difficulties may arise in the case of buckling or in the area of multiple equilibriums, where the convergence problem becomes rather critical and highly depends on an initial point. Here, the number of iterations increases significantly and the computational time becomes non-negligible. To overcome these difficulties, it is proposed to modify the developed iterative scheme and to repeat the computations several times, with slightly modified initial points that are obtained by adding small random noise to $\mathbf{q}_0, \ \boldsymbol{\theta}_0$. Another option is to add small disturbances to $(\boldsymbol{\theta}_i, \mathbf{q}_i)$ at each iteration. These techniques were used in case studies presented in Chap. 5.

The proposed iterative scheme can also be slightly modified to solve the *original problem*, i.e. computing the equilibrium configuration corresponding to the given external loading $\mathbf{F}$(instead of given $\mathbf{t}$). In this case, expressions (6.10), (6.11) are used in the internal loop, while the designed algorithm is supplemented by an external loop, which provides iterative searching for $\mathbf{t}$ corresponding to the given $\mathbf{F}$

$$\mathbf{t}_{i+1} = \mathbf{t}_i + \mathbf{K}_i^{-1} \cdot (\mathbf{F} - \mathbf{F}_i) \tag{6.12}$$

where $\mathbf{t}_i, \ \mathbf{F}_i$ and $\mathbf{K}_i$ are the location, the loading and the stiffness matrix at the i-th iteration respectively. It is worth mentioning that the dual problem is meaningful only if the stiffness matrix $\mathbf{K}_i$ is non-singular. It is obvious that for a separate serial chain with passive joints the matrix $\mathbf{K}_i$ is always singular, while for a serial manipulator, it is usually non-singular (the same as for a parallel manipulator due to the specific assembling of kinematic chains). On the other hand, the dual problem considered in this Section (i.e. computing $\mathbf{F}$ corresponding to $\mathbf{t}$) is always physically meaningful and has at least one solution.

Stiffness Matrix of Serial Chain in the Loaded Mode

The previous sub-section presents a technique that generally allows us to obtain the relation between the elastic deformations and corresponding external force/torque. It is based on sequential computations of loaded equilibriums (and relevant force/torque) for various displacements of the manipulator end-point with respect to its unloaded location. However, though this relation is highly non-linear, common engineering practice operates with the stiffness matrix derived via its linearization.

To compute the desired stiffness matrix, let us consider the neighbourhood of the loaded configuration and assume that the external force and the end-effector location are incremented by some small values $\delta\mathbf{F}$, $\delta\mathbf{t}$. Besides, let us assume that a new configuration also satisfies the equilibrium conditions. Hence, it is necessary to consider simultaneously two equilibriums corresponding to the manipulator state variables $(\mathbf{F}, \mathbf{q}, \boldsymbol{\theta}, \mathbf{t})$ and $(\mathbf{F}+\delta\mathbf{F}, \mathbf{q}+\delta\mathbf{q}, \boldsymbol{\theta}+\delta\boldsymbol{\theta}, \mathbf{t}+\delta\mathbf{t})$. Relevant equations of statistics may be written as

$$\mathbf{J}_\theta^T \cdot \mathbf{F} = \mathbf{K}_\theta \cdot (\boldsymbol{\theta} - \boldsymbol{\theta}_0);\ \mathbf{J}_q^T \cdot \mathbf{F} = 0 \tag{6.13}$$

and

$$\begin{aligned} &(\mathbf{J}_\theta + \delta\mathbf{J}_\theta)^T \cdot (\mathbf{F}+\delta\mathbf{F}) = \mathbf{K}_\theta \cdot (\boldsymbol{\theta} - \boldsymbol{\theta}_0 + \delta\boldsymbol{\theta}); \\ &\left(\mathbf{J}_q + \delta\mathbf{J}_q\right)^T \cdot (\mathbf{F}+\delta\mathbf{F}) = 0 \end{aligned} \tag{6.14}$$

where $\delta\mathbf{J}_q(\mathbf{q}, \boldsymbol{\theta})$ and $\delta\mathbf{J}_\theta(\mathbf{q}, \boldsymbol{\theta})$ are the differentials of the Jacobians due to changes in $(\mathbf{q}, \boldsymbol{\theta})$. Besides, in the neighbourhood of $(\mathbf{q}, \boldsymbol{\theta})$, the kinematic equation may also be presented in the linearized form:

$$\delta\mathbf{t} = \mathbf{J}_\theta(\mathbf{q}, \boldsymbol{\theta}) \cdot \delta\boldsymbol{\theta} + \mathbf{J}_q(\mathbf{q}, \boldsymbol{\theta}) \cdot \delta\mathbf{q} \tag{6.15}$$

Hence, after neglecting the high-order small terms and expanding the differentials via the Hessians of the function $\Psi = \mathbf{g}(\mathbf{q}, \boldsymbol{\theta})^T \cdot \mathbf{F}$, Eqs. (6.13), (6.14) may be rewritten as

$$\begin{aligned} &\mathbf{J}_\theta^T(\mathbf{q}, \theta) \cdot \delta\mathbf{F} + \mathbf{H}_{\theta q}^F(\mathbf{q}, \theta) \cdot \delta\mathbf{q} + \mathbf{H}_{\theta\theta}^F(\mathbf{q}, \theta) \cdot \delta\boldsymbol{\theta} = \mathbf{K}_\theta \cdot \delta\boldsymbol{\theta} \\ &\mathbf{J}_q^T(\mathbf{q}, \theta) \cdot \delta\mathbf{F} + \mathbf{H}_{qq}^F(\mathbf{q}, \theta) \cdot \delta\mathbf{q} + \mathbf{H}_{q\theta}^F(\mathbf{q}, \theta) \cdot \delta\boldsymbol{\theta} = \mathbf{0} \end{aligned} \tag{6.16}$$

where

$$\mathbf{H}_{qq}^F = \partial^2\Psi/\partial\mathbf{q}^2;\ \mathbf{H}_{\theta\theta}^F = \partial^2\Psi/\partial\boldsymbol{\theta}^2;\ \mathbf{H}_{q\theta}^F = \mathbf{H}_{\theta q}^F = \partial^2\Psi/\partial\mathbf{q}\partial\boldsymbol{\theta} \tag{6.17}$$

and the general relation between the increments of the state variables can be presented as

$$\begin{bmatrix} \mathbf{0} & \mathbf{J}_q & \mathbf{J}_\theta \\ \mathbf{J}_q^T & \mathbf{H}_{qq}^F & \mathbf{H}_{q\theta}^F \\ \mathbf{J}_\theta^T & \mathbf{H}_{\theta q}^F & \mathbf{H}_{\theta\theta}^F - \mathbf{K}_\theta \end{bmatrix} \cdot \begin{bmatrix} \delta\mathbf{F} \\ \delta\mathbf{q} \\ \delta\boldsymbol{\theta} \end{bmatrix} = \begin{bmatrix} \delta\mathbf{t} \\ \mathbf{0} \\ \mathbf{0} \end{bmatrix} \tag{6.18}$$

The latter gives a straightforward numerical technique for computing the desired stiffness matrix: direct inversion of the matrix on the left-hand side of (6.18) and extracting from it the upper-left sub-matrix of size 6 × 6. Similarly, the matrices defining linear relations between the end-effector increment $\delta\mathbf{t}$ and the increments of the joint variables $\delta\boldsymbol{\theta}$, $\delta\mathbf{q}$ can be computed, i.e.:

$$\delta\mathbf{F} = \mathbf{K}_C \cdot \delta\mathbf{t};\ \delta\boldsymbol{\theta} = \mathbf{K}_{C\theta} \cdot \delta\mathbf{t};\ \delta\mathbf{q} = \mathbf{K}_{Cq} \cdot \delta\mathbf{t} \tag{6.19}$$

where

$$\begin{bmatrix} \mathbf{K}_C & * & * \\ \mathbf{K}_{Cq} & * & * \\ \mathbf{K}_{C\theta} & * & * \end{bmatrix} = \begin{bmatrix} \mathbf{0} & \mathbf{J}_q & \mathbf{J}_\theta \\ \mathbf{J}_q^T & \mathbf{H}_{qq}^F & \mathbf{H}_{q\theta}^F \\ \mathbf{J}_\theta^T & \mathbf{H}_{\theta q}^F & \mathbf{H}_{\theta\theta}^F - \mathbf{K}_\theta \end{bmatrix}^{-1} \tag{6.20}$$

It is worth mentioning that the internal preloading (expressed by the variable $\boldsymbol{\theta}_0$) is not included in the latter expression explicitly, but it directly influences the variables $(\mathbf{q}, \boldsymbol{\theta})$ describing the equilibrium configuration and corresponding Jacobians and Hessians, which are elements of (6.20). Besides, in contrast to previous works, here it is possible to obtain supplementary matrices $\mathbf{K}_{C\theta}$, $\mathbf{K}_{Cq}$ that give additional measures of the manipulator stiffness which evaluate the sensitivity of the joint coordinates $(\mathbf{q}, \boldsymbol{\theta})$ with respect to the external loading.

In the case when the matrix inverse (6.20) is computationally hard, the variable $\delta\boldsymbol{\theta}$ can be eliminated analytically, using the corresponding static equation: $\delta\boldsymbol{\theta} = \mathbf{k}_\theta^F \cdot \mathbf{J}_\theta^T \cdot \delta\mathbf{F} + \mathbf{k}_\theta^F \cdot \mathbf{H}_{\theta q}^F \cdot \delta\mathbf{q}$, where $\mathbf{k}_\theta^F = \left(\mathbf{K}_\theta - \mathbf{H}_{\theta\theta}^F\right)^{-1}$. This leads to a reduced system of matrix equations with unknowns δF and δq

$$\begin{bmatrix} \mathbf{J}_\theta \cdot \mathbf{k}_\theta^F \cdot \mathbf{J}_\theta^T & \mathbf{J}_q + \mathbf{J}_\theta \cdot \mathbf{k}_\theta^F \cdot \mathbf{H}_{\theta q}^F \\ \mathbf{J}_q^T + \mathbf{H}_{q\theta}^F \cdot \mathbf{k}_\theta^F \cdot \mathbf{J}_\theta^T & \mathbf{H}_{qq}^F + \mathbf{H}_{q\theta}^F \cdot \mathbf{k}_\theta^F \cdot \mathbf{H}_{\theta q}^F \end{bmatrix} \cdot \begin{bmatrix} \delta\mathbf{F} \\ \delta\mathbf{q} \end{bmatrix} = \begin{bmatrix} \delta\mathbf{t} \\ \mathbf{0} \end{bmatrix} \tag{6.21}$$

that may be treated in a similar way, i.e. the desired stiffness matrix is also obtained by direct inversion of the matrix on the left-hand side of (6.21) and extracting from it the upper-left sub-matrix of size 6 × 6:

$$\begin{bmatrix} \mathbf{K}_C & * \\ \mathbf{K}_{Cq} & * \end{bmatrix} = \begin{bmatrix} \mathbf{J}_\theta \cdot \mathbf{k}_\theta^F \cdot \mathbf{J}_\theta^T & \mathbf{J}_q + \mathbf{J}_\theta \cdot \mathbf{k}_\theta^F \cdot \mathbf{H}_{\theta q}^F \\ \mathbf{J}_q^T + \mathbf{H}_{q\theta}^F \cdot \mathbf{k}_\theta^F \cdot \mathbf{J}_\theta^T & \mathbf{H}_{qq}^F + \mathbf{H}_{q\theta}^F \cdot \mathbf{k}_\theta^F \cdot \mathbf{H}_{\theta q}^F \end{bmatrix}^{-1} \tag{6.22}$$

Similar to the linear case, this approach allows us to reduce the dimension of the inverted matrix from $(n+m+6) \times (n+m+6)$ to $(n+6) \times (n+6)$, that in the case of Orthoglide (see Chap. 5) corresponds to 34 × 34 and 10 × 10 respectively.

It is worth mentioning that the structure of the latter matrix is similar to the one obtained for the unloaded manipulator and differs only by Hessians that take into account the influence of the external load. It should also be noted that, because of the presence of the passive joints, the stiffness matrix of a separate serial kinematic chain is always singular, but aggregation of all the chains for a parallel manipulator produces a non-singular stiffness matrix.

Further simplification of (6.22) can be obtained by applying the block matrix inversion technique of Frobenius [29] that yields the following expressions

$$\mathbf{K}_{\mathrm{C}} = \mathbf{K}_{\mathrm{C}}^{0(F)} - \mathbf{K}_{\mathrm{C}}^{0(F)} \cdot \left(\mathbf{J}_{\mathrm{q}} + \mathbf{J}_{\theta} \cdot \mathbf{k}_{\theta}^{F} \cdot H_{\theta q}^{F}\right) \cdot \mathbf{K}_{\mathrm{Cq}} \tag{6.23}$$

where the first term $\mathbf{K}_{C}^{0(F)} = \left(\mathbf{J}_{\theta} \cdot \mathbf{k}_{\theta}^{\mathrm{F}} \cdot \mathbf{J}_{\theta}^{\mathrm{T}}\right)^{-1}$ exactly corresponds to the classical formula defining the stiffness of the kinematic chain without passive joints in the loaded mode [1, 4] and

$$\mathbf{K}_{\mathrm{Cq}} = -\left(\mathbf{H}_{\mathrm{qq}}^{\mathrm{F}} + \mathbf{H}_{\mathrm{q}\theta}^{\mathrm{F}} \cdot \mathbf{k}_{\theta}^{\mathrm{F}} \cdot \mathbf{H}_{\theta\mathrm{q}}^{\mathrm{F}} - \left(\mathbf{J}_{\mathrm{q}}^{\mathrm{T}} + \mathbf{H}_{\mathrm{q}\theta}^{\mathrm{F}} \cdot \mathbf{k}_{\theta}^{\mathrm{F}} \cdot \mathbf{J}_{\theta}^{T}\right) \cdot \mathbf{K}_{\mathrm{C}}^{0(\mathrm{F})} \cdot \left(\mathbf{J}_{\mathrm{q}} + \mathbf{J}_{\theta} \cdot \mathbf{k}_{\theta}^{\mathrm{F}} \cdot \mathbf{H}_{\theta\mathrm{q}}^{\mathrm{F}}\right)\right)^{-1} \cdot \left(\mathbf{J}_{\mathrm{q}}^{\mathrm{T}} + \mathbf{H}_{\mathrm{q}\theta}^{\mathrm{F}} \cdot \mathbf{k}_{\theta}^{\mathrm{F}} \cdot \mathbf{J}_{\theta}^{\mathrm{T}}\right) \cdot \mathbf{K}_{\mathrm{C}}^{0(\mathrm{F})} \tag{6.24}$$

Similarly, the matrix $\mathbf{K}_{\mathrm{C}\theta}$ can be expressed analytically as

$$\mathbf{K}_{\mathrm{C}\theta} = \mathbf{k}_{\theta}^{\mathrm{F}} \cdot \mathbf{J}_{\theta}^{\mathrm{T}} \cdot \mathbf{K}_{\mathrm{C}} + \mathbf{k}_{\theta}^{\mathrm{F}} \cdot \mathbf{H}_{\theta\mathrm{q}}^{\mathrm{F}} \cdot \mathbf{K}_{\mathrm{Cq}} \tag{6.25}$$

Hence, the technique presented here allows us to compute the static equilibrium configuration and Cartesian stiffness matrix for serial chains with external loading applied to the end-point and with preloading in the joints. Further, it will be applied to several case studies that deal with kinematic chains employed in typical parallel manipulators and demonstrate particularities of stiffness analysis of loaded manipulators with passive joints. However, this technique does not cover heavy manipulators, where the own weight of the robot is significant and the external loading is applied to both the end-point and internal points of the kinematic chain. Furthermore, this type of loading will be called an auxiliary one.

6.1.2 Serial Kinematic Chain of Parallel Manipulator with Auxiliary Loadings in Intermediate Nodes

Problem Statement

For stiffness modeling of serial kinematic chain with auxiliary loading let us use the same VJM model as in Sects. 5.2 and 6.1.1, but let us assume that the serial chain has the additional external loadings applied to the internal node points (Fig. 6.2).

These loadings can be caused by gravity forces (generally they are distributed, but in practice, they can be approximated by localized ones) and/or gravity compensators. These forces will be denoted as $\mathbf{G}_j$, where $j = 1, ..., n$ is the node number in the serial chain starting from the fixed base (here, $j = n$ corresponds to the end-point). It should be noted that for computational convenience, it is assumed that the end-point loading consists of two components $\mathbf{G}_n$ and $\mathbf{F}$ of different natures.

It is evident that in general the auxiliary forces $\mathbf{G}_i$ depend on the manipulator configuration. So, let us assume that they are described by the functions

$$\mathbf{G}_j = \mathbf{G}_j(\mathbf{q}, \boldsymbol{\theta}), \tag{6.26}$$

where $\mathbf{q}$ and $\boldsymbol{\theta}$ collect all passive joints and virtual joints coordinates respectively. In contrast, for the external force $\mathbf{F}$, it is assumed that there is no direct relation with the manipulator configuration.

For the serial chains with the auxiliary loadings, it is also required to extend the geometrical model. In particular, in addition to the Eq. (6.1) defining the end-point location, it is necessary to introduce the additional functions

$$\mathbf{t}_j = \mathbf{g}_j(\mathbf{q}, \boldsymbol{\theta}), j = 1, ..., n \tag{6.27}$$

defining locations of the nodes. It is worth mentioning that for the serial chain, the position $\mathbf{t}_j$ depends on a sub-set of the joint coordinates (corresponding to the passive and virtual joints located between the base and the j-th node), but for the purpose of analytical simplicity let us use the whole set of the joint coordinates $(\mathbf{q}, \boldsymbol{\theta})$ as the arguments of the functions $\mathbf{g}_i(\ldots)$.

Using these assumptions and the methodology of Sect. 6.1.1, the problem of stiffness modelling of serial chains with auxiliary loadings can be split into the following sub-problems: (a) deriving the static equilibrium equations for the chain with auxiliary loadings; (b) computing *full-scale "force-deflections relation"* for the end-point and intermediate nodes; (c) *linearization* of the relevant force–deflection relations in the neighbourhood of the equilibrium and computing corresponding stiffness matrix. Let us focus on these sub-problems.

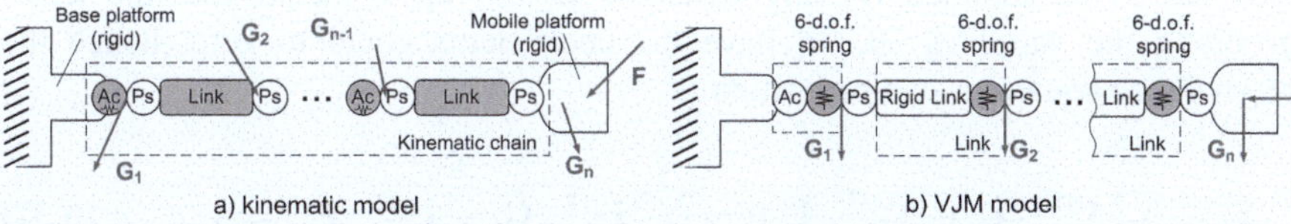

Fig. 6.2 General structure of kinematic chain with auxiliary loading and its VJM model [30]

Static Equilibrium Equations for Serial Chain with Auxiliary Loadings

To obtain a desired stiffness model, let us derive first the static equilibrium equations that differ from the Eq. (6.6) used in the previous Section due to the influence of auxiliary loadings $\mathbf{G}_j$. Let us apply the principle of virtual work and assume that the kinematic chain under external loadings $\mathbf{F}$ and $\mathbf{G} = [\mathbf{G}_1 \ldots \mathbf{G}_n]$ has the configuration $(\mathbf{q},\ \boldsymbol{\theta})$ and the locations of the end-point and the nodes are $\mathbf{t} = \mathbf{g}(\mathbf{q}, \boldsymbol{\theta})$ and $\mathbf{t}_j = \mathbf{g}_j(\mathbf{q}, \boldsymbol{\theta}), j = \overline{1, n}$ respectively.

According to the principle of virtual work, the work of external forces $\mathbf{G},\ \mathbf{F}$ is equal to the work of internal forces $\boldsymbol{\tau}_\theta$ caused by the displacement of the virtual springs $\delta\boldsymbol{\theta}$

$$\sum_{j=1}^{n}\left(\mathbf{G}_j^{\mathrm{T}} \cdot \delta\mathbf{t}_j\right) + \mathbf{F}^{\mathrm{T}} \cdot \delta\mathbf{t} = \boldsymbol{\tau}_\theta^{\mathrm{T}} \cdot \delta\boldsymbol{\theta} \tag{6.28}$$

where the virtual displacements $\delta\mathbf{t}_j$ can be computed from the linearized geometrical model derived from (6.27)

$$\delta\mathbf{t}_j = \mathbf{J}_\theta^{(j)} \cdot \delta\boldsymbol{\theta} + \mathbf{J}_{\mathrm{q}}^{(j)} \cdot \delta\mathbf{q}, j = 1..n \tag{6.29}$$

which includes the Jacobian matrices

$$\mathbf{J}_\theta^{(j)} = \frac{\partial}{\partial\boldsymbol{\theta}}\mathbf{g}_j(\mathbf{q}, \boldsymbol{\theta});\ \mathbf{J}_{\mathrm{q}}^{(j)} = \frac{\partial}{\partial\mathbf{q}}\mathbf{g}_j(\mathbf{q}, \boldsymbol{\theta}) \tag{6.30}$$

with respect to the virtual and passive joint coordinates respectively.

Substituting (6.29) to (6.28) we can obtain the equation

$$\sum_{j=1}^{n}\left(\mathbf{G}_j^{\mathrm{T}} \cdot \mathbf{J}_\theta^{(j)} \cdot \delta\boldsymbol{\theta} + \mathbf{G}_j^{\mathrm{T}} \cdot \mathbf{J}_{\mathrm{q}}^{(j)} \cdot \delta\mathbf{q}\right) + \left(\mathbf{F}^{\mathrm{T}} \cdot \mathbf{J}_\theta^{(n)} \cdot \delta\boldsymbol{\theta} + \mathbf{F}^{\mathrm{T}} \cdot \mathbf{J}_{\mathrm{q}}^{(n)} \cdot \delta\mathbf{q}\right) = \boldsymbol{\tau}_\theta^{\mathrm{T}} \cdot \delta\boldsymbol{\theta} \tag{6.31}$$

which has to be satisfied for any variation of $\delta\boldsymbol{\theta},\ \delta\mathbf{q}$. It means that the terms regrouping the variables $\delta\boldsymbol{\theta},\ \delta\mathbf{q}$ have the coefficients equal to zero, hence the force-balance equations can be written as

$$\boldsymbol{\tau}_\theta = \sum_{j=1}^{n}\mathbf{J}_\theta^{(j)\mathrm{T}} \cdot \mathbf{G}_j + \mathbf{J}_\theta^{(n)\mathrm{T}} \cdot \mathbf{F};\ \mathbf{0} = \sum_{j=1}^{n}\mathbf{J}_{\mathrm{q}}^{(j)\mathrm{T}} \cdot \mathbf{G}_j + \mathbf{J}_{\mathrm{q}}^{(n)\mathrm{T}} \cdot \mathbf{F}, j = 1..n \tag{6.32}$$

Also, these equations can be re-written in block-matrix form as

$$\boldsymbol{\tau}_\theta = \mathbf{J}_\theta^{(\mathrm{G})\mathrm{T}} \cdot \mathbf{G} + \mathbf{J}_\theta^{(\mathrm{F})\mathrm{T}} \cdot \mathbf{F};\ \mathbf{0} = \mathbf{J}_{\mathrm{q}}^{(\mathrm{G})\mathrm{T}} \cdot \mathbf{G} + \mathbf{J}_{\mathrm{q}}^{(\mathrm{F})\mathrm{T}} \cdot \mathbf{F} \tag{6.33}$$

where

$$
\begin{gathered}
\mathbf{J}_\theta^{(F)} = \mathbf{J}_\theta^{(n)};\ \mathbf{J}_\mathrm{q}^{(F)} = \mathbf{J}_\mathrm{q}^{(n)}; \\
\mathbf{J}_\theta^{(G)} = \left[\mathbf{J}_\theta^{(1)\mathrm{T}} \ldots \mathbf{J}_\theta^{(n)\mathrm{T}}\right]^\mathrm{T}; \mathbf{J}_\mathrm{q}^{(G)} = \left[\mathbf{J}_\mathrm{q}^{(1)\mathrm{T}} \ldots \mathbf{J}_\mathrm{q}^{(n)\mathrm{T}}\right]^\mathrm{T}; \mathbf{G} = \left[\mathbf{G}_1^\mathrm{T} \ldots \mathbf{G}_n^\mathrm{T}\right]^\mathrm{T}
\end{gathered}
\tag{6.34}
$$

Finally, taking into account the virtual spring reaction $\boldsymbol{\tau}_\theta = \mathbf{K}_\theta \cdot \left(\boldsymbol{\theta} - \boldsymbol{\theta}^0\right)$, where $\mathbf{K}_\theta = diag\left(\mathbf{K}_{\theta_1}, \ldots, \mathbf{K}_{\theta_n}\right)$, the desired static equilibrium equations can be presented as

$$
\begin{aligned}
&\mathbf{J}_\theta^{(\mathrm{G})\mathrm{T}} \cdot \mathbf{G} + \mathbf{J}_\theta^{(\mathrm{F})\mathrm{T}} \cdot \mathbf{F} = \mathbf{K}_\theta \cdot \left(\boldsymbol{\theta} - \boldsymbol{\theta}^0\right) \\
&\mathbf{J}_\mathrm{q}^{(\mathrm{G})\mathrm{T}} \cdot \mathbf{G} + \mathbf{J}_\mathrm{q}^{(\mathrm{F})\mathrm{T}} \cdot \mathbf{F} = \mathbf{0}
\end{aligned}
\tag{6.35}
$$

It should be noted that compared to previously used equilibrium Eq. (6.6), here there are two additional terms $\mathbf{J}_\mathrm{q}^{(\mathrm{G})\mathrm{T}}\mathbf{G}$ and $\mathbf{J}_\mathrm{q}^{(\mathrm{G})\mathrm{T}}\mathbf{G}$ that take into account the influence of the auxiliary loading $\mathbf{G}$. Further, these equations will be used for computing the static equilibrium configuration and corresponding Cartesian stiffness matrix.

Static Equilibrium Configuration for Serial Chain with Auxiliary Loadings

To obtain a relation between the external loading $\mathbf{F}$ and internal coordinates of the kinematic chain $(\mathbf{q}, \boldsymbol{\theta})$ corresponding to the static equilibrium, Eq. (6.35) should be solved either for the different given values of $\mathbf{F}$ or for the different given values of $\mathbf{t}$. Previously, these problems were referred to as the original and the dual ones respectively, but the dual problem has been discovered to be the most convenient from a computational point of view. Hence, let us solve static equilibrium equations with respect to the manipulator configuration $(\mathbf{q}, \boldsymbol{\theta})$ and external loading $\mathbf{F}$ for given end-effector position $\mathbf{t} = \mathbf{g}(\mathbf{q}, \boldsymbol{\theta})$ and function of auxiliary-loadings $\mathbf{G}(\mathbf{q}, \boldsymbol{\theta})$

$$
\begin{aligned}
&\mathbf{K}_\theta \cdot \left(\boldsymbol{\theta} - \boldsymbol{\theta}^0\right) = \mathbf{J}_\theta^{(\mathrm{G})\mathrm{T}} \cdot \mathbf{G} + \mathbf{J}_\theta^{(\mathrm{F})\mathrm{T}} \cdot \mathbf{F};\ \mathbf{J}_\mathrm{q}^{(\mathrm{G})\mathrm{T}} \cdot \mathbf{G} + \mathbf{J}_\mathrm{q}^{(\mathrm{F})\mathrm{T}} \cdot \mathbf{F} = \mathbf{0} \\
&\mathbf{t} = \mathbf{g}(\mathbf{q}, \boldsymbol{\theta});\ \mathbf{G} = \mathbf{G}(\mathbf{q}, \boldsymbol{\theta})
\end{aligned}
\tag{6.36}
$$

Compared to Sect. 6.1.1, here the unknown variables are the same $(\mathbf{q}, \boldsymbol{\theta}, \mathbf{F})$, but the equation structure is more complicated.

Since this system of equations usually has no analytical solution, an iterative numerical technique can be applied. Similar to the previous Section, the kinematic equations may be linearized in the neighbourhood of the current configuration $(\mathbf{q}_i, \boldsymbol{\theta}_i)$

$$
\mathbf{t}_{i+1} = \mathbf{g}(\mathbf{q}_i, \boldsymbol{\theta}_i) + \mathbf{J}_\theta^{(\mathrm{F})}(\mathbf{q}_i, \boldsymbol{\theta}_i) \cdot (\boldsymbol{\theta}_{i+1} - \boldsymbol{\theta}_i) + \mathbf{J}_\mathrm{q}^{(\mathrm{F})}(\mathbf{q}_i, \boldsymbol{\theta}_i) \cdot (\mathbf{q}_{i+1} - \mathbf{q}_i); \tag{6.37}
$$

where the subscript 'i' indicates the iteration number and the changes in Jacobians $\mathbf{J}_\theta^{(\mathrm{G})}, \mathbf{J}_\theta^{(\mathrm{F})}, \mathbf{J}_\mathrm{q}^{(\mathrm{G})}, \mathbf{J}_\mathrm{q}^{(\mathrm{F})}$ and the auxiliary loadings $\mathbf{G}(\mathbf{q}, \boldsymbol{\theta})$ are assumed to be negligible from iteration to iteration. Correspondingly, the static equilibrium equations in the neighbourhood of $(\mathbf{q}_i, \boldsymbol{\theta}_i)$ may be rewritten as

$$\begin{aligned}
&\mathbf{J}_\theta^{(G)T}(\mathbf{q}_i, \boldsymbol{\theta}_i) \cdot \mathbf{G}(\mathbf{q}_i, \boldsymbol{\theta}_i) + \mathbf{J}_\theta^{(F)T}(\mathbf{q}_i, \boldsymbol{\theta}_i) \cdot \mathbf{F}_{i+1} = \mathbf{K}_\theta \cdot \left(\boldsymbol{\theta}_{i+1} - \boldsymbol{\theta}^0\right) \\
&\mathbf{J}_q^{(G)T}(\mathbf{q}_i, \boldsymbol{\theta}_i) \cdot \mathbf{G}(\mathbf{q}_i, \boldsymbol{\theta}_i) + \mathbf{J}_q^{(F)T}(\mathbf{q}_i, \boldsymbol{\theta}_i) \cdot \mathbf{F}_{i+1} = \mathbf{0}
\end{aligned} \tag{6.38}$$

Thus, combining (6.37) and (6.38), the iterative algorithm for computing the static equilibrium configuration for the given end-effector location can be presented as

$$\begin{bmatrix} \mathbf{F}_{i+1} \\ \mathbf{q}_{i+1} \\ \boldsymbol{\theta}_{i+1} \end{bmatrix} = \begin{bmatrix} \mathbf{0} & \mathbf{J}_q^{(F)}(\mathbf{q}_i, \boldsymbol{\theta}_i) & \mathbf{J}_\theta^{(F)}(\mathbf{q}_i, \boldsymbol{\theta}_i) \\ \mathbf{J}_q^{(F)T}(\mathbf{q}_i, \boldsymbol{\theta}_i) & \mathbf{0} & \mathbf{0} \\ \mathbf{J}_\theta^{(F)T}(\mathbf{q}_i, \boldsymbol{\theta}_i) & \mathbf{0} & -\mathbf{K}_\theta \end{bmatrix}^{-1}
\cdot \begin{bmatrix} \mathbf{t}_{i+1} - \mathbf{g}(\mathbf{q}_i, \boldsymbol{\theta}_i) + \mathbf{J}_\theta^{(F)}(\mathbf{q}_i, \boldsymbol{\theta}_i) \cdot \boldsymbol{\theta}_i + \mathbf{J}_q^{(F)}(\mathbf{q}_i, \boldsymbol{\theta}_i) \cdot \mathbf{q}_i \\ -\mathbf{J}_q^{(G)T}(\mathbf{q}_i, \boldsymbol{\theta}_i) \cdot \mathbf{G}_i \\ -\mathbf{J}_\theta^{(G)T}(\mathbf{q}_i, \boldsymbol{\theta}_i) \cdot \mathbf{G}_i - \mathbf{K}_\theta \cdot \boldsymbol{\theta}^0 \end{bmatrix} \tag{6.39}$$

where $\mathbf{G}_{i+1} = \mathbf{G}(\mathbf{q}_{i+1}, \boldsymbol{\theta}_{i+1})$.

To reduce the size of a matrix to be inverted, the above system can be slightly simplified. In particular, applying the same approach as in the previous case (based on the analytical expression for $\boldsymbol{\theta} = \mathbf{K}_\theta^{-1}\left(\mathbf{J}_\theta^{(G)T} \cdot \mathbf{G} + \mathbf{J}_\theta^{(F)T} \cdot \mathbf{F}\right) + \boldsymbol{\theta}^0$), the unknown variables can be separated into two groups (F, q) and $\boldsymbol{\theta}$. This yields the iterative scheme

$$\begin{aligned}
&\begin{bmatrix} \mathbf{F}_{i+1} \\ \mathbf{q}_{i+1} \end{bmatrix} = \begin{bmatrix} \mathbf{J}_\theta^{(F)}(\mathbf{q}_i, \boldsymbol{\theta}_i) \cdot \mathbf{K}_\theta^{-1} \cdot \mathbf{J}_\theta^{(F)T}(\mathbf{q}_i, \boldsymbol{\theta}_i) & \mathbf{J}_q^{(F)}(\mathbf{q}_i, \boldsymbol{\theta}_i) \\ \mathbf{J}_q^{(F)T}(\mathbf{q}_i, \boldsymbol{\theta}_i) & \mathbf{0} \end{bmatrix}^{-1} \cdot \begin{bmatrix} \boldsymbol{\delta}_{i+1} \\ -\mathbf{J}_q^{(G)T}(\mathbf{q}_i, \boldsymbol{\theta}_i) \cdot \mathbf{G}_i \end{bmatrix} \\
&\boldsymbol{\delta}_{i+1} = \mathbf{t}_{i+1} - \mathbf{g}(\mathbf{q}_i, \boldsymbol{\theta}_i) + \mathbf{J}_\theta^{(F)}(\mathbf{q}_i, \boldsymbol{\theta}_i) \cdot \boldsymbol{\theta}_i + \mathbf{J}_q^{(F)}(\mathbf{q}_i, \boldsymbol{\theta}_i) \cdot \mathbf{q}_i \\
&-\mathbf{J}_\theta^{(F)}(\mathbf{q}_i, \boldsymbol{\theta}_i) \cdot \mathbf{K}_\theta^{-1} \cdot \mathbf{J}_\theta^{(G)T}(\mathbf{q}_i, \boldsymbol{\theta}_i) \cdot \mathbf{G}_i \\
&\boldsymbol{\theta}_{i+1} = \mathbf{K}_\theta^{-1}\left(\mathbf{J}_\theta^{(G)T}(\mathbf{q}_i, \boldsymbol{\theta}_i) \cdot \mathbf{G}_i + \mathbf{J}_\theta^{(F)T}(\mathbf{q}_i, \boldsymbol{\theta}_i) \cdot \mathbf{F}_{i+1}\right) + \boldsymbol{\theta}^0
\end{aligned} \tag{6.40}$$

that requires matrix inversion of the essentially lower order (for example, for the 3-link manipulator with two passive joints and two actuated joints, the size of matrix inversion reduces from 34 to 14).

The proposed algorithm allows us to compute static equilibrium configuration for the serial chains with passive joints and all types of loadings (internal preloading, external loadings applied to any point of the manipulator and loading from the technological process). The convergence properties of this algorithm are similar to the one presented before. It can also be modified in a similar way to solve the problem of computing the equilibrium configuration corresponding to the given external loading

$\mathbf{F}$(instead of the given $\mathbf{t}$). Some examples illustrating its application will be presented in the following Sections.

Stiffness Matrix for a Serial Chain with Auxiliary Loadings

The previous sub-section allows us to obtain the non-linear relation between elastic deflections $\Delta\mathbf{t}$ and external loading $\mathbf{F}$. Since common engineering practice operates with the stiffness matrix, let us linearize this relation in the neighbourhood of the equilibrium. Following the virtual work technique, let us assume that the external force and the end-effector location are incremented by some small values $\delta\mathbf{F}$, $\delta\mathbf{t}$ in the neighbourhood of the current equilibrium configuration. Let us also assume that a new configuration also satisfies the equilibrium conditions. Hence, it is necessary to consider the two equilibriums corresponding to the manipulator state variables $(\mathbf{F}, \mathbf{q}, \boldsymbol{\theta}, \mathbf{t})$ and $(\mathbf{F}+\delta\mathbf{F}, \mathbf{q}+\delta\mathbf{q}, \boldsymbol{\theta}+\delta\boldsymbol{\theta}, \mathbf{t}+\delta\mathbf{t})$ simultaneously. The relevant static equilibrium equations may be written as

$$\begin{aligned}
&\mathbf{t} = \mathbf{g}(\mathbf{q}, \boldsymbol{\theta}) \\
&\mathbf{K}_\theta \cdot \left(\boldsymbol{\theta} - \boldsymbol{\theta}^0\right) = \mathbf{J}_\theta^{(G)T} \cdot \mathbf{G} + \mathbf{J}_\theta^{(F)T} \cdot \mathbf{F} \\
&\mathbf{J}_q^{(G)T} \cdot \mathbf{G} + \mathbf{J}_q^{(F)T} \cdot \mathbf{F} = \mathbf{0}
\end{aligned} \tag{6.41}$$

and

$$\begin{aligned}
&\mathbf{t} + \delta\mathbf{t} = \mathbf{g}(\mathbf{q} + \delta\mathbf{q}, \boldsymbol{\theta} + \delta\boldsymbol{\theta}) \\
&\mathbf{K}_\theta \cdot \left(\boldsymbol{\theta} + \delta\boldsymbol{\theta} - \boldsymbol{\theta}^0\right) = \left(\mathbf{J}_\theta^{(G)} + \delta\mathbf{J}_\theta^{(G)}\right)^T \cdot (\mathbf{G} + \delta\mathbf{G}) + \left(\mathbf{J}_\theta^{(F)} + \delta\mathbf{J}_\theta^{(F)}\right)^T \cdot (\mathbf{F} + \delta\mathbf{F}) \\
&\left(\mathbf{J}_q^{(G)} + \delta\mathbf{J}_q^{(G)}\right)^T \cdot (\mathbf{G} + \delta\mathbf{G}) + \left(\mathbf{J}_q^{(F)} + \delta\mathbf{J}_q^{(F)}\right)^T \cdot (\mathbf{F} + \delta\mathbf{F}) = \mathbf{0}
\end{aligned} \tag{6.42}$$

where the variables $\mathbf{t},\ \mathbf{F},\ \mathbf{G},\ \mathbf{K}_\theta,\ \mathbf{q},\ \boldsymbol{\theta},\ \boldsymbol{\theta}^0$ are assumed to be known.

After linearization of the function $\mathbf{g}(\mathbf{q}, \boldsymbol{\theta})$ in the neighbourhood of loaded equilibrium, the system (6.41), (6.42) is reduced to three equations

$$\begin{aligned}
&\delta\mathbf{t} = \mathbf{J}_\theta^{(F)} \cdot \delta\boldsymbol{\theta} + \mathbf{J}_q^{(F)} \cdot \delta\mathbf{q} \\
&\mathbf{K}_\theta \cdot \delta\boldsymbol{\theta} = \delta\mathbf{J}_\theta^{(G)} \cdot \mathbf{G} + \mathbf{J}_\theta^{(G)} \cdot \delta\mathbf{G} + \delta\mathbf{J}_\theta^{(F)} \cdot \mathbf{F} + \mathbf{J}_\theta^{(F)} \cdot \delta\mathbf{F} \\
&\delta\mathbf{J}_q^{(G)} \cdot \mathbf{G} + \mathbf{J}_q^{(G)} \cdot \delta\mathbf{G} + \delta\mathbf{J}_q^{(F)} \cdot \mathbf{F} + \mathbf{J}_q^{(F)} \cdot \delta\mathbf{F} = \mathbf{0}
\end{aligned} \tag{6.43}$$

which define the desired linear relations between $\delta\mathbf{t}$ and $\delta\mathbf{F}$, $\delta\mathbf{q}$, $\delta\boldsymbol{\theta}$ that are expressed by the stiffness matrices $\mathbf{K}_C$, $\mathbf{K}_{Cq}$, $\mathbf{K}_{C\theta}$ (see Eq. (6.19)). In this system, small variations of Jacobians may be expressed via the second-order derivatives

$$\begin{aligned}\delta\mathbf{J}_{\mathrm{q}}^{(\mathrm{F})} &= \mathbf{H}_{\mathrm{q}\theta}^{(\mathrm{F})}\cdot\delta\boldsymbol{\theta}+\mathbf{H}_{\mathrm{qq}}^{(\mathrm{F})}\cdot\delta\mathbf{q};\;\; \delta\mathbf{J}_{\theta}^{(\mathrm{F})} = \mathbf{H}_{\theta\theta}^{(\mathrm{F})}\cdot\delta\boldsymbol{\theta}+\mathbf{H}_{\theta\mathrm{q}}^{(\mathrm{F})}\cdot\delta\mathbf{q};\\ \delta\mathbf{J}_{\mathrm{q}}^{(\mathrm{G})} &= \mathbf{H}_{\mathrm{q}\theta}^{(\mathrm{G})}\cdot\delta\boldsymbol{\theta}+\mathbf{H}_{\mathrm{qq}}^{(\mathrm{G})}\cdot\delta\mathbf{q};\; \delta\mathbf{J}_{\theta}^{(\mathrm{G})} = \mathbf{H}_{\theta\theta}^{(\mathrm{G})}\cdot\delta\boldsymbol{\theta}+\mathbf{H}_{\theta\mathrm{q}}^{(\mathrm{G})}\cdot\delta\mathbf{q};\end{aligned} \tag{6.44}$$

where

$$\begin{aligned}\mathbf{H}_{\theta\theta}^{(\mathrm{G})} &= \sum_{j=1}^{n}\frac{\partial^2}{\partial\boldsymbol{\theta}^2}\left(\mathbf{g}_j^T(\mathbf{q},\boldsymbol{\theta})\cdot\mathbf{G}_j\right);\; \mathbf{H}_{\mathrm{qq}}^{(\mathrm{G})} = \sum_{j=1}^{n}\frac{\partial^2}{\partial\mathbf{q}^2}\left(\mathbf{g}_j^T(\mathbf{q},\boldsymbol{\theta})\cdot\mathbf{G}_j\right);\\ \mathbf{H}_{\theta\mathrm{q}}^{(\mathrm{G})} &= \sum_{j=1}^{n}\frac{\partial^2}{\partial\boldsymbol{\theta}\partial\mathbf{q}}\left(\mathbf{g}_j^T(\mathbf{q},\boldsymbol{\theta})\cdot\mathbf{G}_j\right);\; \mathbf{H}_{\mathrm{q}\theta}^{(\mathrm{G})} = \sum_{j=1}^{n}\frac{\partial^2}{\partial\mathbf{q}\partial\boldsymbol{\theta}}\left(\mathbf{g}_j^T(\mathbf{q},\boldsymbol{\theta})\cdot\mathbf{G}_j\right);\\ \mathbf{H}_{\theta\theta}^{(\mathrm{F})} &= \frac{\partial^2}{\partial\boldsymbol{\theta}^2}\left(\mathbf{g}^T(\mathbf{q},\boldsymbol{\theta})\cdot\mathbf{F}\right);\; \mathbf{H}_{\mathrm{qq}}^{(\mathrm{F})} = \frac{\partial^2}{\partial\mathbf{q}^2}\left(\mathbf{g}^T(\mathbf{q},\boldsymbol{\theta})\cdot\mathbf{F}\right);\\ \mathbf{H}_{\theta\mathrm{q}}^{(\mathrm{F})} &= \frac{\partial^2}{\partial\boldsymbol{\theta}\partial\mathbf{q}}\left(\mathbf{g}^T(\mathbf{q},\boldsymbol{\theta})\cdot\mathbf{F}\right);\; \mathbf{H}_{\mathrm{q}\theta}^{(\mathrm{F})} = \frac{\partial^2}{\partial\mathbf{q}\partial\boldsymbol{\theta}}\left(\mathbf{g}^T(\mathbf{q},\boldsymbol{\theta})\cdot\mathbf{F}\right);\end{aligned} \tag{6.45}$$

Also, the auxiliary loading $\mathbf{G}$ may be computed via the first-order derivatives as

$$\delta\mathbf{G} = \frac{\partial\mathbf{G}}{\partial\boldsymbol{\theta}}\cdot\delta\boldsymbol{\theta}+\frac{\partial\mathbf{G}}{\partial\mathbf{q}}\cdot\delta\mathbf{q} \tag{6.46}$$

Furthermore, to ensure similarity between the results of this subsection and sub-Sect. 4.2.3, let us introduce the additional notation

$$\begin{aligned}\mathbf{H}_{\theta\theta} &= \mathbf{H}_{\theta\theta}^{(\mathrm{F})}+\mathbf{H}_{\theta\theta}^{(\mathrm{G})}+\mathbf{J}_{\theta}^{(\mathrm{G})\mathrm{T}}\cdot\frac{\partial}{\partial\boldsymbol{\theta}}\mathbf{G};\; \mathbf{H}_{\theta\mathrm{q}} = \mathbf{H}_{\theta\mathrm{q}}^{(\mathrm{G})}+\mathbf{H}_{\theta\mathrm{q}}^{(\mathrm{F})}+\mathbf{J}_{\theta}^{(\mathrm{G})\mathrm{T}}\cdot\frac{\partial}{\partial\mathbf{q}}\mathbf{G};\\ \mathbf{H}_{\mathrm{q}\theta} &= \mathbf{H}_{\mathrm{q}\theta}^{(\mathrm{G})}+\mathbf{H}_{\mathrm{q}\theta}^{(\mathrm{F})}+\mathbf{J}_{\mathrm{q}}^{(\mathrm{G})\mathrm{T}}\cdot\frac{\partial}{\partial\boldsymbol{\theta}}\mathbf{G};\; \mathbf{H}_{\mathrm{qq}} = \mathbf{H}_{\mathrm{qq}}^{(\mathrm{G})}+\mathbf{H}_{\mathrm{qq}}^{(\mathrm{F})}+\mathbf{J}_{\mathrm{q}}^{(\mathrm{G})\mathrm{T}}\cdot\frac{\partial}{\partial\mathbf{q}}\mathbf{G}\end{aligned} \tag{6.47}$$

which allows us to present the system (6.43) in the form

$$\begin{bmatrix}\delta\mathbf{t}\\ \mathbf{0}\\ \mathbf{0}\end{bmatrix} = \begin{bmatrix}\mathbf{0} & \mathbf{J}_{\mathrm{q}}^{(\mathrm{F})} & \mathbf{J}_{\theta}^{(\mathrm{F})}\\ \mathbf{J}_{\mathrm{q}}^{(\mathrm{F})\mathrm{T}} & \mathbf{H}_{\mathrm{qq}} & \mathbf{H}_{\mathrm{q}\theta}\\ \mathbf{J}_{\theta}^{(\mathrm{F})\mathrm{T}} & \mathbf{H}_{\theta q} & -\mathbf{K}_{\theta}+\mathbf{H}_{\theta\theta}\end{bmatrix}\cdot\begin{bmatrix}\delta\mathbf{F}\\ \delta\mathbf{q}\\ \delta\boldsymbol{\theta}\end{bmatrix} \tag{6.48}$$

that has the same structure as (6.18). The desired stiffness matrices can be computed via either a high-order matrix inversion

$$\left[\begin{array}{c|c|c}\mathbf{K}_{\mathrm{C}} & * & *\\ \hline \mathbf{K}_{\mathrm{Cq}} & * & *\\ \hline \mathbf{K}_{\mathrm{C}\theta} & * & *\end{array}\right] = \begin{bmatrix}\mathbf{0} & \mathbf{J}_{\mathrm{q}}^{(\mathrm{F})} & \mathbf{J}_{\theta}^{(\mathrm{F})}\\ \mathbf{J}_{\mathrm{q}}^{(\mathrm{F})\mathrm{T}} & \mathbf{H}_{\mathrm{qq}} & \mathbf{H}_{\mathrm{q}\theta}\\ \mathbf{J}_{\theta}^{(\mathrm{F})\mathrm{T}} & \mathbf{H}_{\theta\mathrm{q}} & -\mathbf{K}_{\theta}+\mathbf{H}_{\theta\theta}\end{bmatrix}^{-1} \tag{6.49}$$

or lower-order inversion

$$\left[\begin{array}{c|c} \mathbf{K}_\mathrm{C} & * \\ \hline \mathbf{K}_\mathrm{Cq} & * \end{array}\right] = \begin{bmatrix} \mathbf{J}_\theta^{(\mathrm{F})} \cdot \mathbf{k}_\theta^\mathrm{F} \cdot \mathbf{J}_\theta^{(\mathrm{F})\mathrm{T}} & \mathbf{J}_\mathrm{q}^{(\mathrm{F})} + \mathbf{J}_\theta^{(\mathrm{F})} \cdot \mathbf{k}_\theta^\mathrm{F} \cdot \mathbf{H}_{\theta\mathrm{q}} \\ \mathbf{J}_\mathrm{q}^{(\mathrm{F})\mathrm{T}} + \mathbf{H}_{\mathrm{q}\theta} \cdot \mathbf{k}_\theta^\mathrm{F} \cdot \mathbf{J}_\theta^{(\mathrm{F})\mathrm{T}} & \mathbf{H}_\mathrm{qq} + \mathbf{H}_{\mathrm{q}\theta} \cdot \mathbf{k}_\theta^\mathrm{F} \cdot \mathbf{H}_{\theta\mathrm{q}} \end{bmatrix}^{-1} \quad (6.50)$$

where $\mathbf{k}_\theta^\mathrm{F}$ denotes the modified joint compliance matrix $\mathbf{k}_\theta^\mathrm{F} = (\mathbf{K}_\theta - \mathbf{H}_{\theta\theta})^{-1}$. It is obvious that, using these notations, the matrices $\mathbf{K}_\mathrm{C}$, $\mathbf{K}_\mathrm{Cq}$, $\mathbf{K}_{\mathrm{C}\theta}$can be also computed analytically in accordance with expressions (6.23), (6.24) and (6.25) respectively.

Thus, in this Section, the non-linear stiffness modelling technique for serial chains was extended for the manipulators with external and internal loading applied to the node-points (auxiliary loading). It allows us to compute values of the internal variables corresponding to the equilibrium, to obtain the non-linear force–deflection relation and to compute the related stiffness matrices. However, the question related to the stability of the equilibrium configuration of the kinematic chain still remains open.

6.2 Non-linear Behaviour of Manipulators Under the Loading

External and internal loadings applied to the serial chains may have an influence on their configurations (i.e. on their shape). Moreover, for the same end-point location, due to the kinematic redundancy, the kinematic chain can have multiple equilibrium configurations, either stable or unstable. For this reason, let us concentrate on the stability analysis of the kinematic chain configuration under the loading. It is worth mentioning that in parallel manipulators, the considered configurations of the chains are equivalent with respect to the location of the end-platform. But in practice, only stable configurations may be observed.

6.2.1 Definition of Configuration Stability for a Serial Chain Under Loading

To investigate possible non-linear effects that may be caused by external and internal loading, it is necessary to extend the notion of stability associated with the stiffness analysis. Traditionally, the stability of compliant mechanical systems (including manipulators) is defined as the *resistance of the end-point location* $\mathbf{t}$ with respect to the "disturbing" effects of an external force $\mathbf{F}$ applied at this point. In such formulation, the stability is completely defined by the stiffness matrix $\mathbf{K}_\mathrm{C}$ that describes the linear relations (6.5) between the force and deflection deviations $\delta\mathbf{F}$, $\delta\mathbf{t}$ with respect to the values $\mathbf{F}$, $\mathbf{t}$. It is obvious that the matrix $\mathbf{K}_\mathrm{C}$ is a positive definite for the stable location $\mathbf{t}$.

However, in the compliant manipulators with passive joints, the equilibrium configuration $(\mathbf{q}, \boldsymbol{\theta})$ corresponding to the same end-point location $\mathbf{t}$ cannot be unique.

Moreover, these configurations may be both "stable" and "unstable" and may correspond to different values of potential energy stored in the virtual springs. From this point of view, it is worth to distinguish stability of the end-point location $\mathbf{t}$ from the stability of the corresponding equilibrium configuration of the kinematic chain $(\mathbf{q}, \boldsymbol{\theta})$, which may be defined as a *resistance of the chain shape* with respect to disturbances in redundant kinematic variables. This issue becomes extremely important for the loaded mode, due to the kinematic redundancy caused by the passive joints and excessive number of virtual springs. As such, small disturbances in $(\mathbf{q}, \boldsymbol{\theta})$ may provoke an essential change of the current equilibrium configuration leading to the reduction of the potential energy and transition to another equilibrium state, while keeping the same end-point location. Hence, it is necessary to evaluate the internal properties of the kinematic chain in the state of the loaded equilibrium that may correspond either to the minimum or maximum of the potential energy for a fixed value of $\mathbf{t}$.

Let us illustrate this notion with the example of a three-link chain (Fig. 6.3a), which includes passive joints at both ends and two virtual torsional springs between the links, which ensure the "straight" configuration for the unloaded mode. It is assumed that both ends of the chain are fixed by the external geometrical constraints while the internal configuration may change without shifting the end-points, in accordance with the redundant parameter value. It is evident that this chain is loaded, but the corresponding value of the force $\mathbf{F}$ depends on a particular configuration. Besides, among a variety of possible configurations (corresponding to given end-point locations), only the equilibrium ones are in the focus of interest.

For this case study, it is convenient to give an energy-based interpretation. The considered kinematic chain has one redundant parameter (rotation angle of any passive joint) and under geometrical constraints may occupy configurations with different shapes. The relevant relation between the energy stored in the virtual springs and the redundant parameter value is presented in Fig. 6.3b. Due to the physical nature of this chain, for each given end-point displacement Δ, the examined plot presents a continuous closed criss-cross curve that has exactly two minimum and

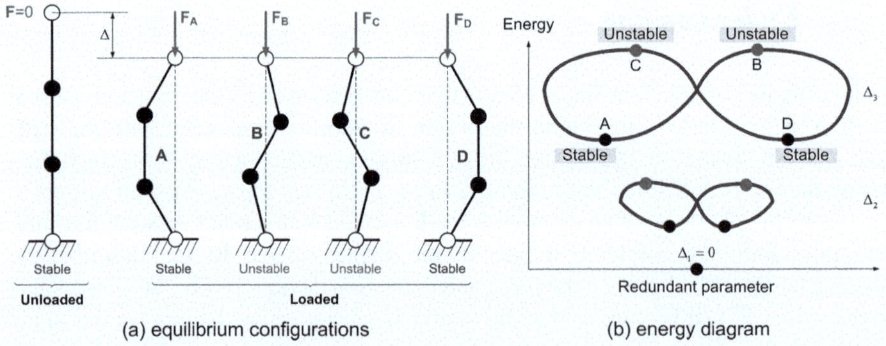

Fig. 6.3 Stable and non-stable configurations of 3-link serial chain and their energy-based interpretation (○—passive join, ●—virtual torsional spring) [31]

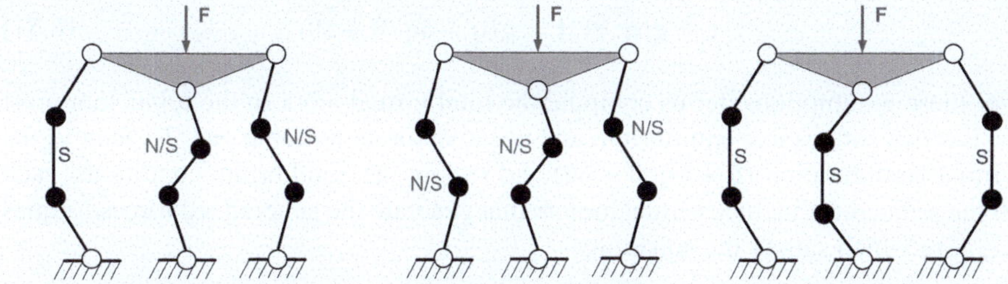

Fig. 6.4 2D parallel manipulators with serial chains in stable and unstable configurations (○—passive join, ●—virtual torsional spring) [31]

maximum points, that correspond to the stable and unstable equilibriums respectively. Hence, the numerical solution of static equilibrium equations is presented in Sects. 6.1.1(b) and 5.2.2(c) may yield both stable and unstable configurations, while in practice only stable ones should be considered. Thus, in addition, to the above-mentioned algorithms, the criterion that allows to distinguish between stable and unstable configurations of the kinematic chain is required.

The physical meaning of this stability notion (related to the kinematic chain shape) is illustrated in Fig. 6.4, which contains several postures of the same parallel manipulator with exactly the same end-platform location. These postures differ in the shapes of serial kinematic chains that may be treated as an internal configuration of the parallel manipulator, which is not "visible" from the end-platform side whose static stability is completely defined by the Cartesian stiffness matrix. In particular, Fig. 6.4a,b present parallel manipulators that include at least one kinematic chain in the unstable configuration that cannot be observed in practice but satisfy the general static equilibrium equation. In contrast, Fig. 6.4c shows the physically realizable posture of the same manipulator (with exactly the same location of the chain end-points) where for all kinematic chains the shapes are stable.

Hence, a full-scale investigation into the stiffness properties of the loaded parallel manipulator must include the stability analysis of the internal kinematic chain configurations as presented in the following sub-section.

6.2.2 *Stability Criterion for Kinematic Chain Configuration*

To evaluate the stability of the static equilibrium configuration $(\mathbf{q}, \boldsymbol{\theta})$ of a separate kinematic chain, let us assume that the end-point is fixed at the point $\mathbf{t} = (\mathbf{p}, \varphi)^{\mathrm{T}}$ corresponding to the external load $\mathbf{F}$, but the joint coordinates are given small virtual displacements $\delta\mathbf{q}$, $\delta\boldsymbol{\theta}$ satisfying the geometrical constraint (6.1), i.e.

$$\mathbf{t} = \mathbf{g}(\mathbf{q}, \boldsymbol{\theta}); \mathbf{t} = \mathbf{g}(\mathbf{q} + \delta\mathbf{q}, \boldsymbol{\theta} + \delta\boldsymbol{\theta}) \tag{6.51}$$

For these assumptions, let us compute the total virtual work in the joints that must be positive for a stable equilibrium and negative for an unstable one. To achieve the virtual configuration $(\mathbf{q} + \delta\mathbf{q}, \boldsymbol{\theta} + \delta\boldsymbol{\theta})$ and restore the equilibrium conditions, each of the joints must include virtual motors that generate the generalized forces/torques $\delta\boldsymbol{\tau}_q, \delta\boldsymbol{\tau}_\theta$ which satisfy the equations:

$$\begin{aligned} &\mathbf{J}_\theta^T \cdot \mathbf{F} = \mathbf{K}_\theta \cdot (\boldsymbol{\theta} - \boldsymbol{\theta}_0); \ (\mathbf{J}_\theta + \delta\mathbf{J}_\theta)^T \cdot \mathbf{F} = \mathbf{K}_\theta \cdot (\boldsymbol{\theta} - \boldsymbol{\theta}_0 + \delta\boldsymbol{\theta}) + \delta\boldsymbol{\tau}_\theta \\ &\mathbf{J}_q^T \cdot \mathbf{F} = 0; \ (\mathbf{J}_q + \delta\mathbf{J}_q)^T \cdot \mathbf{F} = \delta\boldsymbol{\tau}_q \end{aligned} \tag{6.52}$$

After relevant transformations, the virtual torques may be expressed as

$$\delta\boldsymbol{\tau}_\theta = \delta\left(\mathbf{J}_\theta^T \cdot \mathbf{F}\right) - \mathbf{K}_\theta \cdot \delta\boldsymbol{\theta}; \ \delta\boldsymbol{\tau}_q = \delta\left(\mathbf{J}_q^T \cdot \mathbf{F}\right) \tag{6.53}$$

where $\delta(\ldots)$ denotes the differential with respect to $\delta\mathbf{q}, \delta\boldsymbol{\theta}$ that may be expanded via Hessians of the scalar function $\Psi = \mathbf{g}(\mathbf{q}, \boldsymbol{\theta})^T \cdot \mathbf{F}$:

$$\delta\left(\mathbf{J}_\theta^T \cdot \mathbf{F}\right) = \mathbf{H}_{\theta q}^F \cdot \delta\mathbf{q} + \mathbf{H}_{\theta\theta}^F \cdot \delta\boldsymbol{\theta}; \ \delta\left(\mathbf{J}_q^T \cdot \mathbf{F}\right) = \mathbf{H}_{qq}^F \cdot \delta\mathbf{q} + \mathbf{H}_{q\theta}^F \cdot \delta\boldsymbol{\theta} \tag{6.54}$$

provided that

$$\mathbf{H}_{qq}^F = \partial^2\Psi/\partial\mathbf{q}^2; \ \mathbf{H}_{\theta\theta}^F = \partial^2\Psi/\partial\boldsymbol{\theta}^2; \ \mathbf{H}_{q\theta}^F = \mathbf{H}_{\theta q}^{FT} = \partial^2\Psi/\partial\mathbf{q}\partial\boldsymbol{\theta} \tag{6.55}$$

Furthermore, taking into account that the virtual displacement from $(\mathbf{q}, \boldsymbol{\theta})$ to $(\mathbf{q} + \delta\mathbf{q}, \boldsymbol{\theta} + \delta\boldsymbol{\theta})$ leads to a gradual change of the virtual motor torques from $(\mathbf{0}, \mathbf{0})$ to $\left(\delta\boldsymbol{\tau}_q, \delta\boldsymbol{\tau}_\theta\right)$, the virtual work may be computed as half of the corresponding scalar products

$$\delta W = -\frac{1}{2}\left(\delta\boldsymbol{\tau}_\theta^T \cdot \delta\boldsymbol{\theta} + \delta\boldsymbol{\tau}_q^T \cdot \delta\mathbf{q}\right) \tag{6.56}$$

where the minus sign takes into account the adopted conventions for the positive directions of the forces and displacements. Hence, after appropriate substitutions and transformations to the matrix form, the desired stability condition may be written as

$$\delta W = -\frac{1}{2}\left[\delta\boldsymbol{\theta}^T \ \delta\mathbf{q}^T\right] \cdot \begin{bmatrix} \mathbf{H}_{\theta\theta}^F - \mathbf{K}_\theta & \mathbf{H}_{q\theta}^F \\ \mathbf{H}_{\theta q}^F & \mathbf{H}_{qq}^F \end{bmatrix} \cdot \begin{bmatrix} \delta\boldsymbol{\theta} \\ \delta\mathbf{q} \end{bmatrix} > 0 \tag{6.57}$$

where $\delta\mathbf{q}$ and $\delta\boldsymbol{\theta}$ must satisfy the geometrical constraints (6.51).

In order to take into account the relation between $\delta\mathbf{q}$ and $\delta\boldsymbol{\theta}$ that is imposed by (6.51), let us apply the first-order expansion of the function $\mathbf{g}(\boldsymbol{\theta}, \mathbf{q})$ that yields the following linear relation

$$[\mathbf{J}_\theta \ \mathbf{J}_q] \cdot \begin{bmatrix} \delta\boldsymbol{\theta} \\ \delta\mathbf{q} \end{bmatrix} = \mathbf{0} \tag{6.58}$$

Then, applying the SVD- factorization [Strang 1998] of the integrated Jacobian

$$[\mathbf{J}_\theta \ \mathbf{J}_q] = [\mathbf{U}_\theta \ \mathbf{U}_q] \cdot \begin{bmatrix} \mathbf{S}_r & \\ & \mathbf{0} \end{bmatrix} \cdot \begin{bmatrix} \mathbf{V}_\theta^T \\ \mathbf{V}_q^T \end{bmatrix} \tag{6.59}$$

and extracting from $\mathbf{V}_\theta$, $\mathbf{V}_q$ the sub-matrices $\mathbf{V}_\theta^o$, $\mathbf{V}_q^o$ corresponding to zero singular values, a relevant null space of the system (6.58) may be presented as

$$\delta\boldsymbol{\theta} = \mathbf{V}_\theta^o \cdot \delta\boldsymbol{\mu}; \ \delta\mathbf{q} = \mathbf{V}_q^o \cdot \delta\boldsymbol{\mu} \tag{6.60}$$

where $\delta\boldsymbol{\mu}$ is the arbitrary vector of the appropriate dimension (equal to the rank-deficiency of the integrated Jacobian). Hence, the stability condition (6.57) may be rewritten as an inequality

$$\delta W = -\frac{1}{2}\delta\boldsymbol{\mu}^T \cdot \begin{bmatrix} \mathbf{V}_\theta^o \\ \mathbf{V}_q^o \end{bmatrix}^T \cdot \begin{bmatrix} \mathbf{H}_{\theta\theta}^F - \mathbf{K}_\theta & \mathbf{H}_{q\theta}^F \\ \mathbf{H}_{\theta q}^F & \mathbf{H}_{qq}^F \end{bmatrix} \cdot \begin{bmatrix} \mathbf{V}_\theta^o \\ \mathbf{V}_q^o \end{bmatrix} \cdot \delta\boldsymbol{\mu} > 0 \tag{6.61}$$

that must be satisfied for all arbitrary non-zero $\delta\boldsymbol{\mu}$. In other words, the considered static equilibrium configuration $(\mathbf{q}, \boldsymbol{\theta})$ is stable if (and only if) the matrix

$$\begin{bmatrix} \mathbf{V}_\theta^o \\ \mathbf{V}_q^o \end{bmatrix}^T \cdot \begin{bmatrix} \mathbf{H}_{\theta\theta}^F - \mathbf{K}_\theta & \mathbf{H}_{q\theta}^F \\ \mathbf{H}_{\theta q}^F & \mathbf{H}_{qq}^F \end{bmatrix} \cdot \begin{bmatrix} \mathbf{V}_\theta^o \\ \mathbf{V}_q^o \end{bmatrix} < 0 \tag{6.62}$$

is negative-definite. It is worth mentioning that the obtained result is in good agreement with the previous studies Chen 2002], where (for manipulators without passive joints) the stiffness properties were defined by the matrix $\mathbf{K}_\theta - \mathbf{H}_{\theta\theta}^F$ that evidently must be positive-definite for the stable configurations. In Chap. 8 these results are applied for detecting bifurcations and buckling phenomena in typical serial kinematic chains and relevant parallel manipulators.

Thus, proposed stability analysis techniques for serial chains with the passive joints and related matrix stability criterion for the kinematic chain configuration allow estimating the stability of the serial chain configuration under external loading in the case of single and multiple equilibriums.

6.3 Stiffness Model of Parallel Manipulators in Loaded Mode

The non-linear stiffness modelling technique proposed above deals with separate kinematic chains. In order to be applied to the parallel manipulators, it should be extended by appropriate stiffness model aggregation routines, as in the case of linear stiffness modelling. Similar to the unloaded mode, these routines have to be applicable for both perfect and non-perfect kinematic chains. Besides, it is required to develop numerical algorithms for computing both direct and inverse force–deflection relations that are referred below to as non-linear stiffness and compliance models respectively.

6.3.1 Aggregation of Chains Stiffness Models

Let us focus on the aggregation of stiffness models of separate serial chains into the stiffness model of the whole parallel manipulator in the loaded mode. To solve this problem, it is necessary to obtain the non-linear force–deflection relation, which takes into account the elastostatic properties of all kinematic chains, and to compute the corresponding Cartesian stiffness matrix.

Let us assume that the end-points of all kinematic chains are aligned and matched in the same target point $\mathbf{t}_0$, which corresponds to the desired end-platform location. This point is assumed to be known and allows us to compute, from the inverse kinematic model, the actuator and passive joint coordinates defining nominal configurations of the chains a. It is also assumed that the stiffness models of all kinematic chains have been already obtained and are presented in the form of partial non-linear force–deflection relations $\mathbf{F}_i = f_i(\mathbf{t}|\mathbf{t}_0)$ corresponding to the target point $\mathbf{t}_0$.

It is evident that the external loading $\mathbf{F}$ changes the end-platform location $\mathbf{t}_0$, hence it is reasonable to consider the set of locations $\mathbf{t}$ in the neighbourhood of target one. Under the above assumptions, for any given point $\mathbf{t}$ from the neighbourhood of $\mathbf{t}_0$ it is possible to compute both the partial forces $\mathbf{F}_i$ and corresponding equilibrium configurations $(\mathbf{q}_i, \boldsymbol{\theta}_i)$ as well as to evaluate their stability. Then, in accordance with the superposition principle, the desired non-linear force–deflection relation for the whole parallel manipulator can be found by straightforward summation of all partial forces $\mathbf{F}_i$, i.e.

$$\mathbf{F} = \sum_{i=1}^{m} f_i(\mathbf{t}|\mathbf{t}_0) \tag{6.63}$$

where $\mathbf{F}$ denotes the total external loading applied to the end-platform. As a result, corresponding curves can be obtained by multiple repetitions of the above-described procedures for different values of the end-platform location $\mathbf{t}$.

Furthermore, for each given $\mathbf{t}$, the stiffness matrices $\mathbf{K}_C^{(i)}$ of all kinematic chains can be computed using the expression (6.23) or it's modified version that includes the auxiliary loadings $\mathbf{G}_i$ (see previous Section for details). This allows us to compute the Cartesian stiffness matrix $\mathbf{K}_C$ of the whole parallel manipulator as a sum

$$\mathbf{K}_C = \sum_{i=1}^{m} \mathbf{K}_C^{(i)} \tag{6.64}$$

However, the matrices $\mathbf{K}_{Cq}^{(i)}$ and $\mathbf{K}_{Cq}^{(i)}$ defining the "sensitivity" of the chain joint coordinates $(\mathbf{q}_i, \boldsymbol{\theta}_i)$ to the end-platform displacement cannot be aggregated in this way, they should be used separately to evaluate stresses in joints/links and resistance of the chain configurations with respect to external loading $\mathbf{F}$

$$\begin{aligned} \boldsymbol{\tau}_\theta^{(i)} &= \mathbf{J}_\theta^{(i)\mathrm{T}}\, \mathbf{K}_C^{(i)}\, (\mathbf{t} - \mathbf{t}_0) \\ \delta \mathbf{q}_\mathrm{i} &= \mathbf{K}_{Cq}^{(\mathrm{i})}\, (\mathbf{t} - \mathbf{t}_0) \\ \delta \boldsymbol{\theta}_\mathrm{i} &= \mathbf{K}_{C\boldsymbol{\theta}}^{(\mathrm{i})}\, (\mathbf{t} - \mathbf{t}_0) \end{aligned} \tag{6.65}$$

where $\mathbf{J}_\theta^{(i)}$ is the Jacobian matrix of i-th kinematic chain with respect to virtual joint coordinates.

It is worth mentioning that above it was implicitly assumed that the manipulator assembling is equivalent to the aligning and matching of the chain end-frames. To deal with a more general case, when the chains are connected to the different points of the platform, it is necessary to slightly modify the chain geometrical models and to re-compute the forces/torques and the stiffness matrices by adding a virtual rigid link connecting the end-point of the chain and the reference point of the platform (see linear stiffness modelling where these extensions are defined by the vectors $\mathbf{v}_i$). After the relevant transformations that are described in detail in Sect. 5.2.3, the above-presented technique can be applied straightforwardly.

Besides, in contrast to the linear case, here there are no evident differences in stiffness models aggregation of perfect and non-perfect kinematic chains. However, here the chain geometrical errors are implicitly included in the functions $\mathbf{g}_i^\varepsilon(\mathbf{q}_i, \boldsymbol{\theta}_i)$. In particular for non-perfect chains, it is assumed that the nominal values of the joint coordinates $(\mathbf{q}_{0i}, \boldsymbol{\theta}_{0i})$ produce the end-point location vector which differs from $\mathbf{t}_0$:

$$\mathbf{g}_i^\varepsilon(\mathbf{q}_{0i}, \boldsymbol{\theta}_{0i}) = \mathbf{t}_0 + \boldsymbol{\varepsilon}_i \tag{6.66}$$

where $\boldsymbol{\varepsilon}_i$ accumulates influences of all geometrical errors on the end-point location of the i-th chain. As a result, the end-platform cannot be located at the target point $\mathbf{t}_0$ without external loading, i.e.

$$\left. \sum_{i=1}^{m} f_i(\mathbf{t}|\mathbf{t}_0) \right|_{\mathbf{t}=\mathbf{t}_0} \neq \mathbf{0} \tag{6.67}$$

Moreover, without external loading, the end-platform location $\mathbf{t}_\varepsilon$ is different from the target one $\mathbf{t}_0$. The vector $\mathbf{t}_\varepsilon$ can be computed from the equation

$$\sum_{i=1}^{m} f_i(\mathbf{t}_\varepsilon|\mathbf{t}_0) = \mathbf{0} \tag{6.68}$$

which will be considered in the next subsection. Corresponding internal forces $\mathbf{F}_i^\varepsilon$ defining the chain loadings due to the geometrical errors in the chains can be computed by simple substitution $\mathbf{t}_\varepsilon$ to the partial force deflection relations

$$\mathbf{F}_i^\varepsilon = f_i(\mathbf{t}|\mathbf{t}_0)|_{\mathbf{t}=\mathbf{t}_\varepsilon} \tag{6.69}$$

It is obvious that the sum of the $\mathbf{F}_i^\varepsilon$ is equal to zero but they produce stresses in the links and joints if the parallel manipulator is over-constrained.

Hence, the developed aggregation technique allows us to obtain the non-linear force–deflection relation for a parallel manipulator in the loaded mode as well as to compute Cartesian stiffness matrices for any given target point $\mathbf{t}_0$ and given a set of end-point locations $\{\mathbf{t}\}$. This technique is summarized in Fig. 6.5.

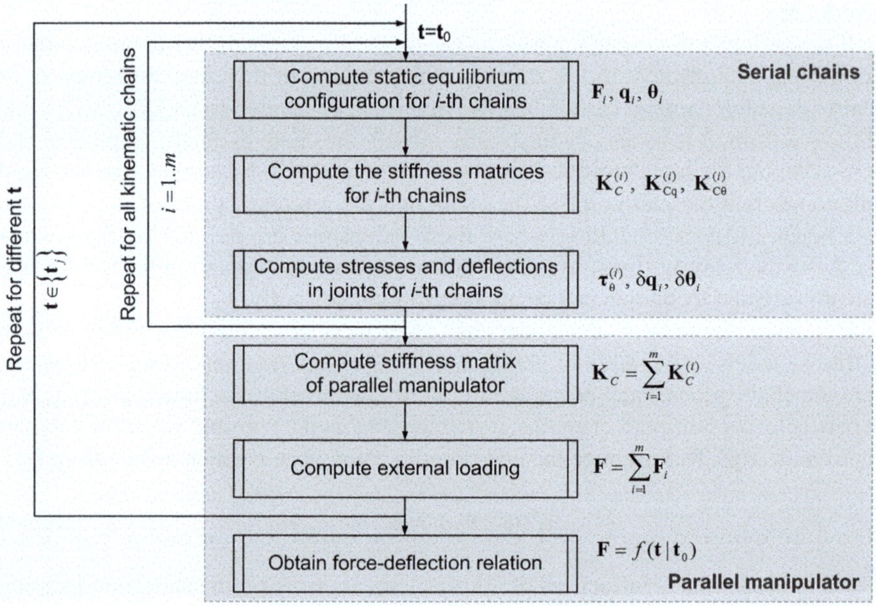

Fig. 6.5 Aggregation of serial chains stiffness models technique [30]

6.3.2 *Compliance Model of Parallel Manipulator*

The non-linear force–deflection relation (6.63) allows us to evaluate the external force/torque **F** required to locate the manipulator in the target point **t** (assuming that the actuated coordinates are computed for the end-platform location $\mathbf{t}_0$ corresponding to the unloaded configuration). However, in practice, it is often necessary to determine the end platform resistance to the external loading, i.e. to compute the deflection caused by the force **F** applied to the end-platform. The desired value can be found in the non-linear compliance model that in the general case is expressed as

$$\mathbf{t} = f^{-1}(\mathbf{F}|\mathbf{t}_0) \tag{6.70}$$

and is defined by the inverse $f^{-1}(\cdots)$ which for parallel manipulators usually exists (due to over-constrained structure). In contrast, for serial chains with passive joints, the function $f^{-1}(\cdots)$ cannot be computed since the corresponding Cartesian stiffness matrix $\mathbf{K}_C^{(i)}$ is singular.

It is obvious that in a general case, the function $f^{-1}(\cdots)$ cannot be expressed analytically. Hence, it is required that a dedicated iterative procedure, which is able to solve the non-linear Eq. (6.70) for **t** (assuming that **F** is given). It is proposed here to apply a modification of the Newton–Raphson technique which iteratively updates the desired value **t** in accordance with the expression

$$\mathbf{t}\prime = \mathbf{t} + \mathbf{K}_C^{-1}(\mathbf{t}|\mathbf{t}_0) \cdot (\mathbf{F} - f(\mathbf{t}|\mathbf{t}_0)) \tag{6.71}$$

where $\mathbf{t}'$ corresponds to the next iteration, $\mathbf{K}_C(\mathbf{t}|\mathbf{t}_0)$ is the Cartesian stiffness matrix computed in the point **t**, and $\mathbf{t}_0$ denotes the unloaded location of the end-platform. For this iterative scheme, $\mathbf{t}_0$ can be also used as the initial value of **t**. Similar to sub-Sect. 4.5.1, within each iterative loop, corresponding configurations $(\mathbf{q}_i, \boldsymbol{\theta}_i)$, the loadings $\mathbf{F}_i$ and stiffness matrices $\mathbf{K}_C^{(i)}$ for each kinematic chain are computed using Eqs. (6.10), (6.20) or (6.39), (6.49), depending on the external loading type.

As it follows from the relevant study, the convergence of this iterative procedure is good enough if the function $f(...)$ is smooth and non-singular in the neighbourhood of $\mathbf{t}_0$. If it is required to improve convergence, it is possible to apply the same technique as in the previous Section, when the force **F** is modified from iteration to iteration in accordance with the expression $\mathbf{F}\prime = \alpha \cdot \mathbf{F}$, where a scalar variable α is monotonically increasing from 0 up to 1. The stopping criterion can be expressed in a straightforward way as

$$\|\mathbf{F} - f(\mathbf{t}|\mathbf{t}_0)\| < \varepsilon_F \tag{6.72}$$

where ε_F is the desired accuracy. A more details presentation of the developed iterative routines is given in Fig. 6.6.

Summarising this sub-section, it is worth mentioning that the developed technique allows to obtain the desired non-linear deflection-force relation describing the

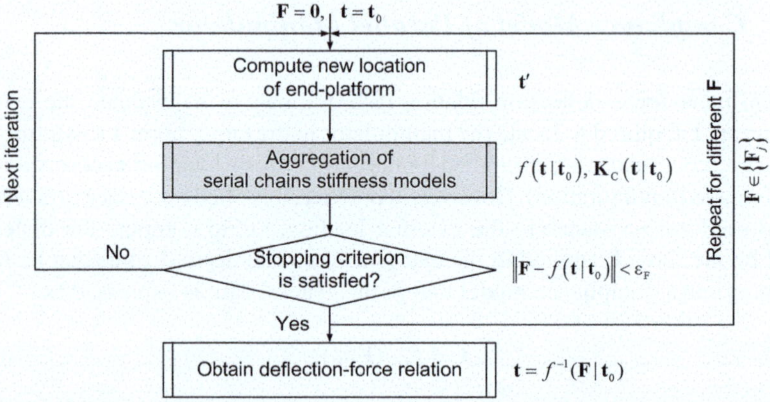

Fig. 6.6 Procedure for obtaining deflection-force relation for loaded parallel manipulator [30]

end-platform resistance with respect to the external force for a given $\mathbf{t}_0$, which corresponds to the unloaded (nominal) manipulator configuration. However, in practice, it is often required to compensate the compliance errors by properly adjusting actuated coordinates. This procedure is equivalent to the modification of $\mathbf{t}_0$. Before considering this problem in detail, let us illustrate the developed aggregation technique by an example.

6.3.3 *Illustrative Example: 2D Translational Manipulator with Preloading in Joints*

Let us apply the proposed techniques to the stiffness analysis of the planar parallel manipulator of the Orthoglide family (Fig. 6.7a). For illustration purposes, let us assume that the only source of the manipulator elasticity is concentrated in actuated drives, while the passive joints may be preloaded by (i) standard linear springs, or (ii) non-linear springs with mechanical stop-limit (see Fig. 6.7 for details).

For this manipulator, the kinematic model includes a single parameter L (the leg length) and the dexterous workspace was defined as the maximum square area that provides the velocity (and force) transmission factors in the range $[0.5, 2.0]$. Using the critical point technique developed for this type of manipulator [33], it was proved that the desired square vertices are located in the points $Q_1(-p,-p)$ and $Q_2(p, p)$, where $p = 0.45$ L. Besides, the square centre $Q_0(0, 0)$ is isotropic with respect to the velocity and force transmission. The parameters of the actuating drives are also assumed identical and their linear stiffness is denoted as K_θ. The auxiliary springs (incorporated in the passive joints) are described by two parameters: the angular stiffness coefficient K_ϑ and the activation angle ϑ_0 that defines the preloading activation point. During simulation, the manipulator end-point has been displaced by value Δ in the direction,

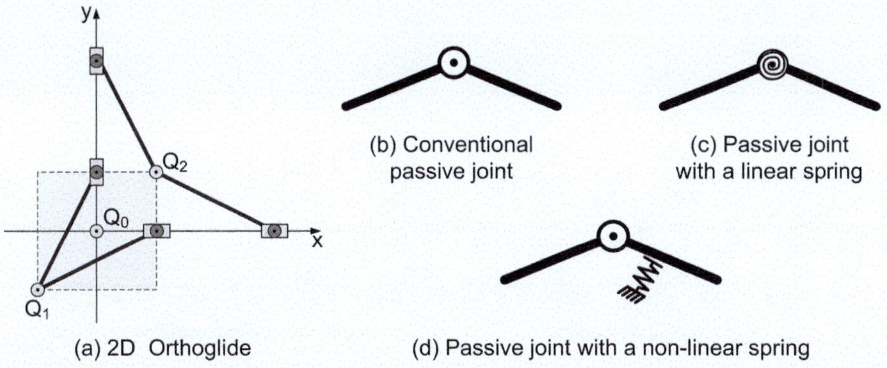

Fig. 6.7 Architecture planar version of the Orthoglide manipulator (**a**), and passive joints and alternatives for them (**b–d**) [32]

Q_0Q_1 or Q_0Q_2, and the corresponding magnitude of the external force F has been computed.

The stiffness analysis results are summarized in Figs. 6.8, 6.9 and Table 6.1. As it follows from them, the original manipulator (without preloading in passive joints) demonstrates rather low stiffness in the neighbourhood of the point Q_2, which is roughly 4 times lower than in the isotropic point Q_0. In contrast, the linear stiffness in point Q_1 is twice as high as in the point Q_0. Besides, in point Q_2, the external loading may provoke the buckling phenomenon that is caused by a local minimum of the force–deflection relation. In this case, the distance-to-singularity is essentially lower than it is estimated from the kinematical model and the manipulator may easily lose its structural stability.

To improve the manipulator stiffness and avoid buckling in the neighbourhood of Q_2, the passive joints were first preloaded by linear springs with an activation angle $\vartheta_0 = 0$. As it follows from Fig. 6.8, the preloading with parameter $K_\vartheta = 0.1 K_\theta L^2$ allows to elimination of the buckling completely and improves the stiffness by a factor of 2.3. On the other hand, the stiffness in points Q_0 and Q_1 change non-essentially,

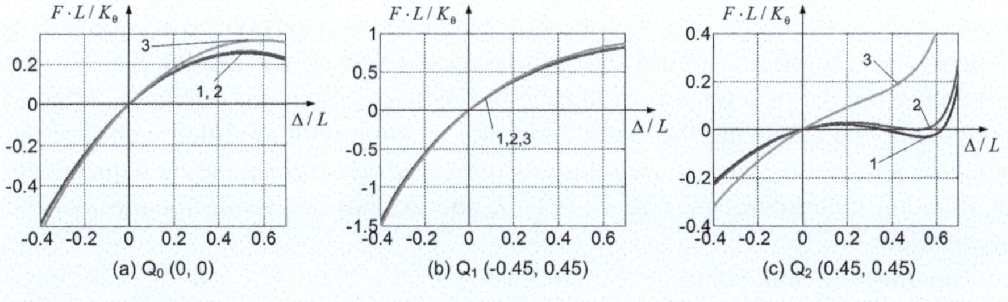

Fig. 6.8 Force–deflection relations $F = f(\Delta/L)$ in critical points: (1) $K_\vartheta = 0$, (2) $K_\vartheta = 0.05\, K_\theta\, L^2$, (3) $K_\vartheta = 0.1\, K_\theta\, L^2$ [32]

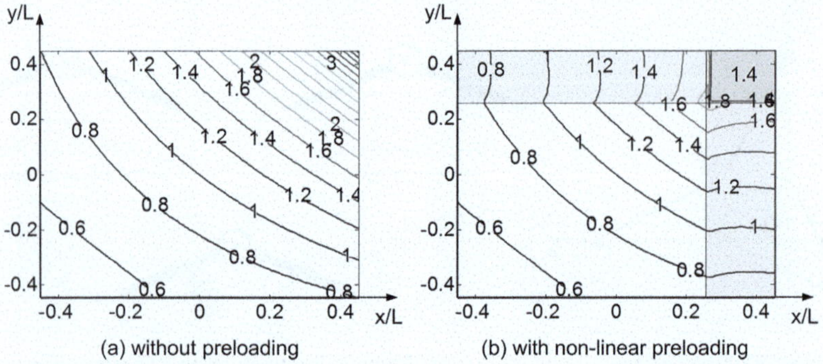

Fig. 6.9 Compliance maps for cases of: **a** manipulator without preloading; **b** manipulator with preloaded nonlinear springs ($K_\vartheta = 0.05\, K_\theta\, L^2$ and $\vartheta_0 = \pi/12$) [32]

Table 6.1 Manipulator stiffness for different linear preloading [32]

Stiffness in preloaded joints	$K_\vartheta = 0$	$0.01\ K_\theta\ L^2$	$0.05\ K_\theta\ L^2$	$0.1\ K_\theta\ L^2$
Point Q0: (isotropic point)				
Actuating joint coordinates ρ	L	L	L	L
Manipulator stiffness $\mathbf{K}_C$	K_θ	$1.01\, K_\theta$	$1.05\, K_\theta$	$1.10\, K_\theta$
Point Q1 (neighbourhood of "bar" singularity)				
Actuating joint coordinates ρ	0.437 L	0.433 L	0.419 L	0.402 L
Manipulator stiffness $\mathbf{K}_C$	$2.276\, K_\theta$	$2.286\, K_\theta$	$2.329\, K_\theta$	$2.382\, K_\theta$
Point Q2: (neighbourhood of "flat" singularity)				
Actuating joint coordinates ρ	1.345 L	1.356 L	1.399 L	1.453 L
Manipulator stiffness $\mathbf{K}_C$	$0.24\, K_\theta$	$0.27\, K_\theta$	$0.39\, K_\theta$	$0.55\, K_\theta$
Critical force $\mathbf{F}_{cr}$	$0.020\, K_\theta\, L$	$0.027\, K_\theta\, L$	–	–

by 10% and 5% respectively. Hence, with respect to the stiffness, such preloading has a positive impact.

The only negative consequence of such preloading is related to the changes in the actuator control strategy. In fact, instead of standard kinematic control, it is necessary to apply the kinetostatic control algorithm (see next Section for details). It allows for compensating the position errors caused by elastic deformations due to the internal preloading and achieving the target end-point location with modified values of the actuated joint coordinates. As it follows from Table 6.1, corresponding adjustments of the joint coordinates may reach $0.1 \cdot L$ and are not negligible for most of the applications.

The most efficient solution that eliminates this problem is to use the non-linear springs with mechanical stop-limits that are activated while approaching Q_2. For instance, as it follows from the dedicated study, the preloading with the parameters $K_\vartheta = 0.05 \cdot K_\theta L^2$, $\vartheta_0 = \pi/12$ provides almost the same improvements in Q_2 as the linear spring while preserving usual control strategies if the preloading is not activated. The efficiency of this approach is illustrated by the compliance maps presented in Fig. 6.9.

Thus, in this Section, the stiffness model aggregation technique for parallel manipulators was extended for the loaded mode. Compared to the related results presented in Sect. 5.2.3, it allows us to obtain a non-linear force–deflection (or deflection-force) relation and to compute the corresponding Cartesian stiffness matrix that depends on the loading magnitude and direction. In practice, this technique can be applied for two purposes: (i) evaluating the external loading impact on the positioning errors and (ii) compensating the compliance errors by proper adjustment of actuated coordinates. The last issue is the focus of the next Section.

6.4 Compensation for Compliance Errors for Parallel Manipulators

In many robotic applications such as machining, grinding, trimming etc., the interaction between the workpiece and technological tool causes essential deflections that significantly decrease the processing accuracy and quality of the final product. To overcome this difficulty, it is possible to modify either the control algorithm or the prescribed trajectory, which is used as the reference input for a control system. This section focuses on the second approach that is considered to be more realistic for the practice. In contrast to the previous works, the proposed technique is based on the non-linear stiffness model that is able to take into account external and internal loadings of high magnitude.

In industrial robotic controllers, the manipulator motions are usually generated using the inverse kinematic model that allows us to compute the input signals for actuators $\boldsymbol{\rho}_0$ corresponding to the desired end-effector location $\mathbf{t}_0$, which is assigned assuming that the compliance errors are negligible. However, if the external loading is essential, the kinematic control becomes non-applicable because of changes in the end-platform location. It can be computed from the non-linear compliance model as

$$\mathbf{t}_{\mathrm{F}} = f^{-1}(\mathbf{F}|\mathbf{t}_0) \tag{6.73}$$

where the subscript ‘F’ refers to the loaded mode.

To compensate for this undeterred end-platform displacement from $\mathbf{t}_0$ to $\mathbf{t}_{\mathrm{F}}$, the target point should be modified in such a way that, under the loading $\mathbf{F}$, the end-platform is located at the desired point $\mathbf{t}_0$. This requirement can be expressed using the stiffness model (6.63) in the following way

$$\mathbf{F} = f\left(\mathbf{t}_0|\mathbf{t}_0^{(\mathrm{F})}\right) \tag{6.74}$$

where $t_0^{(F)}$ denotes the modified target location (which obviously is not attained because of the loading). Hence, the problem is reduced to the solution of the nonlinear

Eq. (6.74) for $\mathbf{t}_0^{(F)}$, while $\mathbf{F}$ and $\mathbf{t}_0$ are assumed to be given. It is worth mentioning that this equation completely differs from the equation $\mathbf{F} = f(\mathbf{t}|\mathbf{t}_0)$ considered in sub-Sect. 4.5.2, where the unknown variable is $\mathbf{t}$. It means that here the compliance model does not allow us to compute the modified target point $\mathbf{t}_0^{(F)}$ straightforwardly, while the linear compensation technique directly operates with the Cartesian compliance matrix [20, 21].

To solve Eq. (6.74) for $\mathbf{t}_0^{(F)}$, the Newton–Raphson technique can be applied. It yields the following iterative scheme

$$\mathbf{t}_0^{(F)\prime} = \mathbf{t}_0^{(F)} + \mathbf{K}_{\text{t.p.}}^{-1}\left(\mathbf{t}_0|\mathbf{t}_0^{(F)}\right) \cdot \left(\mathbf{F} - f\left(\mathbf{t}_0|\mathbf{t}_0^{(F)}\right)\right) \tag{6.75}$$

where the prime corresponds to the next iteration and $\mathbf{K}_{\text{t.p.}}\left(\mathbf{t}_0|\mathbf{t}_0^{(F)}\right)$ is Cartesian stiffness matrix computed with respect to the second argument of the stiffness model (6.63)

$$\mathbf{K}_{\text{t.p.}}(\mathbf{t}|\mathbf{t}_0) = \frac{\partial f(\mathbf{t}|\mathbf{t}_0)}{\partial \mathbf{t}_0} \tag{6.76}$$

which can be interpreted as the target point. Here, the location $\mathbf{t}_0$ can be also used as the initial value of $\mathbf{t}_0^{(F)}$. The stopping criterion can be also expressed as

$$\left\|\mathbf{F} - f\left(\mathbf{t}_0|\mathbf{t}_0^{(F)}\right)\right\| < \varepsilon_F \tag{6.77}$$

where ε_F is the desired accuracy.

To overcome computational difficulties related to the evaluation of the matrix $\mathbf{K}_{\text{t.p.}}\left(\mathbf{t}_0|\mathbf{t}_0^{(F)}\right)$, it is possible to use its simple approximation that does not change from iteration to iteration. In particular, assuming that $\mathbf{t}$ and $\mathbf{t}_0$ are close enough and the stiffness properties do not vary substantially in their neighbourhood, the stiffness model can be approximated by a linear expression $\mathbf{F} = \mathbf{K}_C \cdot (\mathbf{t} - \mathbf{t}_0)$, which gives $\mathbf{K}_{\text{t.p.}} = -\mathbf{K}_C$. Hence, the iterative scheme (6.75) can be modified as

$$\mathbf{t}_0^{(F)\prime} = \mathbf{t}_0^{(F)} - \alpha\mathbf{K}_C^{-1}\left(\mathbf{t}_0|\mathbf{t}_0^{(F)}\right) \cdot \left(\mathbf{F} - f\left(\mathbf{t}_0|\mathbf{t}_0^{(F)}\right)\right) \tag{6.78}$$

where c is the scalar parameter ensuring convergence. Using a non-linear compliance model (6.73), this idea can be also implemented in an iterative algorithm

$$\mathbf{t}_0^{(F)\prime} = \mathbf{t}_0^{(F)} + \alpha\left(\mathbf{t}_0 - f^{-1}\left(\mathbf{F}|\mathbf{t}_0^{(F)}\right)\right) \tag{6.79}$$

which does not include stiffness matrices $\mathbf{K}_C$ or $\mathbf{K}_{\text{t.p.}}$ (see Fig. 6.10 for details). Obviously, this is the most computationally convenient and it will be used in the next sub-section.

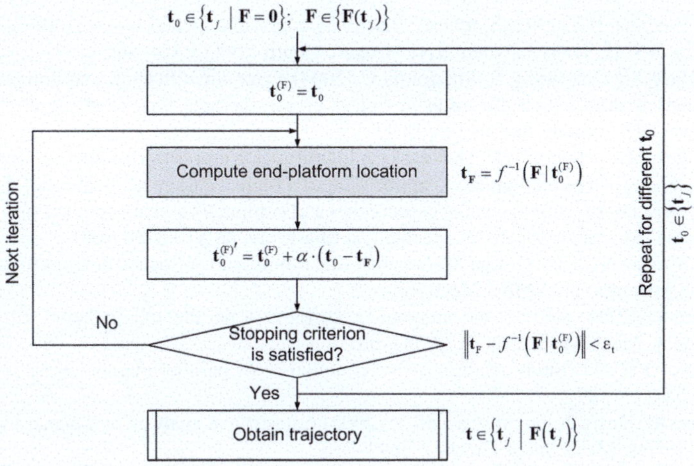

Fig. 6.10 Procedure for compensation of compliance errors in parallel manipulator

Hence, using the computational techniques proposed in this sub-section, it is possible to compensate for the compliance errors by proper adjustment of the reference trajectory that is used as an input for robotic controllers. In this case, the control is based on the inverse kinetostatic model (instead of the kinematic one) that takes into account both the manipulator geometry and elastic properties of its links and joints. The efficiency of this technique is confirmed by an example presented.

References

1. Alici G, Shirinzadeh B (2005) Enhanced stiffness modeling, identification and characterization for robot manipulators. Robot IEEE Trans 21:554–564
2. Shih-Feng C, Imin K Geometrical approach to the conservative congruence transformation (CCT) for robotic stiffness control. In: Robotics and automation, 2002. Proceedings. ICRA '02. IEEE international conference on, vol 541, pp 544–549
3. Yu-Wen L, Jin-Song W, Li-Ping W Stiffness analysis of a Stewart platform-based parallel kinematic machine. In: Proceedings 2002 IEEE international conference on robotics and automation (Cat. No.02CH37292), vol 3674, pp 3672–3677
4. Chen S-F, Kao I (2000) Conservative congruence transformation for joint and Cartesian stiffness matrices of robotic hands and fingers. Int J Robot Res 19:835–847
5. Quennouelle C, Gosselin Cá (2008) Stiffness matrix of compliant parallel mechanisms. In: Advances in robot kinematics: analysis and design. Springer, pp 331–341
6. Merlet J-P, Gosselin C (2008) Parallel mechanisms and robots. In: Siciliano B, Khatib O (eds) Springer handbook of robotics. Springer, Berlin Heidelberg, pp 269–285
7. Quennouelle C, Gosselin CM Instantaneous kinemato-static model of planar compliant parallel mechanisms. In: ASME 2008 international design engineering technical conferences and computers and information in engineering conference. American Society of Mechanical Engineers, pp 163–173
8. Yi B-J, Freeman RA (1993) Geometric analysis of antagonistic stiffness in redundantly actuated parallel mechanisms. J Robot Syst 10:581–603

9. Yi B-J, Freeman RA (1992) Synthesis of actively adjustable springs by antagonistic redundant actuation. ASME Trans J Dynam Syst Measure Control B 114:454–461
10. Meggiolaro M, Dubowsky S, Mavroidis C (2004) Error identification and compensation in large manipulators with application in cancer proton therapy. Sba: Controle & Automação Sociedade Brasileira de Automatica 15:71–77
11. Meggiolaro MA, Mavroidis C, Dubowsky S Identification and compensation of geometric and elastic errors in large manipulators: application to a high accuracy medical robot. In: International design engineering technical conferences and computers and information in engineering conference. American Society of Mechanical Engineers, pp V01AT01A081
12. Wei W, Simaan N (2010) Design of planar parallel robots with preloaded flexures for guaranteed backlash prevention. ASME J Mech Robot 2
13. Chakarov D (2004) Study of the antagonistic stiffness of parallel manipulators with actuation redundancy. Mech Mach Theory 39:583–601
14. Chakarov D (1999) Study of the passive compliance of parallel manipulators. Mech Mach Theory 34:373–389
15. Carricato M, Duffy J, Parenti-Castelli V (2002) Catastrophe analysis of a planar system with flexural pivots. Mech Mach Theory 37:693–716
16. Hines R, Marsh D, Duffy J (1998) Catastrophe analysis of the planar two-spring mechanism. Int J Robot Res 17:89–101
17. Salisbury JK Active stiffness control of a manipulator in Cartesian coordinates. In: Decision and control including the symposium on adaptive processes, 1980 19th IEEE conference on. IEEE, pp 95–100
18. Klimchik A, Pashkevich A, Chablat D, Hovland G (2013) Compliance error compensation technique for parallel robots composed of non-perfect serial chains. Robot Comput Integrat Manufact 29:385–393
19. Surdilović D, Vukobratović M (1996) Deflection compensation for large flexible manipulators. Mech Mach Theory 31:317–329
20. Seo TI (1998) Intégration des effets de déformation d'outil en génération de trajectoire d'usinage. Nantes
21. Gong C, Yuan J, Ni J (2000) Nongeometric error identification and compensation for robotic system by inverse calibration. Int J Mach Tools Manuf 40:2119–2137
22. Jianjun W, Hui Z, Fuhlbrigge T Improving machining accuracy with robot deformation compensation. In: IEEE/RSJ international conference on intelligent robots and systems (IROS 2009), pp 3826–3831
23. Klimchik A, Bondarenko D, Pashkevich A, Briot S, Furet B (2014) Compliance error compensation in robotic-based milling. In: Ferrier JL, Bernard A, Gusikhin O, Madani K (eds) Informatics in control, automation and robotics, vol 283. Springer International Publishing, pp 197–216
24. Kim SH, Min B-K (2020) Joint Compliance error compensation for robot manipulator using body frame. Int J Precis Eng Manuf 21:1017–1023
25. Nguyen VL, Kuo C-H, Lin PT (2022) Compliance error compensation of a robot end-effector with joint stiffness uncertainties for milling: an analytical model. Mech Mach Theory 170:104717
26. Quennouelle C (2009) Modelisation geometrico-statique des mecanismes paralleles compliants. Université Laval
27. Timoshenko S, Goodier JN (1970) Theory of elasticity. McGraw-Hill, Singapore
28. Pashkevich A, Klimchik A, Chablat D (2011) Enhanced stiffness modeling of manipulators with passive joints. Mech Mach Theory 46:662–679
29. Gantmacher FR (1959) The theory of matrices, vol I
30. Klimchik A, Chablat D, Pashkevich A (2014) Stiffness modeling for perfect and non-perfect parallel manipulators under internal and external loadings. Mech Mach Theory 79:1–28
31. Klimchik A, Chablat D, Pashkevich A (2015) Static stability of manipulator configuration: influence of the external loading. Eur J Mech A Solids 51:193–203

32. Pashkevich A, Klimchik A, Chablat D (2010) Stiffness analysis of parallel manipulators with preloaded passive joints. In: Lenarcic J, Stanisic MM (eds) Advances in robot kinematics: motion in man and machine. Springer, Netherlands, pp 465–474
33. Chablat D, Wenger P (2003) Architecture optimization of a 3-DOF translational parallel mechanism for machining applications, the orthoglide. Robot Autom IEEE Trans 19:403–410

Chapter 7
Evaluation of Manipulator Stiffness Model Parameters

7.1 Numerical Computation of Stiffness Parameters for Complex-Shape Links

The stiffness modelling approach developed in this work is based on the VJM-presentation of manipulator links/joints. Its core idea is that each mechanical element with distributed elasticity is replaced by a combination of a rigid body and a lumped virtual spring, whose properties are described by a 6×6 stiffness matrix. Therefore, the accuracy of the manipulator stiffness models considered here directly depends on the accuracy of the stiffness matrices incorporated in them. For this reason, this chapter focuses on the accuracy improvement of the kinetostatic models of manipulator elements and the identification of their parameters.

In robotic literature, there are several ways to obtain the stiffness matrix of a link. The simplest of them assumes that a real link (with rather a complex shape) can be approximated by a simple beam, for which the stiffness matrix can be easily expressed analytically. Another, more accurate, technique deals with multi-beam approximation where the link is presented as a serial chain of rigid bodies separated by virtual springs. For this presentation, the stiffness matrix of the whole link is computed using a common procedure known from stiffness analysis of serial manipulators. However, despite computational convenience and better accuracy, the second approach can be hardly applied to many industrial manipulators (where links may be nonhomogeneous, their shape is quite complex and cross-section is non-constant, etc.). For instance, for the Orthoglide foot (Fig. 7.1), even a four-beam approximation provides an accuracy of only 30–50% [1].

To achieve the desired accuracy, here it is proposed to apply the FEA-based identification methodology that was first introduced in [1] but needs some further development. The corresponding *algorithm* is schematically presented in Fig. 7.2. In the *first step*, a CAD model of the link is created, which properly describes the link shape, cross-sections, distribution and physical properties of the material (density, Young's modulus, Poisson's ratio, etc.). Then, in the *second step*, one of the link ends is fixed

A. Klimchik et al., *Stiffness Modeling of Parallel Robots*, Parallel Robots: Theory and Applications, https://doi.org/10.1007/978-3-031-76320-5_7

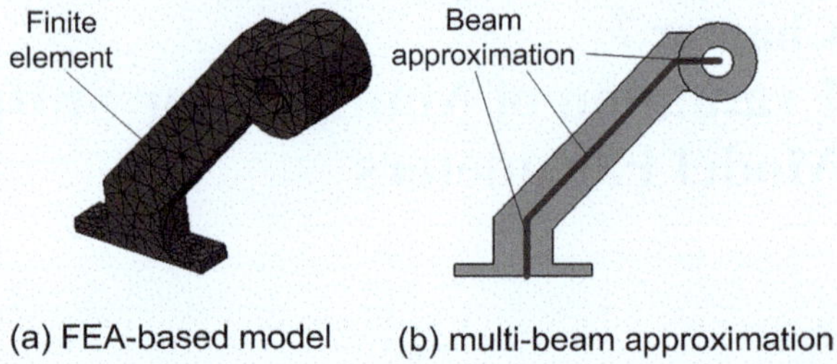

Fig. 7.1 CAD model of the link with complex shape [2]

in accordance with the contact surfaces of an adjacent element (for example by cylindrical surface for the revolute joint). At the second end of the link, a certain distributed (or localized) force/torque is applied in accordance with the contact surfaces of an adjacent element. For these settings, the FEA modelling is performed which yields the deflection field for a huge number of points generated by the meshing procedure. From this set, a subset (so-called '*deflection field*') is extracted corresponding to the neighbourhood of the reference point. This field contains desired information on the link compliance with respect to the applied force/torque. Such virtual experiments are repeated several times, for different directions of forces and torques (the issue of their magnitude is discussed in the following sections). Finally, in the *third step,* the proposed identification procedure that gives the desired stiffness matrix of the link is applied.

It is worth mentioning that the above-described algorithm has some essential advantages, which are not achievable while using any analytical technique. In particular, here it is possible to take into account (straightforwardly and explicitly) the joint elasticity that usually has a number of particularities, such as significant compliance of link/joint areas located close to the contact points or surfaces. For instance, for a

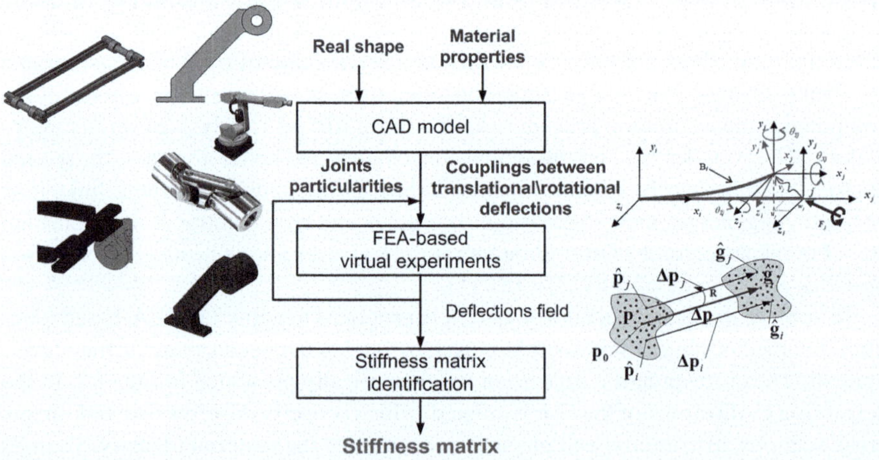

Fig. 7.2 Algorithm for stiffness matrix identification procedure

case study presented in this chapter, some of the stiffness matrix elements are reduced by a factor of 10–12 if the joint particularities are modelled properly. It is evident that existing analytical expressions do not take into account these issues.

Hence, the goal of this chapter is the accuracy improvement of the stiffness matrix identification for manipulator elements. As it follows from the related analysis, particular problems that should be considered here are the following:

(i) Further development of the FEA-based methodology for the link stiffness matrix identification, which allows us to take into account the link shape, coupling between rotational/translational deflections and joint particularities;
(ii) Development of numerical technique, which computes the stiffness matrix from the set of points (displacement field) extracted from the FEA-based virtual experiments;
(iii) Evaluation of this technique's accuracy with respect to virtual experiment settings (meshing type\size, deflection range, fixing and force application method) and size\shape of the virtual sensor;
(iv) Minimization of the identification errors by statistical processing of the experimental data (outlines elimination, determination of the confidence intervals and detecting "zero" elements of the stiffness matrix);
(v) Validation of the developed technique on application examples related to typical parallel manipulators and their comparison with conventional regular-shape approximation models.

7.2 FEA-Based Approach for Identification of Link Stiffness Matrix

Let us start with a detailed description of the FEA-based methodology for the link stiffness matrix identification. It is based on several virtual experiments that are conducted in the CAD environment (CATIA with ANSYS, for instance). Each of these experiments gives some information on the resistance of an elastic body or mechanism to deformations caused by an external force or torque. The desired stiffness model is obtained by integrating data obtained from several different experiments, which differ in the direction of applied forces/torques.

For relatively small deformations, the stiffness properties are defined through the so-called *stiffness matrix* **K**, which defines the linear relation

$$\begin{bmatrix} \mathbf{F} \\ \mathbf{M} \end{bmatrix} = \mathbf{K} \cdot \begin{bmatrix} \mathbf{p} \\ \delta\boldsymbol{\varphi} \end{bmatrix} \tag{7.1}$$

between the three-dimensional translational/rotational displacements $\mathbf{p} = (p_x, p_y, p_z)^{\mathrm{T}}$; $\delta\boldsymbol{\varphi} = (\delta\boldsymbol{\varphi}_x, \delta\boldsymbol{\varphi}_y, \delta\boldsymbol{\varphi}_z)^{\mathrm{T}}$ and the static forces/torques $\mathbf{F} = (F_x, F_y, F_z)^{\mathrm{T}}$, $\mathbf{M} = (M_x, M_y, M_z)^{\mathrm{T}}$ causing this transition. As known from mechanics, **K** is a 6×6 symmetrical semi-definite non-negative matrix, which may include non-diagonal elements to represent the coupling between the translations and rotations [1]. The inverse of **K** is usually called the *compliance matrix* and is denoted as **k**.

It is worth mentioning, that there are some publications, which operate with non-symmetrical stiffness matrices that are derived via differentiation of the force–deflection relation under the assumption that the static loading is essential. However, as it was shown in [3, 4], that non-symmetrical stiffness matrices are obtained for very specific definitions of the force and displacement only (for instance, when the displacement is defined with respect to a moving frame and the scalar product of the force and displacement does not correspond to the work in common sense). On the other hand, for conservative mechanical systems, the stiffness matrix can be defined as the second derivative of the potential energy, which is always symmetrical. Hence, since here we consider only conservative systems and classical definitions of force and displacement, there are no reasons to obtain a non-symmetrical stiffness matrix from the identification procedure presented below.

For robotic manipulators, the matrix **K** can be computed semi-analytically provided that the stiffness matrices of all separate components (links, actuators, etc.) are known with desired precision. However, explicit expressions for the link stiffness matrices can be obtained in simple cases only (truss, beam, etc.). For more sophisticated shapes that are commonly used in robotics, the stiffness matrix is usually estimated via the shape approximation, using a relatively small set of primitives [5]. However, as it follows from the studies [1], the accuracy of this approach is rather low. Hence, in a general case, it is prudent to apply to each link the FEA-based techniques, which hypothetically produce rather accurate results.

Using the FEA, the stiffness matrix **K** (or it's inverse **k**) is evaluated from several numerical experiments, each of which produces the vectors of linear and angular deflections $(\mathbf{p}, \delta\boldsymbol{\varphi})$ corresponding to the applied force and torque $(\mathbf{F}, \mathbf{M})$ (see Fig. 7.3). Then, the desired matrix is computed from the linear system

$$\mathbf{k} = \begin{bmatrix} \mathbf{F}_1 & \ldots & \mathbf{F}_m \\ \mathbf{M}_1 & \ldots & \mathbf{M}_m \end{bmatrix}^{-1} \cdot \begin{bmatrix} \mathbf{p}_1 & \ldots & \mathbf{p}_m \\ \delta\varphi_1 & \ldots & \delta\varphi_m \end{bmatrix} \tag{7.2}$$

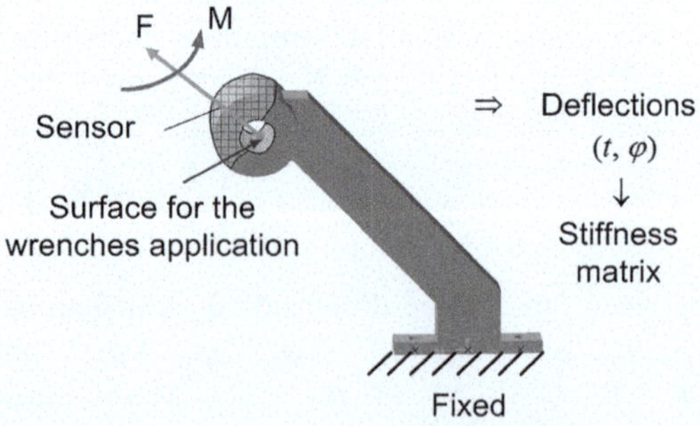

Fig. 7.3 Identification experiment for the element with a complex shape [2]

where m is the number of experiments ($m \geq 6$) and the matrix inverse is replaced by the pseudo-inverse in the case of $m > 6$. It is obvious that the case of $m = 6$ with special arrangement of the forces and torques is numerically attractive

$$\begin{array}{ll} \mathbf{F}_1 = (F_x,\ 0,\ 0)^T\ \mathbf{M}_1 = (0,\ 0,\ 0)^T; & \mathbf{F}_4 = (0,\ 0,\ 0)^T\ \mathbf{M}_4 = (M_z,\ 0,\ 0)^T \\ \mathbf{F}_2 = \left(0,\ F_y,\ 0\right)^T\ \mathbf{M}_2 = (0,\ 0,\ 0)^T; & \mathbf{F}_5 = (0,\ 0,\ 0)^T\ \mathbf{M}_5 = \left(0,\ M_y,\ 0\right)^T. \\ \mathbf{F}_3 = \left(0,\ 0,\ F_y\right)^T\ \mathbf{M}_3 = (0,\ 0,\ 0)^T; & \mathbf{F}_6 = (0,\ 0,\ 0)^T\ \mathbf{M}_6 = (0,\ 0,\ M_z)^T \end{array} \tag{7.3}$$

corresponding to the diagonal structure of the matrix to be inverted. In this case, each FEA experiment produces exactly one column of the compliance matrix

$$\mathbf{k} = \begin{bmatrix} \mathbf{p}_1/F_x & \mathbf{p}_2/F_y & \mathbf{p}_3/F_z & \mathbf{p}_4/M_x & \mathbf{p}_5/M_y & \mathbf{p}_6/M_z \\ \delta\varphi_1/F_x & \delta\varphi_2/F_y & \delta\varphi_3/F_z & \delta\varphi_4/M_z & \delta\varphi_5/M_y & \delta\varphi_6/M_z \end{bmatrix} \tag{7.4}$$

and the values ($\mathbf{p}_\mathrm{i}$, $\delta\boldsymbol{\varphi}_\mathrm{i}$) may be clearly physically interpreted. On the other hand, by increasing the number of experiments ($m > 6$) it is possible to reduce the estimation error. Besides, it is worth mentioning that the classical methodology of the optimal design of an experiment cannot be applied here directly, because it is not possible to include it in the Eq. (7.4) the measurement errors induced by virtual experiments as additive components.

Hence, to obtain the desired stiffness (compliance) matrix, it is required to estimate first the deflections ($\mathbf{p}$, $\delta\boldsymbol{\varphi}$) corresponding to each virtual experiment. This issue is the focus of the following section.

7.3 Numerical Technique for Evaluating the Stiffness Matrix Elements

Usually, in FEA-based experiments, the values ($\mathbf{p}$, $\delta\boldsymbol{\varphi}$) are computed from the spatial location of a single finite element enclosing the reference point. In contrast to this approach, here it is proposed to evaluate ($\mathbf{p}$, $\delta\boldsymbol{\varphi}$) from the set of points (*displacement field*) describing transitions of a rather large number of nodes located in the neighbourhood of the reference point. It is reasonable to assume that such modification will yield a positive impact on the accuracy, since the FEA-modeling errors usually differ from node to node, exposing an almost quasi-stochastic nature.

7.3.1 Related Optimization Problem

To formulate this problem strictly, let us denote the displacement field by a set of vector couples $\{\mathbf{p}_i,\ \Delta\mathbf{p}_i | i = \overline{1, n}\}$ (see Fig. 7.4) where the first component $\mathbf{p}_i$ defines the node's initial location (before applying the force/torque), $\Delta\mathbf{p}_i$ refers to the node's displacement due to the applied force/torque, and n is the number of considered nodes. Then, assuming that all the nodes are close enough to the reference point, this set can be approximated by a "*rigid transformation*"

$$\mathbf{p}_i + \Delta\mathbf{p}_i = \mathbf{R}(\delta\varphi) \cdot \mathbf{p}_i + \mathbf{p}, \quad i = \overline{1, n}, \tag{7.5}$$

that includes as the parameters the linear displacement vector **p** and the orthogonal 3×3 matrix **R** that depends on the rotational displacement $\delta\boldsymbol{\varphi}$. Then, the problem of the deflection estimation can be presented as the best fit of the considered vector field by Eq. (7.5) with respect to the six scalar variables $(p_x, p_y, p_z, \delta\varphi_x, \delta\varphi_y, \delta\varphi_z)$ incorporated in the displacement vector **p** and the rotation matrix **R**.

In practice, the FEA-modeling output provides the deflection vector fields for all nodes referring to all components of the mechanism. So, it is required to select a relevant subset corresponding to the neighbourhood of the reference point $\mathbf{p}_0$. Besides, the node locations $\mathbf{p}_i$ must be expressed relative to this point, i.e. the origin of the coordinate system must be shifted to $\mathbf{p}_0$. The latter is specified by the physical meaning of the deflections in the stiffness analysis.

To estimate the desired deflections $(\mathbf{p}, \delta\boldsymbol{\varphi})$, let us apply the least square technique that leads to the minimization of the sum of squared residuals

$$f = \sum_{i=1}^{n} \|\mathbf{p}_i + \Delta\mathbf{p}_i - \mathbf{R}(\delta\varphi)\,\mathbf{p}_i - \mathbf{p}\|^2 \rightarrow \min_{\mathbf{R},\mathbf{p}} \tag{7.6}$$

with respect to the vector **t** and the orthogonal matrix **R** representing the rotational deflections $\delta\boldsymbol{\varphi}$. The specificities of this problem (that does not allow direct application of the standard methods) are the orthogonal constraint $\mathbf{R}^{\mathrm{T}}\mathbf{R} = \mathbf{I}$ and non-trivial relation between elements of the matrix **R** and the vector $\delta\boldsymbol{\varphi}$. The following subsections present two methods for computing $\mathbf{p}$, $\delta\boldsymbol{\varphi}$, as well as their comparison study.

Fig. 7.4 Field of points from FEA modelling [2]

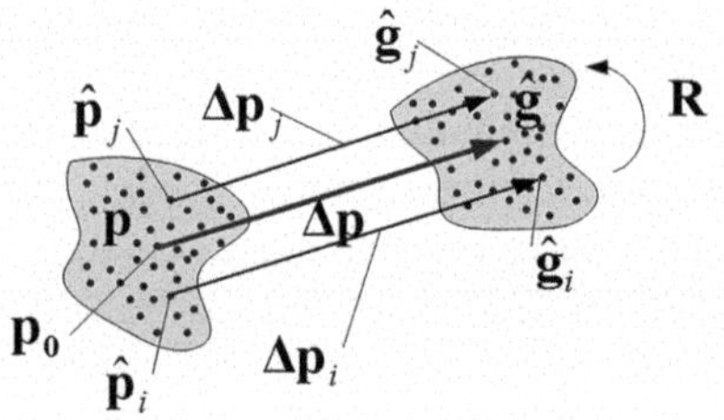

7.3.2 Problem Solution Using SVD-Based Method

For the comparison purposes, first, let us briefly present the exact solution of the optimization problem (7.6). It relies on some results from the matrix algebra referred to as the "Procrustes problem" [6]. The corresponding estimation procedure is decomposed into two steps, which sequentially produce the rotation matrix **R** and the translation vector **p**. Then, the desired vector of rotation angles $\boldsymbol{\delta\varphi}$ is extracted from **R**.

Let us transform the original optimization problem (7.6) to the standard form. First, for the non-constrained variable **p**, straightforward differentiation and equating to zero gives an expression

$$\mathbf{p} = \frac{1}{n}\left(\sum_{i=1}^{n} \Delta\mathbf{p}_i - (\mathbf{R} - \mathbf{I})\sum_{i=1}^{n} \mathbf{p}_i\right) \tag{7.7}$$

Then, after relevant substitution and denoting

$$\begin{aligned} \hat{\mathbf{p}}_i &= \mathbf{p}_i - \frac{1}{n}\sum_{i=1}^{n} \mathbf{p}_i \\ \hat{\mathbf{g}}_i &= \mathbf{p}_i + \Delta\mathbf{p}_i - \frac{1}{n}\sum_{i=1}^{n} (\mathbf{p}_i + \Delta\mathbf{p}_i), \end{aligned} \tag{7.8}$$

the original optimization problem is reduced to the orthogonal Procrustes formulation

$$f = \sum_{i=1}^{n} \|\hat{\mathbf{g}}_i - \mathbf{R}\,\hat{\mathbf{p}}_i\|^2 \to \min_{\mathbf{R}}. \tag{7.9}$$

with the constraint $\mathbf{R}^T\mathbf{R} = \mathbf{I}$. The latter yields the solution [6]

$$\mathbf{R} = \mathbf{V}\mathbf{U}^T \tag{7.10}$$

that is expressed via the singular value decomposition (SVD) of the matrix

$$\sum_{i=1}^{n} \hat{\mathbf{p}}_i \cdot \mathbf{g}_i^T = \mathbf{U}\,\boldsymbol{\Sigma}\,\mathbf{V}^T, \tag{7.11}$$

which requires some computational efforts. Hence, the above expressions (7.7), (7.10) allow to solve the optimization problem (7.6) in terms of variables **R** and **p**.

Further, to evaluate the vector of angular displacements $\boldsymbol{\delta\varphi}$, the orthogonal matrix **R** must be decomposed into a product of elementary rotations around the Cartesian axes x, y, and z. It is obvious that, in a general case, this decomposition is not unique

Table 7.1 Evaluation of the rotation angles from matrix **R** [2]

Method	φ_x	φ_y	φ_z
SVD +	r_{32}	r_{13}	r_{21}
SVD −	$-r_{23}$	$-r_{31}$	$-r_{12}$
SVD ±	$(r_{32}-r_{23})/2$	$(r_{13}-r_{31})/2$	$(r_{21}-r_{12})/2$
SVD + asin	$\operatorname{asin} r_{32}$	$\operatorname{asin} r_{13}$	$\operatorname{asin} r_{21}$
SVD-asin	$-\operatorname{asin} r_{23}$	$-\operatorname{asin} r_{31}$	$-\operatorname{asin} r_{12}$
SVD ± asin	$\operatorname{asin}((r_{32}-r_{23})/2)$	$\operatorname{asin}((r_{13}-r_{31})/2)$	$\operatorname{asin}((r_{21}-r_{12})/2)$

and depends on the rotation order. However, for a small $\delta\boldsymbol{\varphi}$ (that is implicitly assumed for FEA experiments) this matrix may be uniquely presented in differential form

$$\mathbf{R} \cong \begin{bmatrix} 1 & -\delta\varphi_z & \delta\varphi_y \\ \delta\varphi_z & 1 & -\delta\varphi_x \\ -\delta\varphi_y & \delta\varphi_x & 1 \end{bmatrix} \tag{7.12}$$

Using this expression, the desired parameters $\delta\varphi_x,\ \delta\varphi_y,\ \delta\varphi_z$ maybe extracted from $\mathbf{R} = [r_{ij}]$ in several ways (Table 7.1), which are formally equivalent but do not necessarily possess similar robustness with respect to round-off errors. A relevant comparison study will be presented further.

7.3.3 *Efficiency Improvement of the Developed Technique*

To reduce the computational efforts and to avoid the SVD, let us introduce linearization of the rotational matrix **R** at the early stage, using explicit parameterization given by the expression (7.12). This allows rewriting the equation of the 'rigid transformation' (7.5) in the form

$$\Delta\mathbf{p}_i = \mathbf{p}_i \times \delta\boldsymbol{\varphi} + \mathbf{p}; \quad i = \overline{1,n} \tag{7.13}$$

that can be further transformed into a linear system of the following form

$$[\mathbf{I}\ \mathbf{P}_i]\begin{bmatrix} \mathbf{p} \\ \delta\varphi \end{bmatrix} = \Delta\mathbf{p}_i; \quad i = \overline{1,n} \tag{7.14}$$

where $\mathbf{P}_i$ is a skew-symmetric matrix corresponding to the vector $\mathbf{p}_i$:

$$\mathbf{P}_i = \begin{bmatrix} 0 & p_{zi} & -p_{yi} \\ -p_{zi} & 0 & p_{xi} \\ p_{yi} & -p_{xi} & 0 \end{bmatrix} \tag{7.15}$$

Then, applying the standard least-square technique with the objective

$$f = \sum_{i=1}^{n} \|\Delta \mathbf{p}_i - \mathbf{P}_i\, \delta\varphi - \mathbf{p}\|^2 \rightarrow \min_{\varphi,\mathbf{t}} \tag{7.16}$$

one can get the solution

$$\begin{bmatrix} \mathbf{p} \\ \delta\varphi \end{bmatrix} = \begin{bmatrix} n\,\mathbf{I} & \sum_{i=1}^{n} \mathbf{P}_i \\ \sum_{i=1}^{n} \mathbf{P}_i^T & \sum_{i=1}^{n} \mathbf{P}_i^T \mathbf{P}_i \end{bmatrix}^{-1} \begin{bmatrix} \sum_{i=1}^{n} \Delta \mathbf{p}_i \\ \sum_{i=1}^{n} \mathbf{P}_i^T \Delta \mathbf{p}_i \end{bmatrix} \tag{7.17}$$

that employs the 6×6 matrix inversion. This solution can be simplified by shifting the origin of the coordinate system to the point $\mathbf{p}_c = n^{-1} \sum_{i=1}^{n} \mathbf{p}_i$ leading to expression

$$\begin{bmatrix} \mathbf{p} \\ \delta\varphi \end{bmatrix} = \begin{bmatrix} n^{-1}\,\mathbf{I} & \mathbf{0} \\ \mathbf{0} & \left(\sum_{i=1}^{n} \hat{\mathbf{P}}_i^T \hat{\mathbf{P}}_i\right)^{-1} \end{bmatrix} \cdot \begin{bmatrix} \sum_{i=1}^{n} \Delta \mathbf{p}_i \\ \sum_{i=1}^{n} \hat{\mathbf{P}}_i^T \Delta \mathbf{p}_i \end{bmatrix} \tag{7.18}$$

that requires the inversion of the matrix of size 3×3. Here, $\hat{\mathbf{P}}_i$ denotes a skew-symmetric matrix corresponding to the vector $\hat{\mathbf{p}}_i = \mathbf{p}_i - \mathbf{p}_c$.

It is evident that the obtained expression (7.18) is more computationally efficient compared to (7.8) and (7.10). Besides, for some typical cases corresponding to regular meshing patterns of the FEA, expression (7.18) can be further simplified as shown in the following subsection.

7.3.4 Analytical Solutions for Typical Case Studies

In practice, the points $\mathbf{p}_i$ are regularly distributed in the space, in accordance with the meshing options chosen for the FEA modelling. This allows us to inverse the matrix $\sum_{i=1}^{n} \hat{\mathbf{P}}_i^T \hat{\mathbf{P}}_i$ analytically and to obtain very simple expressions for $\mathbf{p}, \delta\boldsymbol{\varphi}$. Let us consider several patterns that are useful for practice and are presented in Fig. 7.5.

Case 1: Symmetrical field (Fig. 7.5a). If the field is symmetrical with respect to its centre $\mathbf{p}_c$, the solution (7.18) can be presented in a compact analytical form as

$$\mathbf{p} = n^{-1} \sum_{i=1}^{n} \Delta \mathbf{p}_i;\ \delta\varphi = \mathbf{D}^{-1} \sum_{i=1}^{n} \hat{\mathbf{P}}_i^T \Delta \mathbf{p}_i \tag{7.19}$$

where the matrix

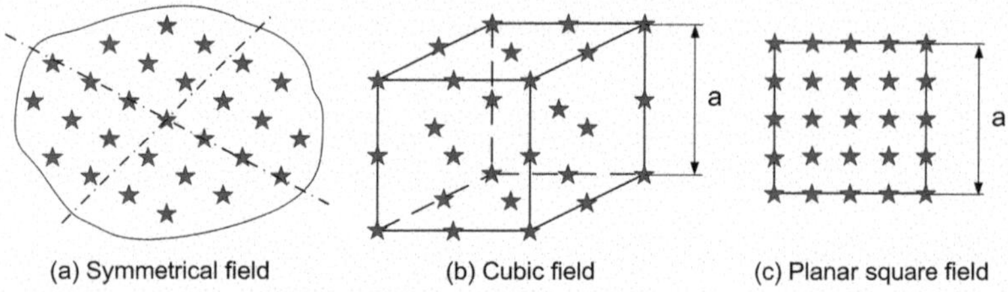

Fig. 7.5 Typical patterns of the deflection field [2]

$$\mathbf{D} = diag\left[\sum_{i=1}^{n} (\hat{p}_{yi}^2 + \hat{p}_{zi}^2) \; \sum_{i=1}^{n} (\hat{p}_{xi}^2 + \hat{p}_{zi}^2) \; \sum_{i=1}^{n} (\hat{p}_{xi}^2 + \hat{p}_{yi}^2) \right] \tag{7.20}$$

is diagonal and easily inverted.

Case 2: Cubic field (Fig. 7.5b). If the field is symmetrical and, in addition, it is produced by uniform meshing of the cubic subspace $a \times a \times a$, the matrix D is expressed as $\mathbf{D} = d \cdot \mathbf{I}$ Where $d = a^2 n(\sqrt{n} - 1)/6(\sqrt{n} + 1)$

Case 3: Planar square field (Fig. 7.5c). For the field produced by uniform meshing of the square $a \times a$ located perpendicular to the x-axis, the expression for the matrix $\mathbf{D} = diag\left[d \; d/2 \; d/2 \right]$.

Summarizing results, it should be noted that the proposed numerical technique is computationally attractive and allows simultaneous estimation of both translational and rotational deflections $\mathbf{p}, \delta\boldsymbol{\varphi}$ from the FEA-produced field. Below it is evaluated from the point of view of precision and robustness.

7.4 Accuracy of Numerical Technique

To be applied in practice, the accuracy of the developed numerical technique should be evaluated with respect to the virtual experiment settings (meshing type\size, deflection range, fixing and force application method) and size\shape of the virtual sensor. Let us consider several case studies corresponding to typical industrial applications.

7.4.1 *Error Sources and Their Impact*

It is obvious that the main source of estimation errors is related to the FEA modelling that highly depends on the size and type of the finite elements, meshing options, incorporated numerical algorithms, computer word length and the round-off principle. Hypothetically, the accuracy can be essentially improved by reducing the mesh size and increasing the number of digits in the presentation of all variables. However, some evident technical constraints do not allow ignoring the FEA limitations.

Another type of error arises from numerical differentiation incorporated into the considered technique. Strictly speaking, the linear relation (7.1) is valid for rather small deflections that may be undetectable against the FEA-modeling defects. On the other side, large deflections may be out of the elasticity range. Hence, it is prudent to find a compromise for the applied forces/torques taking into account both factors.

7.4.2 *Influence of Linearization and Round-Offs*

Since both of the considered algorithms (SVD-based and LIN-based) involve numerous matrix multiplications, they may accumulate round-off errors. Besides, they employ the first-order approximation of the matrix **R** which may create another source of inaccuracy. Hence, it is prudent to obtain numerical assessments corresponding to a typical case study.

For these assessments, there were examined data sets corresponding to the cubic deflection field of size $10 \times 10 \times 10$ mm^3 with the mesh step of 1 mm (1331 points). The deflections have been generated via the 'rigid transformation' (7.5) with the parameters $\mathbf{p} = (a, \ a, \ a)^T$ and $\delta\varphi = (b, \ b, \ b)^T$ presented in Tables 7.2 and 7.3. All calculations have been performed using the double-precision floating-point arithmetic.

Table 7.2 Identification errors for the rotation $\delta\varphi$ [deg] [2]

Method	Rotation amplitude b					
	0.01°	0.05°	0.1°	0.5°	1°	5°
SVD-based technique						
SVD +	$2 \cdot 10^{-6}$	$4 \cdot 10^{-5}$	$2 \cdot 10^{-4}$	$4 \cdot 10^{-3}$	$2 \cdot 10^{-2}$	0.48
SVD −	$2 \cdot 10^{-6}$	$4 \cdot 10^{-5}$	$2 \cdot 10^{-4}$	$4 \cdot 10^{-3}$	$2 \cdot 10^{-2}$	0.48
SVD ±	$9 \cdot 10^{-7}$	$2 \cdot 10^{-5}$	$9 \cdot 10^{-5}$	$2 \cdot 10^{-3}$	$9 \cdot 10^{-3}$	0.24
SVD + asin	$2 \cdot 10^{-6}$	$4 \cdot 10^{-5}$	$2 \cdot 10^{-4}$	$4 \cdot 10^{-3}$	$2 \cdot 10^{-2}$	0.48
SVD-asin	$2 \cdot 10^{-6}$	$4 \cdot 10^{-5}$	$2 \cdot 10^{-4}$	$4 \cdot 10^{-3}$	$2 \cdot 10^{-2}$	0.48
SVD ± ain	$9 \cdot 10^{-7}$	$2 \cdot 10^{-5}$	$9 \cdot 10^{-5}$	$2 \cdot 10^{-3}$	$9 \cdot 10^{-3}$	0.24
LIN-based technique						
LIN	$9 \cdot 10^{-7}$	$2 \cdot 10^{-5}$	$9 \cdot 10^{-5}$	$2 \cdot 10^{-3}$	$9 \cdot 10^{-3}$	0.24
LIN asin	$9 \cdot 10^{-7}$	$2 \cdot 10^{-5}$	$9 \cdot 10^{-5}$	$2 \cdot 10^{-3}$	$9 \cdot 10^{-3}$	0.24

Table 7.3 Identification errors for the translation [mm] [2]

Method	Translation amplitude a			
	0.01 mm	0.1 mm	1.0 mm	10 mm
SVD	10^{-16}	10^{-16}	10^{-16}	$2 \cdot 10^{-14}$
LIN	10^{-16}	10^{-16}	10^{-16}	$3 \cdot 10^{-15}$

As it follows from the analysis, the influence of the linearization and round-offs is negligible for the translation (the induced errors are less than 10^{-14} mm). In contrast, for the rotation, practically acceptable results may be achieved for rather small angular deflections that are less than 1.0° (the errors are up to 0.01°). The latter imposes an essential constraint on the amplitude of the forces/torques in the FEA modelling that must ensure reasonable deflections. Another conclusion concerns the comparison of the SVD-based and LIN-based methods. It justifies the advantages of the proposed LIN-based technique that provides the best robustness and lower computational complexity.

7.4.3 *Influence of FEA-Modelling Errors*

By its general principle, the FEA-modeling is an approximate method that produces some errors caused by discretization. Besides, even for perfect modelling, the deflections in the neighbourhood of the reference point do not exactly obey the Eq. (7.5). Hence, it is reasonable to assume that the 'rigid transformation' (7.5) incorporates some additive random errors

$$\mathbf{p}_i + \Delta\mathbf{p}_i = \mathbf{R}(\delta\varphi)\cdot\mathbf{p}_i + \mathbf{p} + \boldsymbol{\varepsilon}_i; \quad i = \overline{1,n} \tag{7.21}$$

that are supposed to be independent and identically distributed Gaussian random variables with zero mean and standard deviation σ.

In the frame of this assumption, the expression for the deflections (7.18) can be rewritten as

$$\mathbf{p} = \mathbf{p}^{\mathrm{o}} + n^{-1}\sum_{i=1}^{n}\boldsymbol{\varepsilon}_i; \qquad \delta\varphi = \delta\varphi^{\mathrm{o}} + \left(\sum_{i=1}^{n}\hat{\mathbf{P}}_i^T\hat{\mathbf{P}}_i\right)^{-1}\sum_{i=1}^{n}\hat{\mathbf{P}}_i^T\boldsymbol{\varepsilon}_i \tag{7.22}$$

where the superscript 'o' corresponds to the 'true' parameter value. This justifies the usual properties of the adopted point-type estimator (7.18), which is obviously unbiased and consistent. Furthermore, the variance–covariance matrices for $\mathbf{t}$, $\delta\boldsymbol{\varphi}$ may be expressed as

$$\mathrm{cov}[\mathbf{p}] = \frac{\sigma^2}{n}\mathbf{I}; \quad \mathrm{cov}[\delta\varphi] = \sigma^2\left(\sum_{i=1}^{n}\hat{\mathbf{P}}_i^T\hat{\mathbf{P}}_i\right)^{-1} \tag{7.23}$$

allowing us to evaluate the estimation accuracy using common confidence interval techniques.

As it follows from (7.21), for the translational deflection $\mathbf{p}$ the identification accuracy is defined by the standard deviation $\sigma/\sqrt{n}$ and depends on the number of points only. In contrast, for the rotational deflection, the spatial location of the points is a very important issue. In particular, for the cubic field of the size $a \times a \times a$, the standard deviation of the rotation angles may be approximately expressed as $\sigma / a\sqrt{n/6}$.

To evaluate the standard deviation σ describing the random errors **ε**, one may use the residual-based estimator obtained from the expression

$$\mathrm{E}\left(\sum_{i=1}^{n} \|\mathbf{p}_i + \Delta\mathbf{p}_i - \mathbf{R}(\delta\varphi) \cdot \mathbf{p}_i - \mathbf{p}\|^2\right) = (3n - 6)\,\sigma^2. \tag{7.24}$$

The latter may be easily derived taking into account that, for each experiment, the deflection field consists of n three-dimensional vectors that are approximated by the model containing 6 scalar parameters. Moreover, to increase accuracy, it is prudent to aggregate the squared residuals for all FEA experiments and to make relevant estimations using the coefficient $(3\,n - 6)\,m\,\sigma^2$, where m is the experiment number.

In addition, to increase accuracy and robustness, it is reasonable to eliminate outliers in the experimental data. They may appear in the FEA field due to some anomalous causes, such as insufficient meshing of some elements, violation of the boundary conditions in some areas of the mechanical joints, etc. The simplest and most reliable method that is adopted in this research is based on 'data filtering' with respect to the residuals (i.e. eliminating a certain percentage of the points with the highest residual values). Another practical question is related to *detecting zero elements* in the compliance matrix or, in other words, evaluating the statistical significance of the computed values compared to zero. These issues will be studied in the following Section.

7.4.4 Optimal Settings for Identification Experiments

Finally, to evaluate the combined influence of various error sources, let us focus on the deflection identification from the cubic field of size $10 \times 10 \times 10$ mm^3 (1331 points, mesh step 1 mm). In particular, let us contaminate all deflections using the Gaussian noise with the s.t.d. 5×10^{-5} mm which is a typical value discovered from the examined FEA data sets (Table 7.4). Similar to the previous case, all calculations were performed using double-precision floating-point arithmetic (16 decimal digits).

The obtained results confirmed the main theoretical derivations of the previous subsections. The identification errors obey the normal distribution (Fig. 7.6) but their s.t.d. should be evaluated taking into account some additional issues. Thus, the s.t.d. of the translational error is about $1.36 \cdot 10^{-6}$ mm and depends only on the FEA-induced component that is evaluated as $\sigma/\sqrt{n} \approx 1.37 \cdot 10^{-6}$ mm. The influence of the linearization and round-offs is negligible here (this component is less than 10^{-14} mm). Also, this type of error does not depend on the translation amplitude.

Table 7.4 Parameters of the FEA-modeling noise for different mesh types [2]

Meshing options			σ, mm
Type of the mesh	Notation	Size of the finite element (mm)	
Linear mesh	3L	3	$4.10 \cdot 10^{-5}$
	2L	2	$4.59 \cdot 10^{-5}$
	1L	1	$3.87 \cdot 10^{-5}$
Parabolic mesh	5P	5	$6.40 \cdot 10^{-5}$
	3P	3	$5.26 \cdot 10^{-5}$
	2P	2	$5.60 \cdot 10^{-5}$

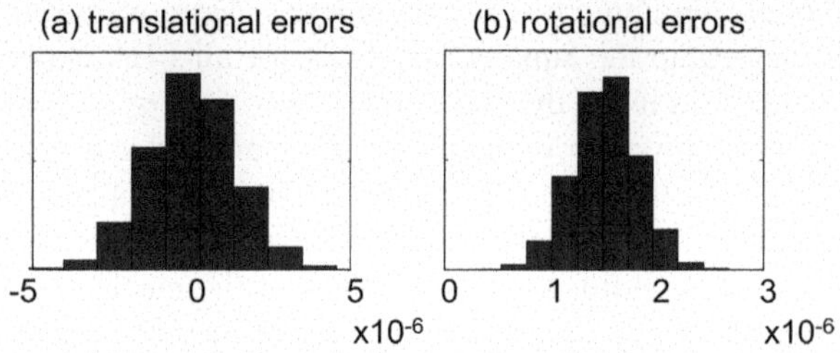

Fig. 7.6 Histograms for the identification errors (a = 1.0 mm, b = 0.1°, σ = 5 × 10^{-5} mm) [2]

In contrast, for the rotational deflections, there exists a strong dependence on the amplitude (Fig. 7.7). In particular, for the angular deflection 0.1°, the s.t.d. of the identification error is about $8.4 \cdot 10^{-5}$ deg, while the FEA-induced component is evaluated as $\sigma / a\sqrt{n/6} \approx 1.8 \cdot 10^{-5}$ deg and the linearization component is about $8.8 \cdot 10^{-5}$ ° (see Table 7.2). Moreover, the simulation results allow us to define preferable values of the angular deflection that may be extracted from the FEA data with the highest accuracy. They show that the deflection angles should be in the range of 0.01 …0.2° to ensure the identification accuracy of about 0.2%

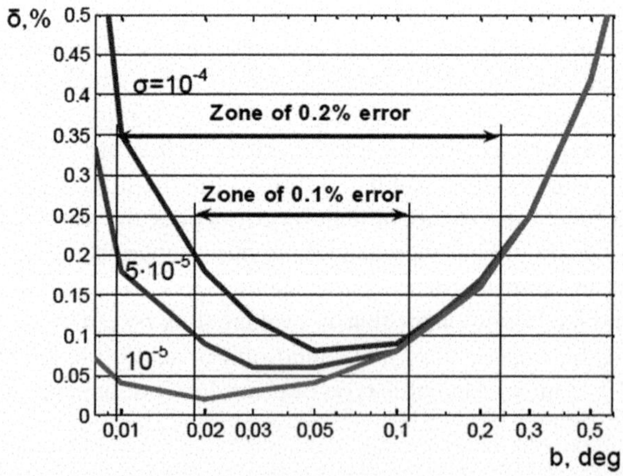

Fig. 7.7 Identification errors for different amplitudes of the rotational deflections [2]

Thus, the case studies considered above confirm that the proposed LIN-based algorithm ensures the same accuracy as the SVD-based one while possessing lower computational complexity. Besides, these results give some practical recommendations for setting the FEA-based experiments and evaluating the identification accuracy.

7.5 Minimization of Identification Errors

To demonstrate the efficiency of the developed technique and to evaluate its applicability to real-life situations, let us consider an example for which the desired compliance matrix can be obtained both numerically and analytically. A comparison of these two solutions provides convenient benchmarks for different FEA-modeling options and also gives some practical recommendations for achieving the required accuracy.

7.5.1 *Benchmark Example*

As an example, let us consider a cantilever beam of size $1000 \times 10 \times 10$ mm^3 (see Fig. 7.8) with the Young's Modulus $E = 2 \cdot 10^5 N/\text{mm}^2$ and the Poisson's Ratio $\nu = 0.266$. These data correspond to the geometry and material properties of a typical robot link studied in this work.

For this element, an analytical expression for the compliance matrix can be presented as

$$\mathbf{k} = \begin{bmatrix} k_{11} & 0 & 0 & 0 & 0 & 0 \\ 0 & k_{22} & 0 & 0 & 0 & k_{26} \\ 0 & 0 & k_{33} & 0 & k_{35} & 0 \\ 0 & 0 & 0 & k_{44} & 0 & 0 \\ 0 & 0 & k_{53} & 0 & k_{55} & 0 \\ 0 & k_{62} & 0 & 0 & 0 & k_{66} \end{bmatrix} \tag{7.25}$$

where non-zero elements are: $k_{35} = k_{53} = -L^2/2EI_y, k_{11} = L/EA, k_{22} = L^3/3EI_z$, $k_{33} = L^3/3EI_y, k_{44} = L/GJ, k_{55} = L/3EI_y, k_{66} = L/3EI_z, k_{26} = k_{62} = L^2/2EI_z$.

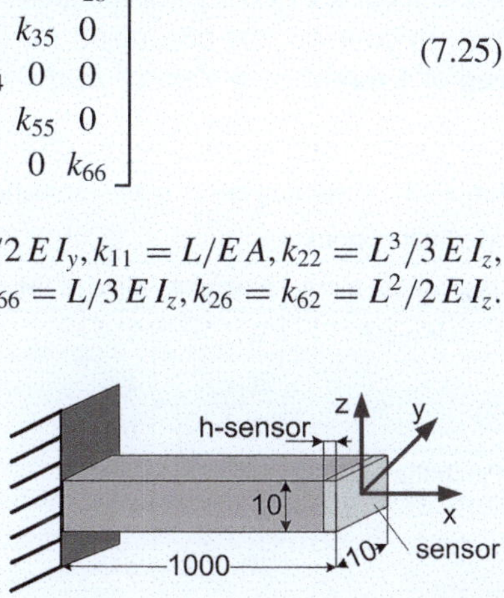

Fig. 7.8 Examined cantilever beam [2]

Here L is the length of the beam, A is its cross-section area, I_y, I_z are the second moments, J is the cross-section property.

The FEA modelling has been performed using the CATIA V5R16 CAD system. During modelling, the loads were applied at one end of the beam with the other end fully clamped. The force/torque amplitudes were determined using the expression (7.23) and the optimal accuracy settings for the deflections 0.1…1.0 mm and 0.01…0.20°, which yielded the following values: $F_x = 1000N$, $F_y = 1N$, $F_z = 1N$, $M_x = 1N \cdot m$, $M_y = 1N \cdot m$, $M_z = 1N \cdot m$. These loads were applied sequentially, providing 6 elementary FEA experiments, each of which produced a single column of the compliance matrix **k**, in accordance with (7.4).

7.5.2 *Optimal Selections of FEA-Modelling Options*

Since the identification errors, which are studied here, essentially depend on the discretization of the FEA-based model and definition of the deflection field, let us focus on the influence of the meshing parameters of the FEA-model and on dimensions of the virtual sensor that specifies this field.

Meshing options. The adopted software provides two basic options for automatic mesh generation: linear and parabolic ones. It is known that, generally, the linear meshing is faster computationally but less accurate. On the other hand, parabolic meshing requires more computational resources but leads to more accurate results.

For the considered case study, both meshing options have been examined and compared with respect to the accuracy of the obtained compliance matrix. The mesh size was gradually reduced from 5 to 1 mm, until achieving the lower limit imposed by the computer memory size. The obtained results (Table 7.5 and Fig. 7.9) clearly demonstrate the advantages of the parabolic mesh, which allows achieving an appropriate accuracy of 0.1% for the mesh step of 2 mm using standard computing capacities. In contrast, the best result for the linear mesh is 12% and corresponds to the step of 1 mm.

Table 7.5 Maximum errors in the estimation of compliance matrix elements [2]

Meshing options			Maximum errors in elements of identified matrix k_{ij} (%)
Type of the mesh	Notation	Size of the finite element (mm)	
Linear mesh	3L	3	27
	2L	2	20
	1L	1	12
Parabolic mesh	5P	5	3.3
	3P	3	0.19
	2P	2	0.10

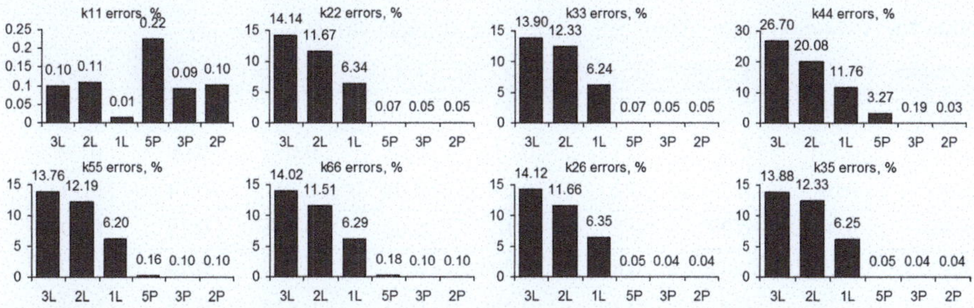

Fig. 7.9 Errors (%) in estimations of non-zero elements of the compliance matrix (3L, 2L, 1L are linear mesh with steps 3, 2, 1 mm; 5P, 3P, 2P are parabolic mesh with steps 5, 3, 2 mm) [2]

The accuracy of these results can also be evaluated by the difference between the deflections computed using analytical expression (7.25) and the identified compliance matrix (assuming that the link is under similar loadings as in virtual experiments). This allows us to simplify the physical meaning of the identification errors and also to obtain a weighted performance measure that neglects some identification failures, whose impact on the reference point deflection is not essential. Relevant results are presented in Fig. 7.10, where there are presented the worst values selected from six virtual experiments. They confirm that the mesh 2P (parabolic with step 2 mm) ensures an accuracy of 0.1% and is preferable for practice. On the other hand, as it follows from a separate study, further reduction of the mesh step does not lead to an essential reduction of the identification errors. Obviously, this conclusion is valid for this case study only, but it can be applied to other cases using proper scaling of dimensions.

Defining the virtual sensor. The developed technique operates with the deflection field bounded by the virtual sensor (see Fig. 7.3), which is located in the neighbourhood of the reference point (RP). As stated above, this neighbourhood should be large enough to neutralize the influence of the FEA-induced errors, but its unreasonable increase may lead to the violation of some essential assumptions and, consequently, to a reduction in the accuracy.

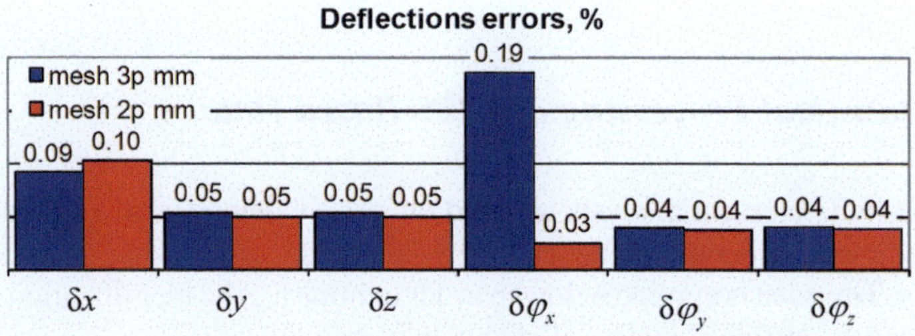

Fig. 7.10 Influence of the stiffness matrix errors on the position accuracy [2]

Table 7.6 Evolution of the identification errors for different virtual sensors [2]

h-sensor	k_{11} (%)	k_{22} (%)	k_{33} (%)	k_{44} (%)	k_{55} (%)	k_{66} (%)	k_{62} (%)	k_{53} (%)
1-layer	0.10	0.05	0.05	0.03	0.10	0.10	0.04	0.04
0.1%	0.11	0.13	0.13	0.01	0.08	0.09	0.09	0.09
0.2%	0.09	0.2	0.2	0.05	0.04	0.04	0.14	0.14
0.3%	0.04	0.28	0.28	0.10	0.02	0.02	0.19	0.19
0.5%	0.06	0.42	0.42	0.20	0.13	0.13	0.29	0.29
1%	0.33	0.8	0.80	0.46	0.41	0.41	0.54	0.54
2%	0.86	1.55	1.55	0.98	0.91	0.92	1.03	1.03

To get a realistic inference concerning this issue, a number of experiments have been carried out, for different definitions of the RP-neighborhood (i.e., the virtual sensor size). The obtained results (Table 7.6) show that the highest accuracy 0.1% is achieved for the one-layer configuration of the deflection field, which is composed of the nodes located on the rare edge of the examined beam. This pattern is very close to the square-type field 10×10 mm^2 studied above (see sub-Sect. 2.3.4). In contrast, increasing the sensor size up to the cubic-type of $10\times10\times10$ mm^3 leads to the identification error of about 0.08%. Hence, in practice, it is reasonable to estimate the deflection values from the field of points corresponding to the square-type virtual sensor.

It is worth mentioning that, in practice, the reference point of the link may be located outside the link material (see Fig. 7.1). Consequently, the RP-neighborhood does not include any finite elements that may be used for creating the deflection field required by the identification procedure. In this case, the link CAD model should be complemented with an additional solid body centred in the reference point and restrained by the surfaces of the corresponding joint (this body should be rigid enough to ensure the correctness of FEA-based simulations). After such modifications, the virtual sensor can be defined in the usual way. As it follows from our experience, the proper definition of the virtual sensor (and additional solid body) as well as applied loading (distributed on the joint surfaces) play a crucial role in conducting virtual experiments.

7.5.3 *Statistical Processing of FEA-Based Data*

Though the finite element analysis is based on strictly deterministic assumptions, it includes tedious computations that may generate some errors, which may be treated statistically. This idea is applied below in order to improve the identification accuracy and to detect the stiffness matrix elements that may be set to zero (in practice, the stiffness matrices with strictly zero elements are rather common).

Eliminating outliers. As noticed above, the FEA-modeling data may include some anomalous samples that do not obey the assumed statistical properties. This phenomenon has been detected in 2 of 6 experiments, (see Fig. 7.11) where the histograms demonstrated the obvious presence of the outliers changing the regular distribution shape (local maximums around the tails). For this reason, a straightforward filtering technique was applied that eliminated 10% of the nodes corresponding to the highest residual values (see Fig. 7.12). This technique essentially improved the identification accuracy, and the maximum error for the compliance matrix elements was reduced from 0.1% to 0.05%.

It is worth mentioning that here, because of the 3-dimensional nature of the problem, each node has been evaluated by three residuals and it was eliminated if any of these residuals was treated as an outlier. In addition, the detailed analysis showed that the outliers were concentrated at the beam edges, which confirms previous assumptions concerning the FEA-induced errors.

Eliminating non-significant elements. According to (7.23), the desired compliance matrix includes a number of zero elements (26 of 36), but the proposed identification procedure may produce some small non-zero values. To evaluate their statistical significance, for each element k_{ij} the confidence interval was computed. Relevant computations were performed using expressions for the variances of the deflections (7.21) and the s.t.d. value of the FEA-modeling noise, which was estimated as $\sigma = 5.6 \cdot 10^{-5}$mm (by averaging for all 6 experiments). Then, the computed confidence intervals were scaled in accordance with (7.4), to be adopted to corresponding elements of the matrix **k** (this procedure involves simple division by the magnitude of the applied force or torque).

Using this approach, the compliance matrix was revised by assigning to zero all non-significant elements. The employed "decision support" algorithm treated an

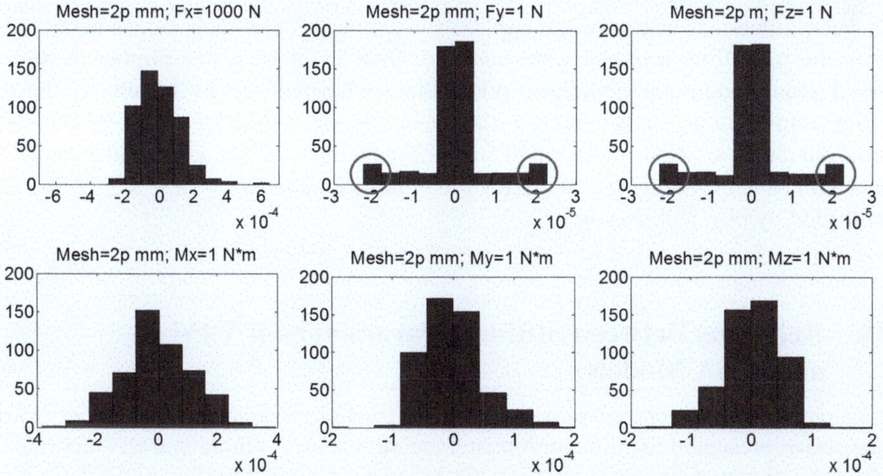

Fig. 7.11 Residuals for stiffness model identification with parabolic mesh of 2 mm [2]

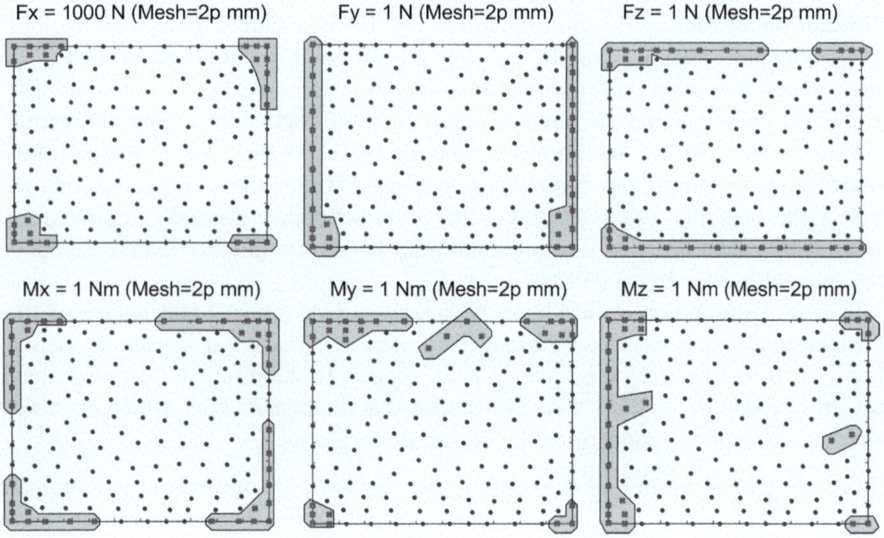

Fig. 7.12 Filtering of the deflection field outliers [2]

element as a non-significant one if its confidence interval included zero. As shown in Table 7.7, this technique allowed us to detect all 26 zero elements mentioned above. It should be also noted that all non-zero elements were evaluated as'significant' ones, with an essential 'safety factor' of 10^2 and higher.

Remarks and comments. The presented illustrative example that deals with a classical element (cantilever beam) confirmed the validity of the developed method but also demonstrated some limitations of the FEA modelling with respect to the stiffness analysis. In particular, some (not very essential but non-negligible) disagreements between numerical values of the applied forces/torques and their values extracted from the modelling protocol were detected. Besides, there are a number of non-trivial issues in defining modelling options that are normally set by default. All these factors contribute to the accuracy, but a practically acceptable level of 0.1% can be achieved rather easily, using standard computing facilities. Some additional enhancement can be achieved by symmetrizing the obtained matrix $\mathbf{k} := (\mathbf{k} + \mathbf{k}^T)/2$ that is motivated by physical reasons.

7.6 Relations Between Stiffness Parameters of VJM and MSA Models

The above-presented identification technique allows us to obtain stiffness matrices of size 6×6, which are common for VJM-based modelling. These matrices describe the link stiffness behaviour that is evaluated assuming that one of the link ends is

Table 7.7 Eliminating non-significant elements [2]

Parameter	Estimated value	CI	Exact value
k_{11}	$5.00 \cdot 10^{-5}$	$\pm 2.3 \cdot 10^{-8}$	$5.00 \cdot 10^{-5}$
k_{22}	$2.00 \cdot 10^{0}$	$\pm 2.4 \cdot 10^{-5}$	$2.00 \cdot 10^{0}$
k_{33}	$2.00 \cdot 10^{0}$	$\pm 2.4 \cdot 10^{-5}$	$2.00 \cdot 10^{0}$
k_{44}	$9.00 \cdot 10^{-6}$	$\pm 6.0 \cdot 10^{-9}$	$9.00 \cdot 10^{-6}$
k_{55}	$6.00 \cdot 10^{-6}$	$\pm 8.2 \cdot 10^{-9}$	$6.00 \cdot 10^{-6}$
k_{66}	$6.00 \cdot 10^{-6}$	$\pm 8.2 \cdot 10^{-9}$	$6.00 \cdot 10^{-6}$
k_{26}	$-3.00 \cdot 10^{-3}$	$\pm 2.4 \cdot 10^{-8}$	$-3.00 \cdot 10^{-3}$
k_{62}	$-3.00 \cdot 10^{-3}$	$\pm 8.5 \cdot 10^{-6}$	$-3.00 \cdot 10^{-3}$
k_{35}	$3.00 \cdot 10^{-3}$	$\pm 2.4 \cdot 10^{-8}$	$3.00 \cdot 10^{-3}$
k_{53}	$3.00 \cdot 10^{-3}$	$\pm 8.5 \cdot 10^{-6}$	$3.00 \cdot 10^{-3}$
k_{31}	$1.8 \cdot 10^{-8}$	$\pm 2.3 \cdot 10^{-8}$	0
k_{32}	$7.8 \cdot 10^{-6}$	$\pm 2.4 \cdot 10^{-5}$	0
k_{23}	$7.8 \cdot 10^{-6}$	$\pm 2.4 \cdot 10^{-5}$	0
k_{45}	$-6.2 \cdot 10^{-10}$	$\pm 8.4 \cdot 10^{-8}$	0
k_{54}	$-5.7 \cdot 10–^{10}$	$\pm 5.6 \cdot 10^{-8}$	0
k_{63}	$-6.7 \cdot 10^{-9}$	$\pm 2.4 \cdot 10^{-8}$	0

fixed and the loading is applied to the second one. In mechanics, such arrangement is known as "cantilever beams" (Fig. 7.13a). Though for kinematic chains studied in the following sections, this assumption (fixing of one of the link ends) is not valid in the strict sense. The stiffness model developed in the following chapters operates with 6×6 stiffness matrices only. This is achieved due to a special procedure of deriving the static equilibrium equations, where the links are considered sequentially, and one-by-one, assuming that each link is fixed in the previous joint and the loading is applied to the next one. This is an essential advantage of the VJM modelling that operates with stiffness matrices of relatively small dimensions.

However, there is an alternative approach (the MSA-modeling) that operates with 12×12 stiffness matrices, which does not assume that any of the link ends is fixed. In contrast, it is assumed here that the links are subject to double-side loading and

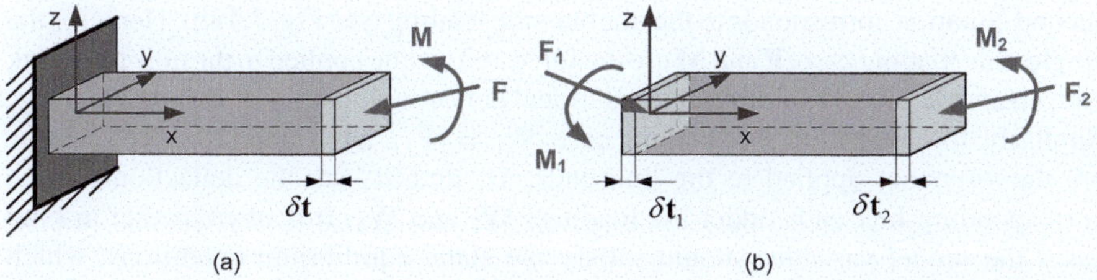

Fig. 7.13 Physical models of the cantilever beam (**a**) and double-side loaded beam (**b**)

deflections are defined for both of these sides (Fig. 7.13b). For such an arrangement, the stiffness matrix should have a size 12 × 12 because it defines the linear relations between two deflection vectors (of size 6 each) and two vectors of loading (of size 6 each). From a physical point of view, it is clear that both stiffness matrices (of size 6 × 6 and 12 × 12) describe the same properties of the link, while the smaller one contains exhaustive information on the stiffness. Hence, it is useful to obtain expressions that allow us to transform a 6 × 6 stiffness matrix into a 12 × 12 one, which is required for the MSA-modeling technique that usually deals with standard and simple link approximations (beams, truss, etc.).

To derive desired expressions, let us denote the above-described stiffness matrices as $\mathbf{K}_{6\times6}$ and $\mathbf{K}_{12\times12}$. Let us also assume that the link geometry and its spatial location are defined by the vector $\mathbf{L} = (l_x,\ l_y,\ l_z)^{\mathrm{T}}$ connecting corresponding nodes. Besides taking into account that both of the matrices are symmetrical, it is convenient to present them in the block form as

$$\mathbf{K}_{6\times6} = \begin{bmatrix} \mathbf{K}_a & \mathbf{K}_b \\ \mathbf{K}_c & \mathbf{K}_d \end{bmatrix} \quad \mathbf{K}_c = \mathbf{K}_b^{\mathrm{T}} \tag{7.26}$$

$$\mathbf{K}_{12\times12} = \begin{bmatrix} \mathbf{K}_{11} & \mathbf{K}_{12} \\ \mathbf{K}_{21} & \mathbf{K}_{22} \end{bmatrix}, \quad \mathbf{K}_{21} = \mathbf{K}_{12}^{\mathrm{T}},$$

$$\mathbf{K}_{ij} = \begin{bmatrix} \mathbf{K}_{ija} & \mathbf{K}_{ijb} \\ \mathbf{K}_{ijc} & \mathbf{K}_{ijd} \end{bmatrix} \quad \mathbf{K}_{ijc} = \mathbf{K}_{ijb}^{\mathrm{T}} \tag{7.27}$$

where $i = 1, 2$ and $j = 1, 2$.

Let us consider simultaneously two static equilibrium equations

$$\begin{bmatrix} \mathbf{F} \\ \mathbf{M} \end{bmatrix} = \begin{bmatrix} \mathbf{K}_a & \mathbf{K}_b \\ \mathbf{K}_c & \mathbf{K}_d \end{bmatrix} \cdot \begin{bmatrix} \delta\mathbf{p} \\ \delta\boldsymbol{\varphi} \end{bmatrix} \tag{7.28}$$

and

$$\begin{bmatrix} \mathbf{W}_1 \\ \mathbf{W}_2 \end{bmatrix} = \begin{bmatrix} \mathbf{K}_{11} & \mathbf{K}_{12} \\ \mathbf{K}_{21} & \mathbf{K}_{22} \end{bmatrix} \cdot \begin{bmatrix} \delta\mathbf{t}_1 \\ \delta\mathbf{t}_2 \end{bmatrix} \tag{7.29}$$

where the first one corresponds to the single-end loading (see Fig. 7.13a) and the second equation corresponds to the double-side loading (see Fig. 7.13b). Here, for the single-side loading case, **F** and **M** are the force and torque applied to the non-fixed link end, $\delta\mathbf{p}$ and $\delta\boldsymbol{\varphi}$ are translational and rotational deflections caused by external loading. Similarly, for the double-side loading case, $\mathbf{W}_1 = (\mathbf{F}_1,\ \mathbf{M}_1)^{\mathrm{T}}$ and $\mathbf{W}_2 = (\mathbf{F}_2,\ \mathbf{M}_2)^{\mathrm{T}}$ are the wrenches applied to the link ends, $\delta\mathbf{t}_1$ and $\delta\mathbf{t}_2$ are the deflections of the corresponding link ends under the loadings $\mathbf{W}_1$ and $\mathbf{W}_2$. It is obvious that in both cases the model variables should satisfy the static equilibrium conditions, which eliminate some redundancy in notations for the second case.

To obtain desired matrices, let us assume first that for the double-side model $\delta\mathbf{t}_1 = \mathbf{0}$. This allows us straightforwardly apply the Eq. (7.28) and static equilibrium constrain that yield

$$\mathbf{W}_1 = \begin{bmatrix} -\mathbf{F} \\ -\mathbf{M} + \mathbf{F} \times \mathbf{L} \end{bmatrix}; \quad \mathbf{W}_2 = \begin{bmatrix} \mathbf{F} \\ \mathbf{M} \end{bmatrix}; \quad \delta\mathbf{t}_2 = \begin{bmatrix} \delta\mathbf{t} \\ \delta\varphi \end{bmatrix} \tag{7.30}$$

That leads to the following presentation of the Eq. (7.29)

$$\begin{bmatrix} -\mathbf{F} \\ -\mathbf{M} + \mathbf{F} \times \mathbf{L} \\ \mathbf{F} \\ \mathbf{M} \end{bmatrix} = \begin{bmatrix} \mathbf{K}_{11} & \mathbf{K}_{12} \\ \mathbf{K}_{21} & \mathbf{K}_{22} \end{bmatrix} \cdot \begin{bmatrix} \mathbf{0} \\ \mathbf{0} \\ \delta\mathbf{p} \\ \delta\varphi \end{bmatrix} \tag{7.31}$$

which leads to

$$\mathbf{K}_{22} = \mathbf{K}_{6\times 6} \tag{7.32}$$

Further, taking into account that $\mathbf{F} \times \mathbf{L} = -(\mathbf{L}\times) \cdot \mathbf{F}$ and using symmetrical properties of stiffness matrix,

$$\begin{aligned} \mathbf{K}_{12} &= \begin{bmatrix} -\mathbf{K}_a & -\mathbf{K}_b \\ -\mathbf{K}_c - (\mathbf{L}\times) \cdot \mathbf{K}_a & -\mathbf{K}_d - (\mathbf{L}\times) \cdot \mathbf{K}_b \end{bmatrix}; \\ \mathbf{K}_{21} &= \mathbf{K}_{12}^{\mathrm{T}} = \begin{bmatrix} -\mathbf{K}_a & -\mathbf{K}_b + \mathbf{K}_a \cdot (\mathbf{L}\times) \\ -\mathbf{K}_c & -\mathbf{K}_d + \mathbf{K}_c \cdot (\mathbf{L}\times) \end{bmatrix} \end{aligned} \tag{7.33}$$

where

$$(\mathbf{L}\times) = \begin{bmatrix} 0 & -l_z & l_y \\ l_z & 0 & -l_x \\ -l_y & l_x & 0 \end{bmatrix} \tag{7.34}$$

Similarly, assuming that $\delta\mathbf{t}_2 = \mathbf{0}$, the model (7.29) can be re-written using expressions

$$\mathbf{W}_1 = \begin{bmatrix} \mathbf{F} \\ \mathbf{M} \end{bmatrix}; \quad \mathbf{W}_2 = \begin{bmatrix} -\mathbf{F} \\ -\mathbf{M} + \mathbf{F} \times \mathbf{L} \end{bmatrix}; \quad \delta\mathbf{t}_1 = \begin{bmatrix} \delta\mathbf{t} \\ \delta\varphi \end{bmatrix} \tag{7.35}$$

and presented in the form

$$\begin{bmatrix} \mathbf{F} \\ \mathbf{M} \\ -\mathbf{F} \\ -\mathbf{M}+\mathbf{F}\times\mathbf{L} \end{bmatrix} = \begin{bmatrix} \mathbf{K}_{11} & \mathbf{K}_{12} \\ \mathbf{K}_{21} & \mathbf{K}_{22} \end{bmatrix} \cdot \begin{bmatrix} \delta\mathbf{p}_1 \\ \delta\varphi_1 \\ \mathbf{0} \\ \mathbf{0} \end{bmatrix} \tag{7.36}$$

that after the substitution of $\mathbf{K}_{12}$, $\mathbf{K}_{21}$, $\mathbf{K}_{22}$ and relevant transformations give the following expression

$$\mathbf{K}_{11} = \begin{bmatrix} \mathbf{K}_a & \mathbf{K}_b - \mathbf{K}_a \cdot (\mathbf{L}\times) \\ \mathbf{K}_c + (\mathbf{L}\times)\cdot\mathbf{K}_a & \mathbf{K}_d - \mathbf{K}_c\cdot(\mathbf{L}\times) + (\mathbf{L}\times)\cdot\mathbf{K}_b - (\mathbf{L}\times)\cdot\mathbf{K}_a\cdot(\mathbf{L}\times) \end{bmatrix} \tag{7.37}$$

Finally, integrating all sub-matrices, the desired matrix of size 12 × 12 can be presented in the form

$$\mathbf{K}_{12\times12} = \begin{bmatrix} \mathbf{K}_{AA} & \mathbf{K}_{AB} \\ \mathbf{K}_{BA} & \mathbf{K}_{BB} \end{bmatrix} \tag{7.38}$$

with

$$\begin{aligned}
\mathbf{K}_{AA} &= \begin{bmatrix} \mathbf{K}_a & \mathbf{K}_b - \mathbf{K}_a \cdot (\mathbf{L}\times) \\ \mathbf{K}_c + (\mathbf{L}\times)\cdot\mathbf{K}_a & \mathbf{K}_d - \mathbf{K}_c\cdot(\mathbf{L}\times) + (\mathbf{L}\times)\cdot\mathbf{K}_b - (\mathbf{L}\times)\cdot\mathbf{K}_a\cdot(\mathbf{L}\times) \end{bmatrix} \\
\mathbf{K}_{AB} &= \begin{bmatrix} -\mathbf{K}_a & -\mathbf{K}_b \\ -\mathbf{K}_c - (\mathbf{L}\times)\cdot\mathbf{K}_a & -\mathbf{K}_d - (\mathbf{L}\times)\cdot\mathbf{K}_b \end{bmatrix} \\
\mathbf{K}_{BA} &= \begin{bmatrix} -\mathbf{K}_a & -\mathbf{K}_b + \mathbf{K}_a\cdot(\mathbf{L}\times) \\ -\mathbf{K}_c & -\mathbf{K}_d + \mathbf{K}_c\cdot(\mathbf{L}\times) \end{bmatrix} \\
\mathbf{K}_{BB} &= \begin{bmatrix} \mathbf{K}_a & \mathbf{K}_b \\ \mathbf{K}_c & \mathbf{K}_d \end{bmatrix}
\end{aligned} \tag{7.39}$$

which includes elements of $\mathbf{K}_{6\times6}$ and the link's geometrical parameters $(\mathbf{L}\times)$ only.

Therefore, the stiffness matrix identification technique proposed before can also produce 12 × 12 matrices required for the manipulator stiffness modelling based on the MSA approach. This also enhances previous results of [7, 8] where the manipulator topology was quite general (parallel and with cross-linkage), but the links' stiffness properties were described rather roughly.

7.7 Illustrative Example: Stiffness Parameters of Orthoglide Manipulator Links

Let us apply the proposed methodology to the identification of the link stiffness matrices for a parallel manipulator of the Orthoglide family [9]. The principal components of the manipulator are presented in Fig. 7.14, where the elements (b, e, f) are treated as flexible ones and the element (c) is assumed to be rigid. In previous works, relevant stiffness matrices have been obtained either via single-beam approximation [5] or by using an FEA-based identification procedure with a linear meshing option [1]. Besides, joint particularities have not been taken into account and their influence on the elements of the stiffness matrix was not studied. Hence, it is quite possible that some stiffness matrix elements have been identified with essential errors.

The FEA-based virtual experiments were performed using the parabolic mesh of size 2 mm (2P in Fig. 7.9) and the test forces/torques of 1 N and 1 N·m respectively, which, for considering link size and material properties, correspond to the area of linear force–deflection relation. As it follows from Table 7.5, such settings ensure an accuracy of about 0.1%. Identification results are presented in Table 7.8, where the obtained matrices include a number of zero elements that were detected using the technique presented above. It is also worth mentioning that these matrices are symmetrical, which confirms the validity of the developed method.

For comparison purposes, similar matrices have been computed using other methods (single- and multi-beam approximations, the FEA modelling with linear meshing). They are presented in Table 7.9 and show essential dissimilarity in the evaluation of some matrix elements by different methods. For instance, for the parallelogram bar (i.e., link (f) in Fig. 7.14), the torsional compliance defined by the element k_{44} differs by a factor of 13. The main reason for this is that the developed technique takes into account the joint particularities that define the force distribution rule for the applied loadings (previous results assumed that the loading was localized in the reference point). However, the final conclusion concerning the accuracy of the

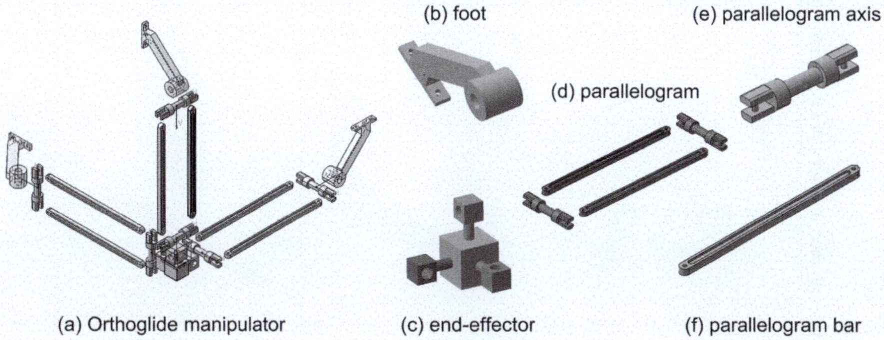

Fig. 7.14 CAD model of Orthoglide and its principal components: Orthoglide (**a**), foot (**b**), end-effector (**c**), parallelogram (**d**), parallelogram axis (**e**) and bar (**f**) [2]

Table 7.8 Compliance matrices of Orhoglide links [2]

Link		Compliance matrix
Foot		$k_{Foot} = \begin{bmatrix} 2.77\cdot10^{-4} & -3.28\cdot10^{-4} & 0 & 0 & 0 & -4.03\cdot10^{-6} \\ -3.28\cdot10^{-4} & 4.14\cdot10^{-4} & 0 & 0 & 0 & 5.41.10^{-6} \\ 0 & 0 & 1.94\cdot10^{-3} & 1.12\cdot10^{-5} & -1.49\cdot10^{-5} & 0 \\ 0 & 0 & 1.12\cdot10^{-5} & 2.29\cdot10^{-7} & 0 & 0 \\ 0 & 0 & -1.49\cdot10^{-5} & 0 & 2.30\cdot10^{-7} & 0 \\ -4.03\cdot10^{-6} & 5.41\cdot10^{-6} & 0 & 0 & 0 & 8.42.10^{-8} \end{bmatrix}$
Axis	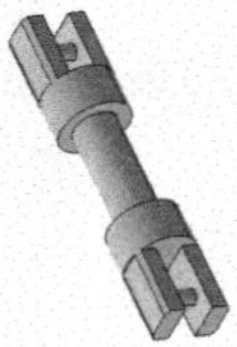	$k_{Axis} = \begin{bmatrix} 6.23\cdot10^{-6} & 0 & 0 & 0 & 0 & 0 \\ 0 & 2.83\cdot10^{-5} & 0 & 0 & 0 & 1.40\cdot10^{-7} \\ 0 & 0 & 2.59\cdot10^{-5} & 0 & -9.65\cdot10^{-7} & 0 \\ 0 & 0 & 0 & 2.77\cdot10^{-7} & 0 & 0 \\ 0 & 0 & -9.65\cdot10^{-7} & 0 & 4.84\cdot10^{-7} & 0 \\ 0 & 1.40\cdot10^{-7} & 0 & 0 & 0 & 1.20.10^{-7} \end{bmatrix}$
Bar	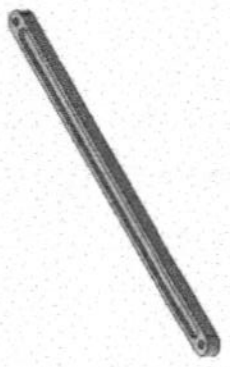	$k_{Bar} = \begin{bmatrix} 4.55\cdot10^{-5} & 0 & 0 & 0 & 0 & 0 \\ 0 & 2.33\cdot10^{-1} & 0 & 0 & 0 & 1.13\cdot10^{-4} \\ 0 & 0 & 5.08\cdot10^{-2} & 0 & -2.39\cdot10^{-4} & 0 \\ 0 & 0 & 0 & 2.88\cdot10^{-5} & 0 & 0 \\ 0 & 0 & -2.39\cdot10^{-4} & 0 & 1.50\cdot10^{-6} & 0 \\ 0 & 1.13\cdot10^{-3} & 0 & 0 & 0 & 7.19\cdot10^{-6} \end{bmatrix}$

obtained stiffness matrices can be obtained after integration of these matrices in the stiffness model of the entire manipulator and comparing it with a straightforward FEA model of the manipulator assembly.

Hence, the presented examples confirm the advantages of the developed stiffness matrix identification technique and demonstrate its ability to take into account complex link shapes as well as joint particularities related to the force/torque distribution. In addition, it produces symmetrical matrices with some zero elements in accordance with the physical properties of the considered manipulator components.

Table 7.9 Comparison of the compliance matrix elements obtained from different methods [2]

Method	Compliance matrix elements					
	k_{11} mm/N	k_{22} mm/N	k_{33} mm/N	k_{44} rad/N•mm	k_{55} rad/N · mm	k_{66} rad/N · mm
Link (b): foot						
Single-beam approximation	$3.45·10^{-4}$	$3.45·10^{-4}$	$18.1·10^{-4}$	$2.10·10^{-7}$	$2.10·10^{-7}$	$0.91·10^{-7}$
Four-beam approximation	$2.77·10^{-4}$	$4.34·10^{-4}$	$17.9·10^{-4}$	$2.11·10^{-7}$	$1.95·10^{-7}$	$0.91·10^{-7}$
FEA-based evaluation (linear mesh)	$2.45·10^{-4}$	$3.24·10^{-4}$	$15.9·10^{-4}$	$2.07·10^{-7}$	$2.06·10^{-7}$	$1.71·10^{-7}$
FEA-based evaluation (parabolic mesh)	$2.77·10^{-4}$	$4.15·10^{-4}$	$19.4·10^{-4}$	$2.29·10^{-7}$	$2.30·10^{-7}$	$0.84·10^{-7}$
Link (e): parallelogram axis						
Single beam approximation	$1.34·10^{-6}$	$2.65·10^{-5}$	$2.65·10^{-5}$	$4.29·10^{-8}$	$3.18·10^{-8}$	$3.18·10^{-8}$
FEA-based evaluation (linear mesh)	$1.99·10^{-6}$	$1.29·10^{-5}$	$1.50·10^{-5}$	$6.81·10^{-6}$	$8.23·10^{-6}$	$2.67·10^{-6}$
FEA-based evaluation (parabolic mesh)	$6.23·10^{-6}$	$2.83·10^{-5}$	$2.59·10^{-5}$	$2.77·10^{-7}$	$4.84·10^{-7}$	$1.20·10^{-7}$
Link (f): parallelogram bar						
Single-beam approximation	$3.75·10^{-5}$	$4.38·10^{-2}$	$1.09·10^{-1}$	$3.96·10^{-6}$	$3.40·10^{-6}$	$1.37·10^{-6}$
FEA-based evaluation (linear mesh)	$4.50·10^{-5}$	$3.64·10^{-2}$	$8.01·10^{-2}$	$3.76·10^{-6}$	$2.65·10^{-6}$	$1.09·10^{-6}$
FEA-based evaluation (parabolic mesh)	$4.55·10^{-5}$	$5.08·10^{-2}$	$2.33·10^{-1}$	$2.88·10^{-5}$	$7.19·10^{-6}$	$1.50·10^{-6}$

In the following chapters, these matrices will be used for the VJM-based stiffness modelling of parallel manipulators.

References

1. Pashkevich A, Chablat D, Wenger P (2009) Stiffness analysis of overconstrained parallel manipulators. Mech Mach Theory 44:966–982
2. Klimchik A, Pashkevich A, Chablat D (2013) CAD-based approach for identification of elasto-static parameters of robotic manipulators. Finite Elem Anal Des 75:19–30
3. Zefran M, Kumar V (1998) A geometrical approach to the study of the Cartesian stiffness matrix. J Mech Des 124:30–38
4. Kövecses J, Angeles J (2007) The stiffness matrix in elastically articulated rigid-body systems. Multibody SysDyn 18:169–184
5. Majou F, Gosselin C, Wenger P, Chablat D (2007) Parametric stiffness analysis of the Orthoglide. Mech Mach Theory 42:296–311
6. Gower JC, Dijksterhuis GB (2004) Procrustes problems. OUP Oxford
7. Deblaise D, Hernot X, Maurine P. A systematic analytical method for PKM stiffness matrix calculation. In: IEEE international conference on robotics and automation (ICRA 2006). IEEE, pp 4213–4219
8. Klimchik A, Pashkevich A, Chablat D (2019) Fundamentals of manipulator stiffness modeling using matrix structural analysis. Mech Mach Theory 133:365–394
9. Chablat D, Wenger P (2003) Architecture optimization of a 3-DOF translational parallel mechanism for machining applications, the orthoglide. Rob Automat IEEE Trans 19:403–410

Chapter 8
Stiffness Modelling of Typical Manipulator Linkages and Their Assemblies

8.1 Stiffness Modelling of Quasi-Serial Chains

8.1.1 Stiffness Model of a Typical Serial Manipulator

Let us obtain first the reduced stiffness model of a typical serial industrial manipulator. As it was mentioned above, the desired model takes into account the actuator compliances only, which usually are the most important elastic components of the entire manipulator. In the case of the serial manipulator, the robot arm contains two actuated revolute joints $q_2.q_3$ and two links of lengths $l_2,\ l_3$ (Fig. 8.1). It should be mentioned that in some robots' link #3 may contain two geometrical parameters l_{3x}, l_{3y} (see Fig. 8.1). Nevertheless, these two parameters can be easily presented as a single link length $l_3 = \sqrt{l_{3x}^2 + l_{3y}^2}$ and joint offset $\Delta q_3 = atan2(l_{3y}, l_{3x})$. Such kinematics can be found in the industrial robots produced by the majority of robot manufacturers (Fig. 8.2).

The considered manipulator stiffness model contains two elastic parameters k_2 and k_3 describing compliances of corresponding joints. For this geometry, the Cartesian stiffness matrix $\mathbf{K}_C$ describing manipulator resistance to the external force $\mathbf{F}$ applied to the manipulator end-effector can be expressed as [1–3]

$$\mathbf{K}_C = \left(\mathbf{J}_\text{S}\, \mathbf{k}_\theta\, \mathbf{J}_\text{S}^T\right)^{-1} \tag{8.1}$$

where the diagonal matrix $\mathbf{k}_\text{S} = diag(k_2,\ k_3)$ collects the joint compliances and $\mathbf{J}_\text{S}$ is kinematic Jacobian. For the serial arm, the Jacobian matrix can be computed as

$$\mathbf{J}_\text{S} = \begin{bmatrix} -l_2 \cos\ q_2 - l_3 \cos\ (q_2 + q_3) & -l_3 \cos\ (q_2 + q_3) \\ l_2 \sin\ q_2 + l_3 \sin(q_2 + q_3) & l_3 \sin(q_2 + q_3) \end{bmatrix} \tag{8.2}$$

A. Klimchik et al., *Stiffness Modeling of Parallel Robots*, Parallel Robots: Theory and Applications, https://doi.org/10.1007/978-3-031-76320-5_8

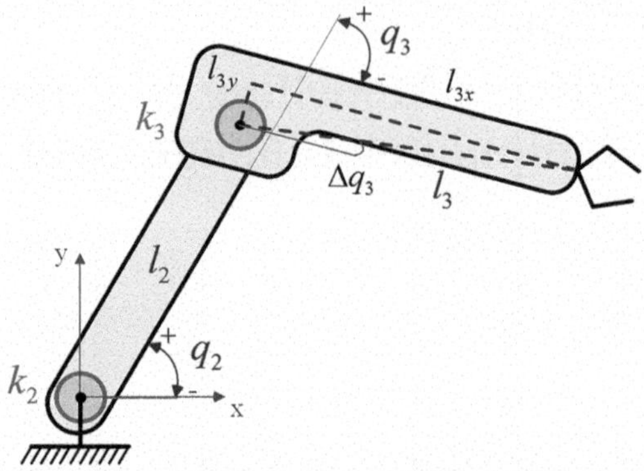

Fig. 8.1 Kinematic structure and basic parameters of serial robot arm [4]

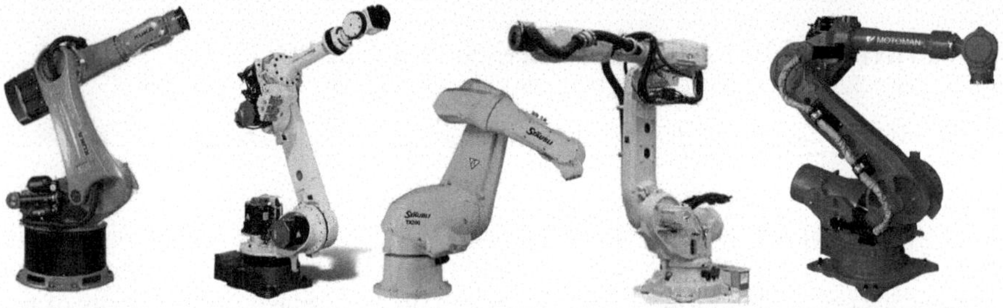

Fig. 8.2 Industrial Robots of Serial Architecture [4]

where l_2. l_3 are the link lengths, q_2, q_3 are the actuated joint coordinates. After substituting (8.2) in (8.1) and several mathematical transformations one can get the following expression for the serial manipulator compliance $\mathbf{k}_C = \mathbf{K}_C^{-1}$

$$\begin{aligned}
\mathbf{k}_{CS} &= \begin{bmatrix} k_{11} & k_{12} \\ k_{21} & k_{22} \end{bmatrix} \\
k_{11} &= k_2(l_2 c_2 + l_3 c_{23})^2 + k_3 l_3^2 c_{23}^2 \\
k_{22} &= k_2(l_2 s_2 + l_3 s_{23})^2 + k_3 l_3^2 s_{23}^2 \\
k_{12} &= k_{21} = -k_2(l_2^2 c_2 s_2 + l_2 l_3 c_3 + l_3^2 c_{23} s_{23}) - k_3 l_3^2 c_{23} s_{23}
\end{aligned} \tag{8.3}$$

where $c_2 = \cos q_2$, $s_2 = \sin q_2$, $c_3 = \cos q_3$, $s_3 = \sin q_3$, $c_{23} = \cos(q_2 + q_3)$, $s_{23} = \sin(q_2 + q_3)$. As follows from our experience, the joint compliance k_2, k_3 are quite close to each other and may be assumed to be equal, i.e. $k_2 = k_3 = k$. Besides, it is reasonable to introduce a coefficient $\mu = l_3/l_2$ defining the ratio between the link lengths. This allows us to present the compliance matrix in a more compact form

$$\mathbf{k}_{CS} = kl_2^2 \begin{bmatrix} (c_2 + \mu c_{23})^2 + \mu^2 c_{23}^2 & -c_2 s_2 - \mu c_3 - 2\mu^2 c_{23} s_{23} \\ -c_2 s_2 - \mu c_3 - 2\mu^2 c_{23} s_{23} & (s_2 + \mu s_{23})^2 + \mu^2 s_{23}^2 \end{bmatrix} \tag{8.4}$$

which is convenient for the comparison analysis.

8.1.2 Stiffness Model of Typical Quasi-Serial Manipulator

Then, applying a similar technique, let us obtain the reduced stiffness model of a typical quasi-serial industrial manipulator. In contrast to its serial counterpart, the robot arm of the quasi-serial manipulator contains a kinematic parallelogram whose adjacent sides can be treated as links #2 and #3 (Fig. 8.3). Besides, here the actuators are usually located on the same axis (corresponding to joint #2 of the serial manipulator, see Fig. 8.1). Such kinematics is quite common in heavy industrial robots (Fig. 8.4) since it allows for reducing the robot arm inertia and improving its dynamic properties. For comparison purposes, the geometric and elastic parameters of the quasi-serial arm are assumed to be the same as in the serial case (see Fig. 8.3).

For the quasi-serial geometry, the desired stiffness matrix can be computed using the expression (8.1), where the Jacobian should be replaced by

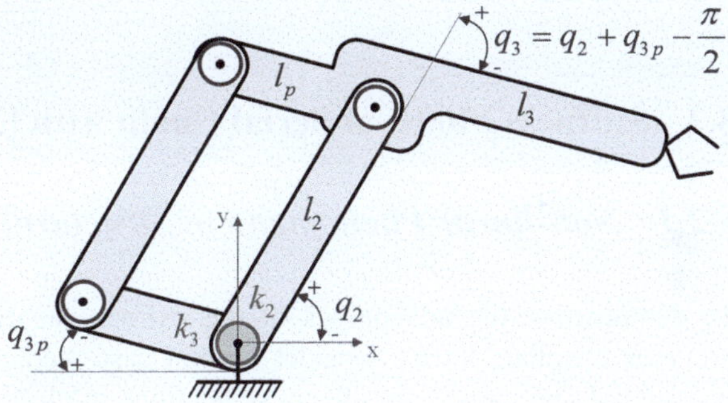

Fig. 8.3 Kinematic structure and basic parameters of quasi-serial robot arm [4]

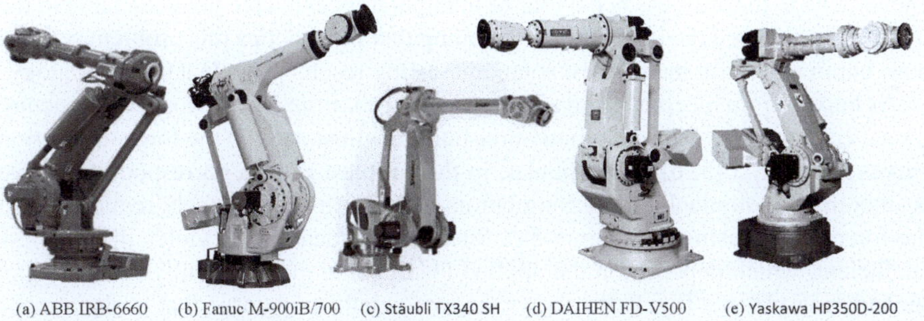

(a) ABB IRB-6660 (b) Fanuc M-900iB/700 (c) Stäubli TX340 SH (d) DAIHEN FD-V500 (e) Yaskawa HP350D-200

Fig. 8.4 Industrial robots of quasi-serial architecture [4]

$$\mathbf{J}_\mathrm{P} = \begin{bmatrix} -l_2 \cos\, q_2 & -l_3 \cos\, q_{3p} \\ l_2 \sin\, q_2 & l_3 \sin q_{3p} \end{bmatrix} \tag{8.5}$$

where q_{3p} is the coordinate of the actuated joint #3, which differs from q_3 in serial manipulator. Using these notations, the compliance matrix of the quasi-serial manipulator can be expressed as

$$\mathbf{k}_\mathrm{CP} = \begin{bmatrix} k_2 l_2^2 \mathrm{c}_2^2 + k_3 l_3^2 \mathrm{c}_{3p}^2 & -k_2 l_2^2 \mathrm{c}_2 \mathrm{s}_2 - k_3 l_3^2 \mathrm{c}_{3p} \mathrm{s}_{3p} \\ -k_2 l_2^2 \mathrm{c}_2 \mathrm{s}_2 - k_3 l_3^2 \mathrm{c}_{3p} \mathrm{s}_{3p} & k_2 l_2^2 \mathrm{s}_2^2 + k_3 l_3^2 \mathrm{s}_{3p}^2 \end{bmatrix} \tag{8.6}$$

where $\mathrm{c}_{3p} = \cos q_{3p}$, $s_{3p} = \sin\, q_{3p}$. Following the assumptions concerning the stiffness coefficients, i.e. $k_2 = k_3 = k$ and introducing the link length ratio $\mu = l_3/l_2$, the above matrix can be presented in the compact form as

$$\mathbf{k}_\mathrm{CP} = k l_2^2 \begin{bmatrix} \mathrm{c}_2^2 + \mu^2 \mathrm{c}_3^2 & -\mathrm{c}_2 \mathrm{s}_2 - \mu^2 \mathrm{c}_3 \mathrm{s}_3 \\ -\mathrm{c}_2 \mathrm{s}_2 - \mu^2 \mathrm{c}_3 \mathrm{s}_3 & \mathrm{s}_2^2 + \mu^2 \mathrm{s}_3^2 \end{bmatrix} \tag{8.7}$$

Using the compliance matrix $\mathbf{k}_\mathrm{CP}$, it is possible to estimate manipulator end-effector deflections under the given external loading and for a given robot configuration and also to evaluate the influence of the link-length ratio on the manipulator stiffness behaviour.

8.2 Stiffness Model of Serial Chain with End-Point Loading

8.2.1 Non-linear Phenomena in Structural Mechanics

To demonstrate the advantages of the stiffness analysis technique developed in previous chapters, let us consider several case studies assuming that the loading is essential and causes significant deflections in elastic elements of the manipulators. In contrast to previous works, it allows us to investigate potential non-linear phenomena in the stiffness behaviour of manipulators that were not detectable in the frame of linear models. It is worth mentioning that for robotics this problem is fairly new, but in structural mechanics, some interesting results provide a useful analogy.

As known from mechanics, in a general case the stiffness behaviour of a structure (or a mechanism) may be essentially non-linear as illustrated by the force–deflection curves in Fig. 8.5 [5, 6]. In particular, in the simplest case, a corresponding curve is smooth, monotonically increasing but its inclination progressively reduces while the force is increasing (see Fig. 8.5a). Such a non-linear shape yields the stiffness coefficient, which depends on the applied loading and monotonically reduces. The second example (see Fig. 8.5b) shows the force–deflection curve that is also monotonically increasing but contains a salient point that corresponds to a sudden reduction of the stiffness coefficient (buckling). Finally, the third example (see Fig. 8.5c)

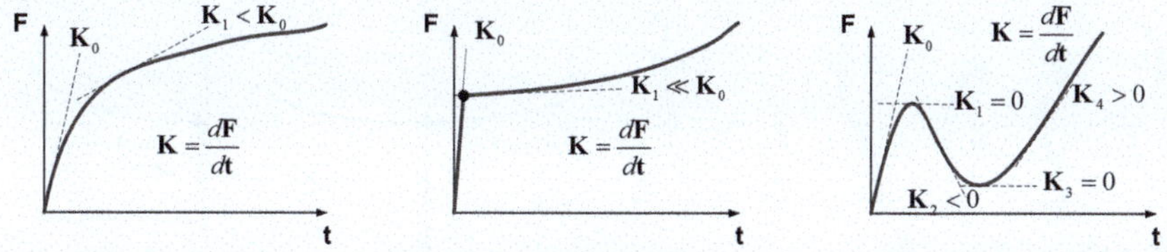

(a) non-linear force-deflection curve (b) non-linear force-deflection curve with buckling (c) non-linear force-deflection curve with negative stiffness

Fig. 8.5 Examples of non-linear force–deflection curves

presents a non-monotonic (but smooth) force–deflection curve with local maximum and minimum. In this case, the curve includes a region with negative inclination (i.e. negative stiffness coefficient) and the system behaviour under the loading is highly non-linear. Here, deflection increases continuously at the beginning, but then it changes abruptly, while the loading achieves its critical value (it is also called buckling). Analyzing these examples, we can conclude that in robotic systems, similar phenomena may exist.

In the following sections, we will pay special attention to the buckling phenomena since they are considered potentially very dangerous for mechanical structures and mechanisms and may provoke failures. In the general theory of elasticity, several types of buckling are usually distinguished:

- ***local buckling*** of a structural element (rod, beam, etc.), observed as a sudden change of the element shape and significant reduction of its stiffness;
- ***geometrical buckling*** of a mechanism or structure (arches, shells, etc.), observed as a sudden change of mechanism/structure geometrical configuration;
- ***contact buckling*** associated with boundary conditions for the mechanism/structure elements that can be observed similar to either local or geometrical buckling.

Below, it will be shown that similar buckling phenomena also exist in robotic systems, but have never been studied in detail.

For local buckling, a classical example is usually related to the *axially compressed column* [6], which is also referred to as "Euler buckling". As shown in Fig. 8.6, depending on end-point constraints, the loading may provoke different changes in the column shape that appear after achieving a critical value for the compressing force. Besides, here the column end-point can abruptly be moved either in axial or lateral direction. It is obvious that in application to the robotic manipulator, similar behaviour may be observed both for a separate link and for a serial chain of links. However, it is unlikely that the VJM model can detect the local buckling for a separate link due to its simplified approximation (rigid rod and virtual spring at the end).

For the geometrical buckling, a representative example is a *two-rod arch* with vertical loading presented in Fig. 8.7. Here, when the loading exceeds its critical value, the elastostatic equilibrium becomes unstable and the system suddenly changes its configuration. Making the analogy with robotic manipulators, it is clear that similar

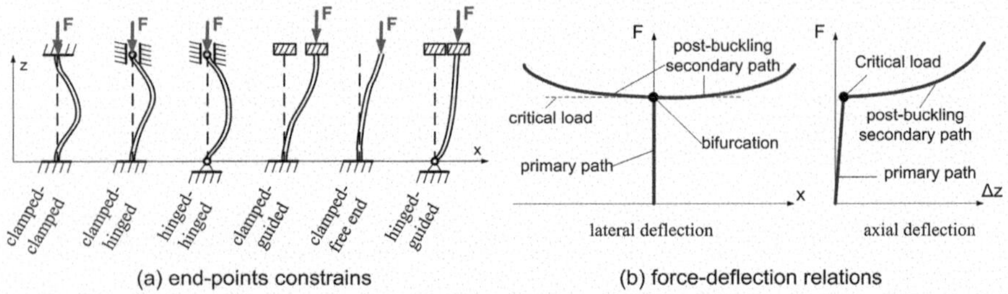

Fig. 8.6 Local buckling in axially compressed column

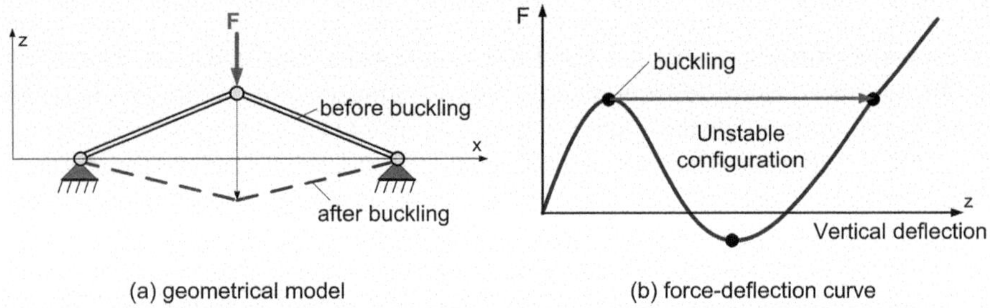

Fig. 8.7 Geometrical buckling in vertically loaded arch

behaviour can be observed in parallel architectures with end-point loading and in serial chains with auxiliary loading applied in the intermediate nodes.

It is worth mentioning that in previous works devoted to the manipulator stiffness analysis, much attention has not been paid to the buckling phenomena. The authors mainly concentrated on the computing of the stiffness matrices implicitly assuming that the force–deflection relation is monotonic and its second-order approximation is good enough. Obviously, this approach does not allow us to detect salient points or local maximums/minimums that are essential for full-scale non-linear force–deflection analysis. To our knowledge, the only works that contain some comprehensive results on non-linear stiffness analysis are [7, 8] where several case studies have been considered using the catastrophe theory. However, it is unlikely that such a technique can be applied in a general case.

8.2.2 Kinetostatic Model of Serial Kinematic Chain and Its Parameterization

To apply the developed technique, first, it is required to derive the models that describe the geometry of the considered serial chain and the stiffness properties of its elements.

For all sub-cases, similar parameterization will be used that allows us to present results in a comparable form.

Geometric model. The geometry of the examined kinematic chain (Fig. 8.8) can be defined as $U_pR_aR_aS_p$ where R, U and S denote respectively the rotational, universal and spherical joints, and the subscripts '*p*' and '*a*' refer to the passive and active joints respectively. Using the homogeneous matrix transformations, it can be described by the equation

$$\mathbf{T} = \mathbf{R}_u(\mathbf{q}_0)\,\mathbf{T}_x(L)\,\mathbf{T}_s(\boldsymbol{\theta}_1)\,\mathbf{R}_z(q_{a1})\,\mathbf{T}_x(L)\,\mathbf{T}_s(\boldsymbol{\theta}_2)\,\mathbf{R}_z(q_{a2})\,\mathbf{T}_x(L)\,\mathbf{T}_s(\boldsymbol{\theta}_3)\,\mathbf{R}_s(\mathbf{q}_t) \tag{8.8}$$

where $\mathbf{R}_z(\ldots)$ and $\mathbf{T}_x(\ldots)$ are the elementary rotation/translation matrices around/along the *z*- and *x*-axes, $\mathbf{R}_u(\ldots)$ is the homogeneous rotation matrix of the universal joint (incorporating two elementary rotations), $\mathbf{R}_s(\ldots)$ is the homogeneous rotation matrix of the universal joint (incorporating three elementary rotations), $q_{a1},\ q_{a2}$ are the coordinates of the actuated joints, *L* is the length of the links, $\mathbf{q}_0$ is the coordinate vector of the universal passive joint located at the robot base, $\mathbf{q}_t$ is the coordinate vector corresponding to the passive spherical joint at the end-platform, $\mathbf{T}_s(\ldots)$ is the homogenous matrix function describing elastic deformations in the links and actuators (they are represented by the virtual coordinates incorporated in the vectors $\boldsymbol{\theta}_1,\ \boldsymbol{\theta}_2,\ \boldsymbol{\theta}_3$). It is obvious that this model can be easily transformed into the form $\mathbf{t} = \mathbf{g}(\mathbf{q},\ \boldsymbol{\theta})$ used in the frame of the developed technique.

To investigate the particularities of this kinematic chain with respect to the external loading, let us also consider three typical postures that differ in values of the actuated coordinates:

- ***S-configuration***: the links are located along a straight line (Fig. 8.8a), the actuated coordinates are $q_{a1} = q_{q2} = 0$
- ***Π-configuration***: the chain takes a trapezoid shape (Fig. 8.8b), the actuated coordinates are $q_{a1} = q_{q2} = -30°$
- ***Z-configuration***: the chain takes a zig-zag shape (Fig. 8.8c), the actuated coordinates are $q_{a1} = -q_{a2} = 30°$

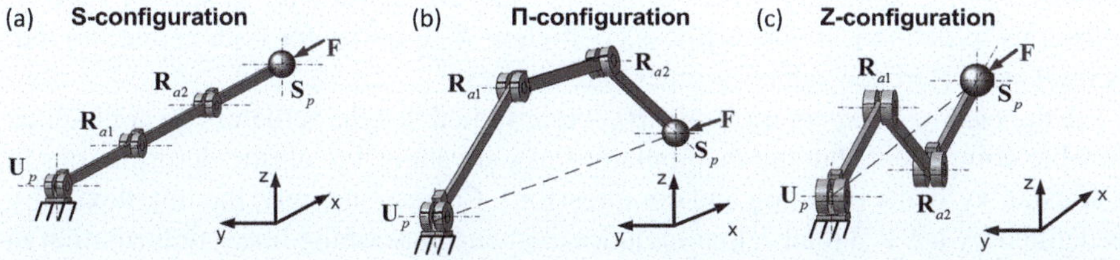

Fig. 8.8 Examined kinematical chain and its typical configurations (U_p—passive universal joint, R_{a1}, R_{a2}—actuated rotating joints, S_p—passive spherical joint) [9]

For presentation convenience, let us also assume that the coordinates $\mathbf{q}_0$ of the universal passive joint are computed to ensure the location of the end-effector on the Cartesian axis x.

Stiffness models. In order to investigate possible non-linear effects in the stiffness behaviour of this chain, let us consider several cases that differ in stiffness models of the links and actuated joints:

Case of 1D-springs (Model A): the flexible elements are localized in the actuating drives while the links are considered as strictly rigid. It allows without loss of generality to reduce the original $U_pR_aR_aS_p$ model down to $R_pR_aR_aR_p$ and define a single stiffness parameter K_θ (similar for both actuators) that will be used as a reference value for further analysis. Besides, it is possible to ignore the end-effector orientation and consider a single passive joint coordinate q (at the base) and two virtual joint coordinates θ_1, θ_2 (at actuators).

Case of 2D springs (Model B): the actuators do not include flexible components but the manipulator links are subject to non-negligible elastic deformations in Cartesian xy-plane (bending and compression). Correspondingly, the link flexibility is defined by a 3×3 matrix that includes elements describing deformation in x- and y-directions and rotational deformation with respect to the z-axis.

Case of 3D springs (Model C): the actuators are strictly rigid but the link flexibility is described by a full-scale 3D model that incorporates all deflections along and around x-,y-, and z-axes of the three-dimensional Cartesian space. The relevant stiffness matrix of the links has dimension 6×6. The kinematics of this model corresponds to the general expression $U_pR_aR_aS_p$, it includes two passive joints $(\mathbf{q}, \mathbf{q}_t)$ incorporating in total five passive coordinates and three virtual springs with 18 virtual coordinates in total (six for each link).

Parameterization of stiffness models for serial kinematic chains. In order to simplify the comparison of numerical values derived from the stiffness analysis of examined serial kinematic chains, the compliance matrices corresponding to all considered case studies have been parameterized similarly. Relevant expressions are presented below.

In the *case of 1D-springs (Model A),* it is assumed that all flexible elements are localized in the actuating drives while the links are considered as strictly rigid. It allows us to define a single stiffness coefficient K_θ (similar for both actuators) that is used as a reference value for other case studies.

In the *case of 2D springs (Model B),* it is assumed that the actuators do not include flexible components but the manipulator links are subject to elastic deformations in Cartesian xy-plane (bending and compression). Correspondingly, the link flexibility is defined by a 3×3 matrix that includes elements describing linear deformation in x- and y-directions and rotational deformation with respect to the z-axis. Relevant stiffness matrix may be written as [10]

$$\mathbf{K} = \frac{E}{L^3} \cdot \begin{bmatrix} A \cdot L^2 & 0 & 0 \\ 0 & 12 \cdot I & -6 \cdot I \cdot L \\ 0 & -6 \cdot I \cdot L & 4 \cdot I \cdot L^2 \end{bmatrix} \tag{8.9}$$

where L is the length of the links, I and A are respectively its second moment and area of the cross-section, E is the Young module. To ensure comparability with model A, the element $k_{3,3}$ (corresponding to z-rotation) of the compliant matrix $\mathbf{k} = \mathbf{K}^{-1}$ was denoted as $1/K_\theta$. It allows to elimination of the Young module and presents the desired matrix as

$$\mathbf{K} = K_\theta \cdot \begin{bmatrix} I/A & 0 & 0 \\ 0 & L^2/3 & L/2 \\ 0 & L/2 & 1 \end{bmatrix}^{-1} \tag{8.10}$$

where, for the rectangular cross-section $a \times b$, the remaining parameters are expressed as $A = ab$ and $I = ab^3/12$.

In the *case of 3D springs (Model C)*, the actuators are strictly rigid but the link flexibility is described by a full-scale 3D model that incorporates all deflections along and around x-,y-, and z-axes of the three-dimensional Cartesian space. The relevant stiffness matrix of the links has the dimension 6×6 and may be expressed as [10]

$$\mathbf{K} = \frac{E}{L^3} \cdot \begin{bmatrix} A \cdot L^2 & 0 & 0 & 0 & 0 & 0 \\ 0 & 12 \cdot I_z & 0 & 0 & 0 & -6 \cdot I_z \cdot L \\ 0 & 0 & 12 \cdot I_y & 0 & 6 \cdot I_y \cdot L & 0 \\ 0 & 0 & 0 & G \cdot J \cdot L^2/E & 0 & 0 \\ 0 & 0 & 6 \cdot I_y \cdot L & 0 & 4 \cdot I_y \cdot L^2 & 0 \\ 0 & -6 \cdot I_z \cdot L & 0 & 0 & 0 & 4 \cdot I_z \cdot L^2 \end{bmatrix} \tag{8.11}$$

where A, I_y, I_z are the area and the second moments of the link cross-section, J is the polar moment, and E and G are the Young and Coulomb modules of the material. For the rectangular cross-section $a \times b$, the required parameters may be computed as $A = ab$ and $I_y = a^3b/12$, $I_z = ab^3/12$.Similar to the previous case, the parameterization is performed by defining the compliance with respect to the z-axis as $1/K_\theta$ (here, it is the element $k_{6,6}$ of the compliant matrix $\mathbf{k} = \mathbf{K}^{-1}$). This leads to the expression:

$$\mathbf{K} = K_\theta \cdot \begin{bmatrix} I_z/A & 0 & 0 & 0 & 0 & 0 \\ 0 & L^2/3 & 0 & 0 & 0 & L/2 \\ 0 & 0 & k_I \cdot L^2/3 & 0 & k_I \cdot L/2 & 0 \\ 0 & 0 & 0 & k_J \cdot I_z/(2 \cdot L \cdot (1+\nu)) & 0 & 0 \\ 0 & 0 & k_I \cdot L/2 & 0 & k_I & 0 \\ 0 & L/2 & 0 & 0 & 0 & 1 \end{bmatrix}^{-1} \tag{8.12}$$

where the coefficient k_J depends on the cross-section shape, $k_I = I_y/I_z$, and ν is the Poisson ratio coefficient.

Let us use these parameterizations of flexible elements sequentially and compare the stiffness analysis results for S, Π and Z configurations of the kinematic chain.

8.2.3 Stiffness Analysis for Serial Chain with 1D-Springs

In this case, the model includes a minimum number of flexible elements (two 1D virtual springs in the actuated joints) and may be tackled analytically. However, despite its simplicity, it is potentially capable of detecting the buckling phenomena at least for S-configuration, because of the evident mechanical analogy to straight columns' behaviour under axial compression. It is also useful to evaluate other initial configurations with respect to the multiple loaded equilibriums, and their stability and to compare with numerical results provided by the developed technique.

In the frame of this model, the end-effector motions are restricted to Cartesian xy-plane and the geometry is defined by equations

$$
\begin{aligned}
x &= L \cdot \cos q + L \cdot \cos q_{12} + L \cdot \cos q_{13}, \\
y &= L \cdot \sin q + L \cdot \sin q_{12} + L \cdot \sin q_{13}
\end{aligned}
\tag{8.13}
$$

where L is the length of links, q is the passive joint coordinate, θ_1, θ_2 are the virtual spring coordinates, $q_{12} = q + \theta_1$ and $q_{13} = q + \theta_1 + \theta_2$. It is obvious that this model includes exactly one redundant variable. Without loss of generality, one can assume that the redundant variable is the passive joint coordinate q while the manipulator end-effector is initially located at the point $(x, y) = (3L - \delta,\ 0)$, where δ is a linear displacement along the x-axis.

Then, assuming that the initial values of the actuating coordinates (i.e. before the loading) are denoted as θ_1^0, θ_2^0, the potential energy stored in the virtual springs may be expressed as the following function of the redundant variable

$$
E(q) = \frac{1}{2} K_\theta \cdot \left(\theta_1(q) - \theta_1^0\right)^2 + \frac{1}{2} K_\theta \cdot \left(\theta_2(q) - \theta_2^0\right)^2 \tag{8.14}
$$

where K_θ is the stiffness coefficient, and θ_1, θ_2 are computed via the inverse kinematics. Using these equations, the desired equilibriums may be computed from the extremum of $E(q)$. In particular, stable equilibriums correspond to minima of this function, and unstable ones correspond to maxima.

To illustrate this approach, Fig. 8.9 presents a case study for the initial S-configuration. It allows for comparing 12 different shapes of the deformed chain and selecting the best /worst cases with respect to the energy. As it follows from these results, here there are two symmetrical maximal and two minimal, i.e. two stable and two unstable equilibriums. Besides, the stable equilibriums correspond to Π-shaped deformed postures, and the unstable ones correspond to Z-shaped postures, as shown in Fig. 8.10. A more detailed analysis allows deriving analytical expressions for the force and energy for small values of δ:

$$\text{Stable equilibrium: } E_{\min} \approx \delta \cdot K_\theta/L;\ F_S \approx K_\theta/L$$

$$\text{Unstable equilibrium: } E_{\max} \approx \delta \cdot 3K_\theta/L;\ F_N \approx 3K_\theta$$

Hence, the external force $F \leq K_\theta/L$ cannot change the manipulator shape, similar to small compressing of straight columns that cannot cause lateral deflections. Consequently, in this case, the straight configuration is stable. Further, for $K_\theta/L < F \leq 3K_\theta/L$, the straight configuration may be hypothetically restored but becomes unstable, so any small disturbance will cause sudden reshaping in the direction of the stable trapezoid-type posture. And finally, for $F > 3K_\theta/L$, there may exist two types of unstable equilibriums: the trivial straight type and a more complicated zig-zag one.

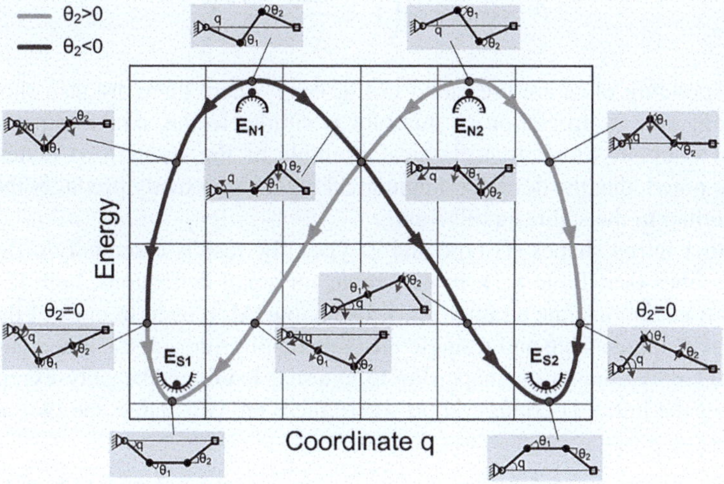

Fig. 8.9 Potential energy $E(q)$ and geometric postures for different values of q (Model A, case of initial S-configuration, $\delta = L/10$) [11]

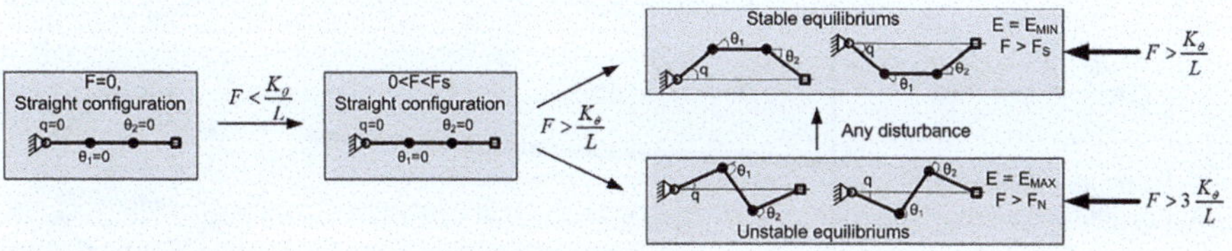

Fig. 8.10 Evolution of the S-configuration under external loading [9]

If the assumption concerning small values of δ is released, analytical solutions for the non-trivial equilibriums may be still derived. In particular, for a stable equilibrium, one can get

$$F_S(\Delta) = \frac{K_\theta}{L} \cdot \frac{\varphi}{\sin\varphi} \tag{8.15}$$

where $\varphi = \pm\arccos(1 - \Delta/2)$. For the unstable equilibrium, a similar equation may be written as

$$F_N(\Delta) = \frac{K_\theta}{L} \cdot \frac{\cos(q+\theta) + 2\cdot\cos q}{\sin\theta} \cdot \theta \tag{8.16}$$

where

$$q = \pm\arccos\left(\frac{12 - 6\Delta + \Delta^2}{12 - 4\Delta}\right);$$
$$\theta = \mp\arccos\left(1 - \frac{3\Delta}{2} + \frac{\Delta^2}{4}\right) \tag{8.17}$$

Corresponding plots are presented in Fig. 8.11 where there are also shown the bifurcation. The interpretation of this plot is similar to the axial compression of a straight column, which is a classical example of the strength of materials. It should be noted, that the developed numerical algorithm exactly produces the curve corresponding to the stable equilibrium.

For other initial shapes (Π-type and Z-type), the results essentially differ from the above ones (see Table 8.1). In particular, for small deflections, both Π-and Z-configuration demonstrate a rather linear behaviour. Moreover, in most of the cases there exist a single stable and a single unstable equilibrium, so the kinematic chain cannot suddenly change its shape due to external loading. The only exception is the case of the initial Π-configuration where there are two stable and two unstable

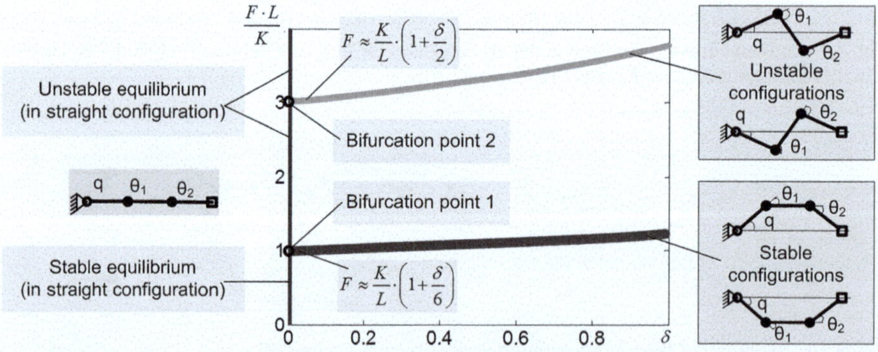

Fig. 8.11 Model A: Force–deflection relations and bifurcations for the initial S-configuration [11]

equilibriums, and here there exists a bifurcation of the stable equilibriums corresponding to the cuspidal point of the function $F(\Delta)$ where the stiffness reduces sharply. Another conclusion concerns the profile of the force–deflection plots that are highly nonlinear in all cases.

A more detailed analysis shows that Π-configuration demonstrates a good analogy with an axially compressed imperfect column where the deflection starts from the beginning of the loading and there is no sudden buckling, but the stiffness essentially reduces while the loading increases. In particular, for the examined kinematic chain the stiffness coefficient is about 1.78 K_θ/L^2 at the beginning and 0.43 K_θ/L^2 at the end of the curve $F(\Delta)$.

However, for Z-configuration that corresponds to the unloaded zig-zag shape, the stiffness behaviour demonstrates the buckling that leads to sudden transformation from a symmetrical to a non-symmetrical posture as shown in Fig. 8.12. Here, there exist two stable equilibriums that differ in the values of the potential energy, so the

Table 8.1 Force–deflection relations for different elasticity models of serial chain with passive joints [9]

	S-configuration	Π-configuration	Z-configuration
Elasticity models	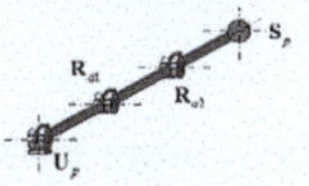 $\theta_1^0 = \theta_2^0 = 0$	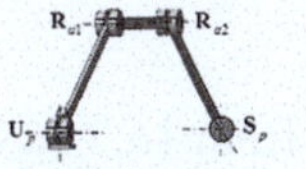 $\theta_1^0 = \theta_2^0 = -30^\circ$	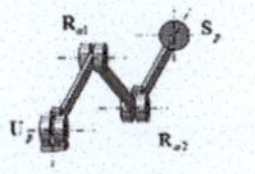 $\theta_1^0 = -30^\circ, \theta_2^0 = 30^\circ$
Model A: 1D-springs	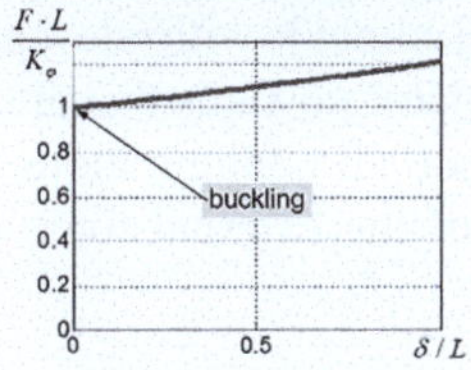	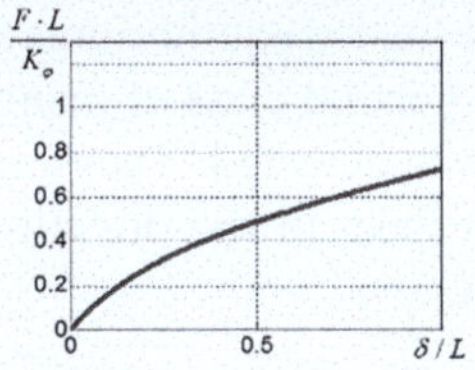	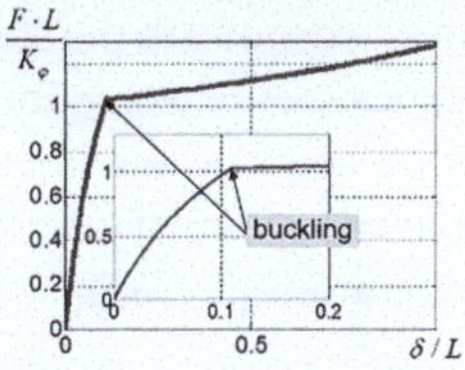
Model B: 2D-springs	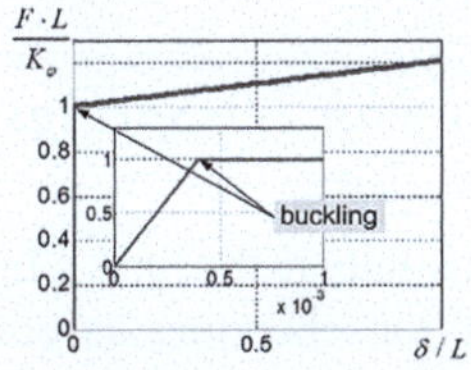	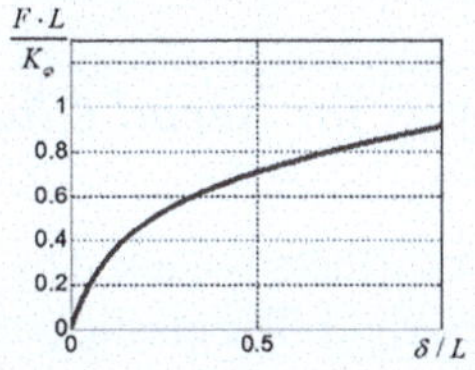	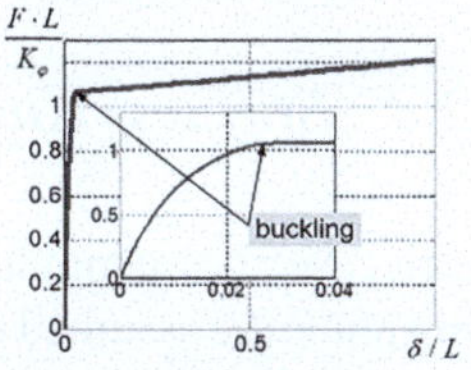
Model C: 3D-springs	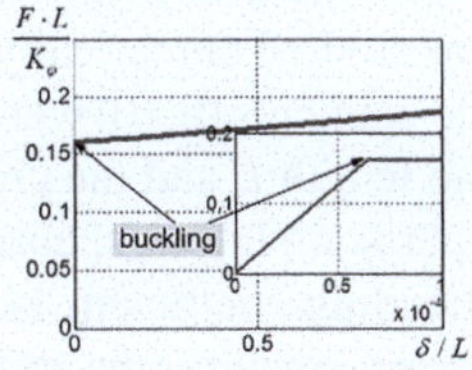	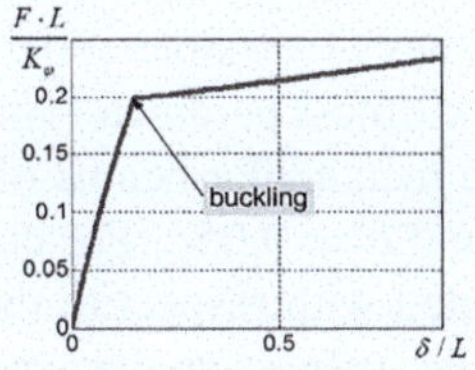	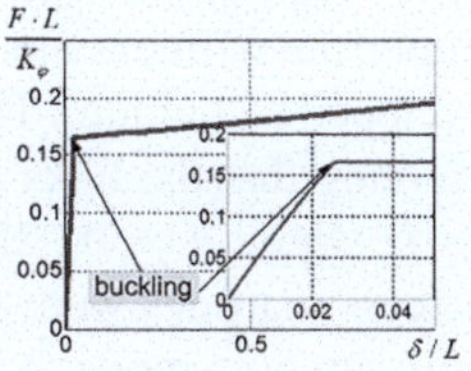

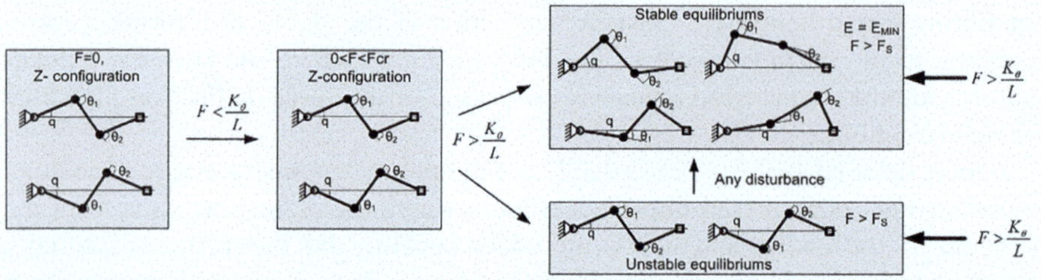

Fig. 8.12 Evolution of the Z-configuration under external loading [9]

Table 8.2 Summary of stiffness analysis for serial chain with passive joints and 1D-springs [11]

Configuration	Critical force	Stiffness for unloaded mode	Stiffness around the buckling		Stiffness for large deformations ($\delta \approx L$)
			$F < F_{cr}$	$F > F_{cr}$	
S-configuration	$\frac{K_\theta}{L}$	∞	∞	$0.20\ \frac{K_\theta}{L}$	$0.22\ \frac{K_\theta}{L}$
Π-configuration	–	$1.78\ \frac{K_\theta}{L}$	–	–	$0.43\ \frac{K_\theta}{L}$
Z-configuration	$1.03\ \frac{K_\theta}{L}$	$16.7\ \frac{K_\theta}{L}$	$5.50\ \frac{K_\theta}{L}$	$0.20\ \frac{K_\theta}{L}$	$0.39\ \frac{K_\theta}{L}$

stiffness coefficient reduces from 16.7 K_θ/L^2 at the beginning to 0.39 K_θ/L^2 at the end of the curve $F(\Delta)$ (see Table 8.2).

Hence, in the case of model A, the developed numerical technique provides results that correspond to the stable loaded equilibriums and coincide with relevant analytical solutions. Besides, a more general class of manipulator postures that are dangerous with respect to buckling was detected. They include all configurations possessing an axial symmetry with respect to the direction of the external force (S- and Z-configurations, for instance).

8.2.4 *Stiffness Analysis for Serial Chain with 2D-Springs*

In this case, the manipulator stiffness is caused by elasticity in the links while the actuating joints are assumed to be rigid. The elastic deflections (bending and compression) are still restricted by the Cartesian xy-plane but each link includes three virtual joints. Totally, the stiffness model has 11 variables, so it was studied numerically (using the developed technique). The stiffness parameters of the elements (8.10) were evaluated assuming that the links are rectangular beams of the length L and the cross-section $a \times b$, where $a = 0.02L$ and $b = 0.05L$. For comparison purposes, corresponding stiffness matrices were scaled with respect to the bending coefficient $1/K_\theta$ to keep similarity with model A. The stiffness analysis was performed for the above-defined S-, Π- and Z-configurations, assuming that the external force is directed along the

Table 8.3 Summary of stiffness analysis for serial chain with passive joints and 2D-springs [157]

Configuration	Critical force	Stiffness for unloaded mode	Stiffness around the buckling		Stiffness for large deformations ($\delta \approx L$)
			$F < F_{cr}$	$F > F_{cr}$	
S-configuration	$\frac{K_\theta}{L}$	$2500\ \frac{K_\theta}{L^2}$	$2500\ \frac{K_\theta}{L^2}$	$0.20\ \frac{K_\theta}{L^2}$	$0.22\ \frac{K_\theta}{L^2}$
Π-configuration	–	$5.31\ \frac{K_\theta}{L^2}$	–	–	$0.34\ \frac{K_\theta}{L^2}$
Z-configuration	$1.07\ \frac{K_\theta}{L}$	$100\ \frac{K_\theta}{L^2}$	$0.13\ \frac{K_\theta}{L^2}$	$0.13\ \frac{K_\theta}{L^2}$	$0.16\ \frac{K_\theta}{L^2}$

x-axis. A summary of the modelling results is presented in Table 8.1 and Table 8.3 and are briefly described.

For *S-configuration*, here there is still a very strong analogy with the compression of the straight column. In particular, at the beginning of the loading, the links are subject to axial compression and the stiffness is very high (about 2500 K_θ/L^2, for the assumed link proportions). Then, after the buckling, the manipulator changes its shape to become non-symmetrical and the stiffness falls down to 0.20 K_θ/L^2. The critical force may be also computed using the previous results, as $F_0 = K_\theta/L$.

For *Π-configuration*, the stiffness properties are also qualitatively equivalent to the case of Model A but the stiffness coefficient is slightly lower (in the frame of the adopted parameterization); it varies from 5.31 K_θ/L^2 to 0.34 K_θ/L^2. For *Z-configuration*, it also detected the buckling that occurs if the loading approaches the critical value $F_0 = 1.07\ K_\theta/L$. At this point, the stiffness falls down from $100 \cdot K_\theta/L^2$ to $0.13 \cdot K_\theta/L^2$, which essentially differs from model A.

Hence, the stiffness analysis of Model B demonstrates qualitative similarity but some quantitative differences compared to Model A. The latter is caused by different arrangements of the elastic elements in the virtual joints that correspond to other physical assumptions.

8.2.5 Stiffness Analysis for Serial Chain with 3D-Springs

Finally, for model C, the link elasticity is described in 3D space and corresponding stiffness matrices have dimension 6 × 6 (8.12) (the actuating joints are assumed perfect and rigid, similar to model B). It is also assumed that the links are rectangular beams of the length L with the cross-section $a \times b$, where $a = 0.02L$, $b = 0.05L$ and the smaller value a corresponds to z-direction, which was not studied above. To ensure comparability of all examined cases, the link stiffness matrices were parameterized with respect to the bending coefficient of the z-axis K_θ. In total, the stiffness model includes 23 variables (five for passive joints and 18 for the virtual springs of three links) and it was studied numerically. The stiffness analysis was performed for the same initial configurations (S-, Π- and Z-type) and for the same direction of the external force as for models A and B. A Summary of the modelling results is presented in Tables 8.1 and 8.4 and are briefly described.

For *S-configuration*, the results are qualitatively similar to the ones obtained for model B. Besides, the numerical value of the stiffness for the non-loaded case is the same, 2500 K_θ/L^2. However, here the buckling occurs for an essentially lower critical force, 0.16 K_θ/L, that corresponds to sudden lateral deflection in the z-direction. For comparison, according to the Euler formula, the local buckling of the links occurs for a compressing force of 0.40 K_θ/L, which is 2.5 times higher than the loading that provokes the geometric buckling of the examined mechanism. Then, after the buckling, the stiffness falls down to 0.20 K_θ/L^2. It is worth mentioning that the axial deflection corresponding to the critical force is very low, it is equal to $7 \cdot 10^{-5} \cdot \delta/L$.

In contrast, *Π-configuration* buckling which does not exist in models A and B was detected. In particular, if the external force exceeds the critical value 0.20 K_θ/L the stiffness suddenly reduces from 1.03 K_θ/L^2 to 0.04 K_θ/L^2 (for comparison, the stiffness coefficient for unloaded mode is 1.70 K_θ/L^2). Physically it is also explained by sudden deflection in the z-direction, which was beyond the capabilities of previous models. It is also worth mentioning that, in this case study, the stiffness of manipulator links in the z-direction is essentially lower than in the y-direction. Another interpretation of this buckling phenomenon may be presented as a sudden loss of symmetry with respect to the xy-plane.

For *Z-configuration*, the results remain qualitatively the same as above, but corresponding numerical values are changed. Thus, manipulator stiffness for the unloaded mode is 7.52 K_θ/L, then it gradually reduces to 5.88 K_θ/L and, after buckling, falls down to 0.03 K_θ/L. The corresponding value of the critical force is 0.17 K_θ/L and it also corresponds to the z-deflection.

Summarized results concerning all models A, B, and C are presented in Tables 8.1, 8.2, 8.3 and 8.4. As it follows from them, a full-scale 3D stiffness analysis yields essentially lower values of critical force compared to models A and B. Besides, for model C, all examined postures demonstrated buckling related to sudden deflections in the z-direction. This presents another source of potential structural instability of kinematic chains that possess symmetry with respect to a plane.

Table 8.4 Summary of stiffness analysis for serial chain with passive joints and 3D-springs [11]

Configuration	Critical force	Stiffness for unloaded mode	Stiffness around the buckling		Stiffness for large deformations ($\delta \approx L$)
			$F < F_{cr}$	$F > F_{cr}$	
S-configuration	$0.16\ \frac{K_\theta}{L}$	$2500\ \frac{K_\theta}{L^2}$	$2500\ \frac{K_\theta}{L^2}$	$0.023\ \frac{K_\theta}{L^2}$	$0.029\ \frac{K_\theta}{L^2}$
Π-configuration	$0.20\ \frac{K_\theta}{L}$	$1.70\ \frac{K_\theta}{L^2}$	$1.03\ \frac{K_\theta}{L^2}$	$0.043\ \frac{K_\theta}{L^2}$	$0.13\ \frac{K_\theta}{L^2}$
Z-configuration	$0.17\ \frac{K_\theta}{L}$	$7.52\ \frac{K_\theta}{L^2}$	$5.88\ \frac{K_\theta}{L^2}$	$0.027\ \frac{K_\theta}{L^2}$	$0.035\ \frac{K_\theta}{L^2}$

8.3 Stiffness Model of Serial Chain with Auxiliary Loadings in Intermediate Points

Let us now focus on the non-linear stiffness analysis of a serial chain under auxiliary loadings applied to an intermediate node. It is assumed that the considered chain consists of two links (either rigid or flexible) separated by a flexible joint. The relevant analysis includes evaluating stiffness variation due to the loading, detecting buckling and computing corresponding critical forces, as well as analysis of the auxiliary spring influence on the chain stiffness.

8.3.1 *Serial Chain with Torsional Springs*

Let us consider first a 2-link manipulator with a compliant actuator between the links and two passive joints at both ends. It is assumed that the left passive joint is fixed, while the right one can be moved along x direction (Fig. 8.13a). Besides, here both rigid links have the same length L and the actuator stiffness is K_θ. The positions of the end-effector and the actuator for such a manipulator are defined as

$$\begin{aligned} x_e &= 2L \cdot \cos\alpha; \quad x_a = L \cdot \cos\alpha \\ y_e &= 0; \quad y_a = L \cdot \sin\alpha \end{aligned} \tag{8.18}$$

where α is the angle between the first link and x-direction, and indices 'e' and 'a' correspond to the end-effector and actuator positions respectively.

Let us assume that the initial configuration (i.e. for $M_\theta = 0$) of the manipulator corresponds to $\theta_0 = -\pi/6$, where $\theta = -2 \cdot \alpha$ is the coordinate of the actuated joint. It is also assumed that the external loading G is applied to the intermediate node (Fig. 8.13a) and it is required to apply the external loading (F_x, F_y) at the end-point to compensate for the auxiliary loading G. Since this example is quite simple, it is possible to obtain the force–deflection relation and the stiffness coefficient both analytically and numerically.

Let us start with the analytical stiffness analysis taking into account static forces/torques presented in Fig. 8.13b. The static equilibrium equation of this system can

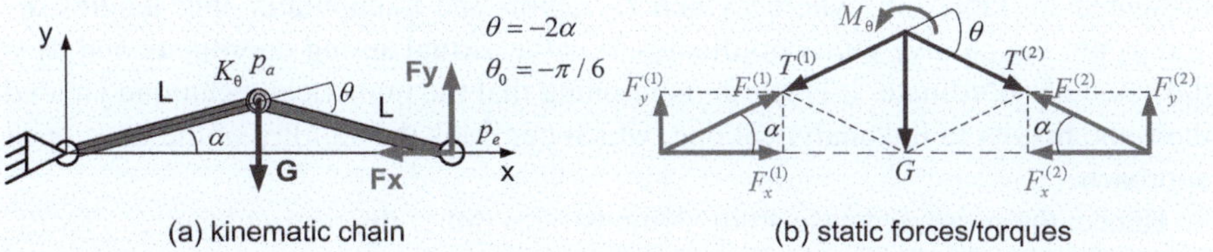

Fig. 8.13 Kinematic chain with compliant actuator between two links and its static forces/torques [12]

be written as

$$M_\theta + M_{Fx} + M_{Fy} = 0 \tag{8.19}$$

where

$$M_\theta = K_\theta \cdot (\theta - \theta_0), \quad M_{Fx} = -F_x \cdot L \cdot \sin\alpha, \quad M_{Fy} = F_y \cdot L \cdot \cos\alpha \tag{8.20}$$

Taking into account that $\theta = -2\alpha$, $\theta_0 = -2\alpha_0$ and $F_y = -G/2$, the force–displacement relation can be expressed in a parametric form as

$$\begin{array}{c} F_x = -\frac{G}{2} \cdot \frac{\cos\alpha}{\sin\alpha} - 2 \cdot \frac{K_\theta}{L} \cdot \frac{\alpha - \alpha_0}{\sin\alpha}; \quad F_y = -\frac{G}{2}; \\ x = 2 \cdot L \cdot \cos\alpha; \quad y = 0 \end{array} \tag{8.21}$$

and stiffness of the manipulator can be presented in a parametric form as

$$K_x = -\frac{G}{4L} \cdot \frac{1}{\sin^3\alpha} - \frac{K_\theta}{L^2} \cdot \frac{(\alpha - \alpha_0) \cdot \cos\alpha - \sin\alpha}{\sin^3\alpha}; \quad K_y = 0 \tag{8.22}$$

where $\alpha \in (-\pi/2;\ \pi/2)$ is treated as a parameter.

As follows from the expression (8.22), the stiffness coefficient K_x essentially depends on the auxiliary loading G. In particular, for the initial configuration (when $M_\theta = 0$), the coefficient K_x can be both positive and negative or even equal to zero when the auxiliary loading is equal to its critical value $G^* = 4K_\theta/L \cdot \sin\alpha_0$. It is evident that the case $G > G^*$ is very dangerous from a practical point of view, since the chain configuration is unstable.

Summarized results for this case study are presented in Fig. 8.14 and in Table 8.5 that contain the force–deflection relations and values of translational stiffness K_x respectively. They show that the auxiliary loading G significantly reduces the stiffness of the serial chain. For example, in an initial configuration ($\Delta x = 0$), for $G = 0$ the stiffness is $14.9 \cdot K_\theta/L^2$, while for $G = 0.5 \cdot G^*$ it reduces to $8.67 \cdot K_\theta/L^2$. Further increasing the auxiliary loading up to $G = 1.5 \cdot G^*$ leads to an unstable configuration with negative stiffness $-7.46 \cdot K_\theta/L^2$. Moreover, in the neighbourhood of the critical value of $G \approx G^*$ (Fig. 8.15), the force–deflection curves have extremum points which may provoke buckling.

Some details concerning functions and matrices used in relevant expressions are presented in Table 8.6, where L_1 and L_2 denote the manipulator link lengths, q_1 and q_2 are the passive joint coordinates, θ is the virtual spring coordinate and θ_0 is the actuator coordinate. It is worth mentioning that the numerical technique yielded the same results as the analytical one, which confirms the validity of the developed approach.

Hence, the presented case study demonstrates rather interesting features of stiffness behaviour for kinematic chains under auxiliary loading that were not studied previously (negative stiffness, non-monotonic force–deflection curves, etc.). This

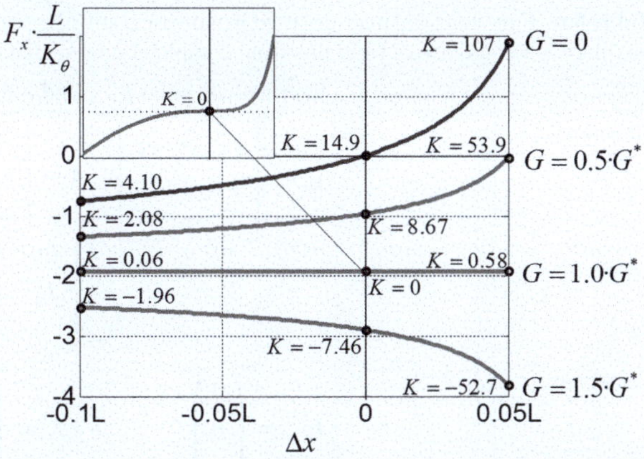

Fig. 8.14 Force–deflection relations for different values of auxiliary loading G: chain with torsional spring ($G^* = 4K_\theta/L \cdot \sin\alpha_0$, $K_x = K \cdot K_\theta/L^2$) [12]

Table 8.5 Translational stiffness K_x for different values of auxiliary loading G and different displacements Δx ($G^* = 4K_\theta/L \cdot \sin\alpha_0$) [12]

Auxiliary loading G	Translational stiffness K_x			
	$\Delta x = -0.1L$ 	$\Delta x = -0.05L$ 	$\Delta x = 0$ 	$\Delta x = 0.05L$ 
0	$4.10 \cdot \frac{K_\theta}{L^2}$	$6.69 \cdot \frac{K_\theta}{L^2}$	$14.9 \cdot \frac{K_\theta}{L^2}$	$107 \cdot \frac{K_\theta}{L^2}$
$0.5 \cdot G^*$	$2.08 \cdot \frac{K_\theta}{L^2}$	$3.36 \cdot \frac{K_\theta}{L^2}$	$8.67 \cdot \frac{K_\theta}{L^2}$	$53.9 \cdot \frac{K_\theta}{L^2}$
$1.0 \cdot G^*$	$0.06 \cdot \frac{K_\theta}{L^2}$	$0.03 \cdot \frac{K_\theta}{L^2}$	0	$0.58 \cdot \frac{K_\theta}{L^2}$
$1.5 \cdot G^*$	$-1.96 \cdot \frac{K_\theta}{L^2}$	$-3.31 \cdot \frac{K_\theta}{L^2}$	$-7.46 \cdot \frac{K_\theta}{L^2}$	$-52.7 \cdot \frac{K_\theta}{L^2}$

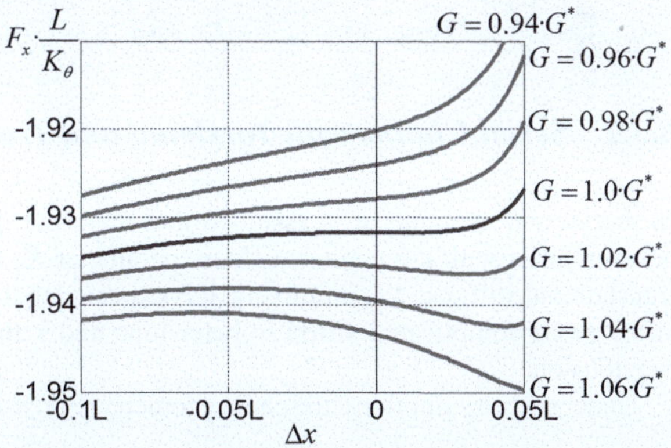

Fig. 8.15 Force-deflections relations for different values of auxiliary loading G in the neighbourhood of critical value $G = G^*$:

motivates the consideration of more sophisticated examples, with more complicated compliant elements.

Table 8.6 Functions and matrices used in numerical stiffness analysis of two-link manipulator with auxiliary loading (case of rigid links and compliant intermediate joint)

		Intermediate point p_a	End-effector p_e
Position		$\mathbf{g}_a = \begin{bmatrix} L_1 \cdot \cos q_1 \\ L_1 \cdot \sin q_1 \\ q_1 \end{bmatrix}$	$\mathbf{g} = \begin{bmatrix} x_e \\ y_e \\ \varphi_e \end{bmatrix} = \begin{bmatrix} L_1 \cdot \cos q_1 + L_2 \cdot \cos(q_1 + \theta) \\ L_1 \cdot \sin q_1 + L_2 \cdot \sin(q_1 + \theta) \\ q_1 + \theta + q_2 \end{bmatrix}$
Jacobians	$\mathbf{J}_\theta$	$\mathbf{J}_\theta^{(G)} = \begin{pmatrix} 0 \\ 0 \\ 1 \end{pmatrix}$	$\mathbf{J}_\theta^{(F)} = \begin{pmatrix} -L_2 \cdot \sin(q_1 + \theta) \\ -L_2 \cdot \cos(q_1 + \theta) \\ 1 \end{pmatrix}$
	$\mathbf{J}_q$	$\mathbf{J}_q^{(G)} = \begin{pmatrix} -L_1 \cdot \sin q_1 & 0 \\ L_1 \cdot \cos q_1 & 0 \\ 1 & 1 \end{pmatrix}$	$\mathbf{J}_q^{(F)} = \begin{pmatrix} -L_1 \cdot \sin q_1 - L_2 \cdot \sin(q_1 + \theta) & 0 \\ L_1 \cdot \cos q_1 + L_2 \cdot \cos(q_1 + \theta) & 0 \\ 1 & 1 \end{pmatrix}$
Hessians	$\mathbf{H}_{\theta\theta}$	$\mathbf{H}_{\theta\theta}^{(G)} = [0]$	$\mathbf{H}_{\theta\theta}^{(F)} = [h_2]$
	$\mathbf{H}_{\theta q}$	$\mathbf{H}_{\theta q}^{(G)} = \begin{bmatrix} 0 & 0 \end{bmatrix}$	$\mathbf{H}_{\theta q}^{(F)} = \begin{bmatrix} h_2 & 0 \end{bmatrix}$
	$\mathbf{H}_{q\theta}$	$\mathbf{H}_{q\theta}^{(G)} = \begin{bmatrix} 0 & 0 \end{bmatrix}^T$	$\mathbf{H}_{q\theta}^{(F)} = \begin{bmatrix} h_2 & 0 \end{bmatrix}^T$
	$\mathbf{H}_{qq}$	$\mathbf{H}_{qq}^{(G)} = \begin{bmatrix} h_1 & 0 \\ 0 & 0 \end{bmatrix}$	$\mathbf{H}_{qq}^{(F)} = \begin{bmatrix} h_3 & 0 \\ 0 & 0 \end{bmatrix}$

$h_1 = -L_1 \cdot \sin q_1 \cdot G_y,\;\; h_2 = -L_2 \cdot \cos(q_1 + \theta) \cdot F_x - L_2 \cdot \sin(q_1 + \theta) \cdot F_y h_3 = -(L_1 \cdot \cos q_1 + L_2 \cdot \cos(q_1 + \theta)) \cdot F_x - (L_1 \cdot \sin q_1 + L_2 \cdot \sin(q_1 + \theta)) \cdot F_y$

8.3.2 Serial Chains with Torsional and Translational Springs

In the second example, it is assumed that there are three compliant elements: an actuated joint with a torsional stiffness parameter K_θ and two non-rigid links with translational stiffness K_L. Intuitively, it is expected that such systems should demonstrate more complicated stiffness behaviour under the loadings compared to the previous one.

Under such assumptions, the chain geometry (Fig. 8.16a) is described by equations

$$
\begin{aligned}
x_e &= 2(L - \Delta L) \cdot \cos\alpha; \quad x_a = (L - \Delta L) \cdot \cos\alpha \\
y_e &= 0; \quad y_a = (L - \Delta L) \cdot \sin\alpha
\end{aligned}
\tag{8.23}
$$

where ΔL is the axial deformation of the links (assumed to be similar to both of them) and the remaining notation corresponds to the previous Section. The static

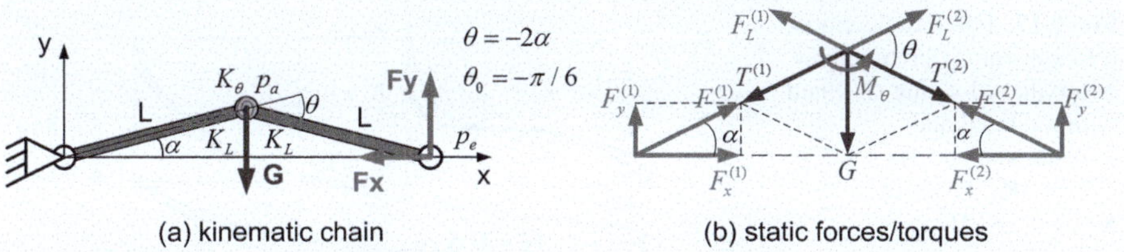

Fig. 8.16 Kinematic chain with compliant links and actuator and its static forces/torques

equilibrium of this system is defined by the balance of forces

$$F_x \cdot \cos\alpha + F_y \cdot \sin\alpha = -K_L \cdot \Delta L \tag{8.24}$$

and the balance of torques (8.19) where L is replaced by $(L - \Delta L)$. Furthermore, applying a similar methodology as above, the force–displacement relation can be expressed in a parametric form as

$$F_x = -\frac{G}{2} \cdot \frac{\cos\alpha}{\sin\alpha} - \\ - \frac{8K_L \cdot K_\theta \cdot (\alpha - \alpha_0)}{2 \cdot L \cdot K_L \cdot \sin\alpha - G + \sqrt{(G - 2 \cdot K_L \cdot L \cdot \sin\alpha)^2 - 32 \cdot K_L \cdot K_\theta \cdot (\alpha - \alpha_0) \cdot \sin\alpha}} \tag{8.25}$$

$$x = \frac{8K_L \cdot \sin\alpha \cdot \cos\alpha}{2 \cdot L \cdot K_L \cdot \sin\alpha - G + \sqrt{(G - 2 \cdot K_L \cdot L \cdot \sin\alpha)^2 - 32 \cdot K_L \cdot K_\theta \cdot (\alpha - \alpha_0) \cdot \sin\alpha}} \tag{8.26}$$

where

$$\Delta L = \left(\frac{G}{4K_L} \cdot \frac{1}{sin\alpha} + \frac{L}{2}\right) - \sqrt{\left(\frac{G}{4K_L} \cdot \frac{1}{sin\alpha} - \frac{L}{2}\right)^2 - 2 \cdot \frac{K_\theta}{K_L} \cdot \frac{\alpha - \alpha_0}{\sin\alpha}} \tag{8.27}$$

This expression also allows us to compute the stiffness coefficient K_x (via the parametric differentiation), but it is not presented here due to its cumbersome form.

The force–deflection relations corresponding to expressions (8.25) are presented in Fig. 8.17. Compared to the previous case, here for $G = 0...G^*$ there is only a quantitative difference (i.e. the shape of the examined curves remains almost the same). However, for $G > G^*$ the chain may not only be unstable with respect to end-effector loading F_x, but also the chain configuration may become unstable. Geometrically, the qualitative difference of the latter is observed similar to buckling in a vertically loaded arch (see Sect. 8.2). A summary of different chain configurations and their stiffness behaviour is presented in Fig. 8.18.

Similar results have also been obtained numerically using the expression presented in Table 8.7, which corresponds to a more general case, where the link lengths are

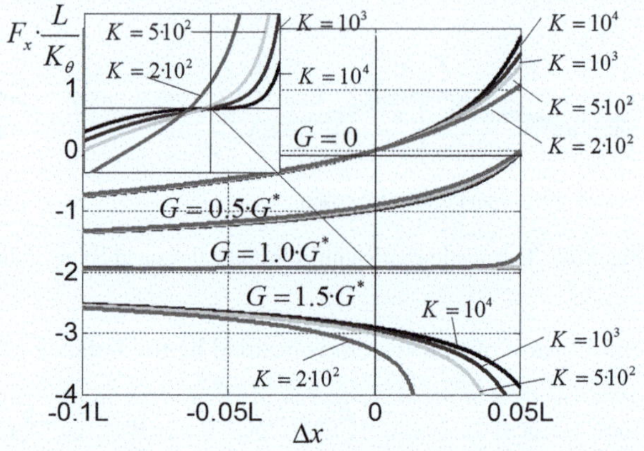

Fig. 8.17 Force-deflections relations for different values of auxiliary loading G: chain with torsional and translational springs ($G^* = 4K_\theta/L \cdot \sin\alpha_0$, $K_L = K \cdot K_\theta/L^2$)

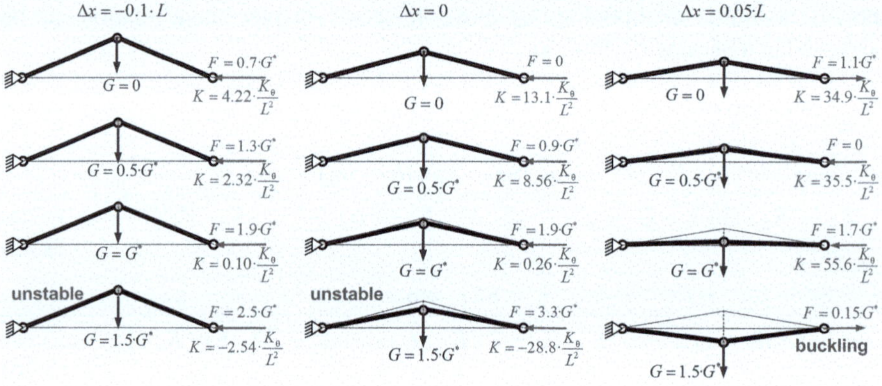

Fig. 8.18 Configuration of kinematic chain with auxiliary loading: case of torsional and translational springs ($G^* = 4K_\theta/L \cdot \sin\alpha_0, K_L = 2 \cdot 10^2 \cdot K_\theta/L^2$)

not assumed to be the same and the auxiliary loading can be oriented arbitrarily and includes the torque component. This more general, model can be useful for stiffness analysis of serial manipulators with auxiliary loading caused by gravity compensators where G is not oriented strictly vertically.

Summarizing, it should be concluded that auxiliary loading essentially influences the stiffness behaviour of robotic manipulators. This may reduce the stiffness coefficient and also provoke undesirable phenomena (such as buckling) that must be taken into account by designers. This justifies previous results and gives perspectives for practical applications.

Table 8.7 Functions and matrices used in numerical stiffness analysis of two-link manipulator with auxiliary loading (case of compliant links and compliant intermediate joint)

		Internal point p_a	End-effector p_e
Position		$x_a=(L_1+\theta_1)\cdot\cos q_1$ $y_a=(L_1+\theta_1)\cdot\sin q_1$ $\varphi_a=q_1$	$\mathbf{g}=\begin{bmatrix}(L_1+\theta_1)\cdot\cos q_1+(L_2+\theta_3)\cdot\cos(q_1+\theta_2)\\(L_1+\theta_1)\cdot\sin q_1+(L_2+\theta_3)\cdot\sin(q_1+\theta_2)\\q_1+\theta_2+q_2\end{bmatrix}$
Jacobians	$\mathbf{J}_\theta$	$\mathbf{J}_\theta^{(F)}=\begin{pmatrix}j_1&0&0\\j_2&0&0\\0&0&0\end{pmatrix}$	$\mathbf{J}_\theta^{(F)}=\begin{pmatrix}j_1&j_5&j_7\\j_2&j_6&j_8\\0&1&0\end{pmatrix}$
	$\mathbf{J}_q$	$\mathbf{J}_q^{(G)}=\begin{pmatrix}j_3&0\\j_4&0\\1&1\end{pmatrix}$	$\mathbf{J}_q^{(F)}=\begin{pmatrix}j_3+j_5&0\\j_4+j_6&0\\1&1\end{pmatrix}$
Hessians	$\mathbf{H}_{\theta\theta}$	$\mathbf{H}_{\theta\theta}^{(G)}=\begin{bmatrix}0&0&0\\0&0&0\\0&0&0\end{bmatrix}$	$\mathbf{H}_{\theta\theta}^{(F)}=\begin{bmatrix}0&0&0\\0&h_3&h_4\\0&h_4&0\end{bmatrix}$
	$\mathbf{H}_{\theta q}$	$\mathbf{H}_{\theta q}^{(G)}=\begin{bmatrix}h_1&0\\0&0\\0&0\end{bmatrix}$	$\mathbf{H}_{\theta q}^{(F)}=\begin{bmatrix}h_5&0\\h_3&0\\h_4&0\end{bmatrix}$
	$\mathbf{H}_{q\theta}$	$\mathbf{H}_{q\theta}^{(G)}=\begin{bmatrix}h_1&0&0\\0&0&0\end{bmatrix}$	$\mathbf{H}_{q\theta}^{(F)}=\begin{bmatrix}h_5&h_3&h_4\\0&0&0\end{bmatrix}$
	$\mathbf{H}_{qq}$	$\mathbf{H}_{qq}^{(G)}=\begin{bmatrix}h_2&0\\0&0\end{bmatrix}$	$\mathbf{H}_{qq}^{(F)}=\begin{bmatrix}h_6&0\\0&0\end{bmatrix}$

$j_1=\cos q_1, j_2=\sin q_1, j_3=-(L_1+\theta_1)\cdot\sin q_1, j_4=(L_1+\theta_1)\cdot\cos q_1,$
$j_5=-(L_2+\theta_3)\cdot\sin(q_1+\theta_2), j_6=(L_2+\theta_3)\cdot\cos(q_1+\theta_2), j_7=\cos(q_1+\theta_2),$
$j_8=\sin(q_1+\theta_2)$
$h_1=-\sin q_1\cdot G_x+\cos q_1\cdot G_y,\ h_2=-(L_1+\theta_1)\cdot\cos q_1\cdot G_x-(L_1+\theta_1)\cdot\sin q_1\cdot G_y,$
$h_3=-(L_2+\theta_3)\cdot\cos(q_1+\theta_2)\cdot F_x-(L_2+\theta_3)\cdot\sin(q_1+\theta)\cdot F_y,$
$h_4=-\sin(q_1+\theta_2)\cdot F_x+\cos(q_1+\theta)\cdot F_y$
$h_5=-\sin q_1\cdot F_x+\cos q_1\cdot F_y, h_6=h_3-(L_1+\theta_1)\cdot\cos q_1\cdot F_x-(L_1+\theta_1)\cdot\sin q_1\cdot F_y$

8.4 Non-linear Stiffness Analysis of Parallelogram-Based Linkages

Let us now consider several examples that deal with the stiffness analysis of 3-d.o.f. translational manipulators of the Orthoglide family and their kinematic chains that include parallelogram-based linkages. Each of these manipulators consists of three identical kinematic chains actuated by mutually orthogonal linear drives, which are arranged in a special way to ensure almost isotropic workspace kinematic properties and to restrict the end-effector motions in translation only. This architecture ensures the velocity transmission factors close to 1.0 (similar to the conventional Cartesian-type machines) and provides high stiffness in the unloaded mode [13]. However, the nonlinear stiffness properties of such manipulators and potential buckling phenomena have not been studied in previous works yet.

8.4.1 Kinetostatic Models of Translational Parallel Manipulators

For the Orthoglide manipulator, each kinematic chain includes a foot, a kinematic parallelogram with two axes and two bars, and an end-effector (Fig. 8.19). To convert this architecture into a set of strictly serial chains two types of approximations were used:

3-PUU model, where the legs are comprised of equivalent limbs with U-joints at the ends (here, the limb stiffness is described as a 6-dof virtual spring corresponding to the parallelogram bars with doubled cross-section area);

3-PU_cU_c model, where there is an explicit kinematic constraint describing parallelism of the parallelogram axes (in this case, the limb stiffness is described by equivalent 5-d.o.f. virtual spring). More details concerning these approximations can be found in [14].

Under these assumptions, the kinematic model of the PUU-chain is expressed as

$$\mathbf{T} = \mathbf{T}_{\text{base}}\ \mathbf{T}_x(\theta_{\text{a}})\ \mathbf{V}(\boldsymbol{\theta}_{\text{Foot}})\ \mathbf{R}_y(q_1)\ \mathbf{R}_z(q_2)\ \mathbf{T}_x(L)\ \mathbf{V}(\boldsymbol{\theta}_{\text{Link}})\ \mathbf{R}_z(q_3)\ \mathbf{R}_y(q_4)\ \mathbf{T}_{\text{Tool}} \tag{8.28}$$

where $\mathbf{T}_{\text{base}}$, $\mathbf{T}_{\text{Tool}}$ are the constant transformations matrices (the matrix $\mathbf{T}_{\text{base}}$ includes also the "foot" transformation); θ_{a}, $\boldsymbol{\theta}_{\text{Foot}}$, $\boldsymbol{\theta}_{\text{Link}}$ are the virtual joint coordinates of

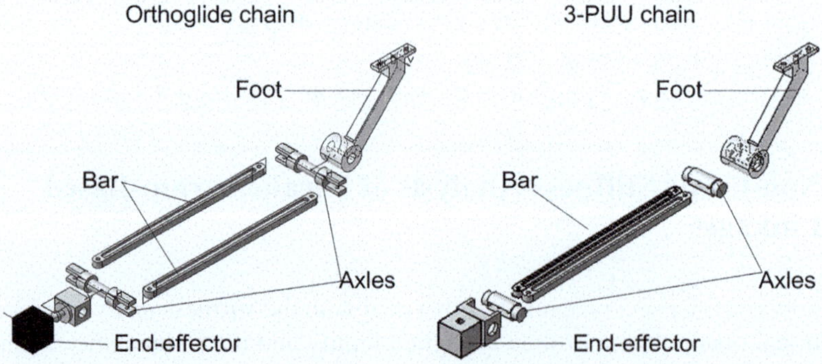

Fig. 8.19 Kinematic chains of the Orthoglide manipulator and its 3-PUU counterpart

the actuator, the "foot" and the "link" respectively; $\mathbf{V}(\ldots)$ is the homogeneous matrix function of the virtual springs, and $\mathbf{T}_x,\ \mathbf{T}_y,\ \mathbf{T}_z,\ \mathbf{R}_x,\ \mathbf{R}_y,\ \mathbf{R}_z$ are elementary homogeneous transformation matrices. In the case of PU_cU_c-chain, the third passive joint is eliminated by adding the kinematic constraint $q_3 + q_2 = 0$). It is also assumed that for the PUU-chain the matrix $\mathbf{V}(\ldots)$ is expressed via the multiplication of six elementary transformations:

$$\mathbf{V}(\theta_x, \theta_y, \theta_z, \theta_{\varphi x}, \theta_{\varphi y}, \theta_{\theta z}) = \mathbf{T}_x(\theta_x)\ \mathbf{T}_y(\theta_y)\ \mathbf{T}_z(\theta_z)\ \mathbf{R}_x(\theta_{\varphi x})\ \mathbf{R}_y(\theta_{\varphi y})\ \mathbf{R}_z(\theta_{\varphi z}) \quad (8.29)$$

These presentations allow computing the Jacobians and Hessians semi-analytically, via the straightforward differentiation of relevant products of homogenous matrices.

The elasticity of the manipulator elements has been evaluated using the FEA-based methodology and special accuracy improvement tools presented in Sect. 7.1. As it follows from our study, the rigidity of the parallelogram axes is high compared to the bar and the foot. For the remaining components, the compliance matrices can be estimated from virtual experiments (see Sect. 7.1).

In the case of the PU_cU_c chain, the parallelogram stiffness matrix depends on the joint variable q_2 and is computed from the analytical expression [14].

$$\mathbf{K}_{\text{plg}}(q_2) = 2 \cdot \begin{bmatrix} K_{11} & 0 & 0 & 0 & 0 & 0 \\ 0 & K_{22} & 0 & 0 & 0 & K_{26} \\ 0 & 0 & 0 & 0 & 0 & 0 \\ 0 & 0 & 0 & K_{44} + \frac{d^2 c_2^2 K_{22}}{4} & 0 & \frac{d^2 s_{22} K_{22}}{8} \\ 0 & 0 & 0 & 0 & \frac{d^2 c_2^2 K_{11}}{4} & 0 \\ 0 & K_{26} & 0 & K_{44} + \frac{d^2 s_{22} K_{22}}{8} & 0 & K_{66} + \frac{d^2 s_2^2 K_{22}}{4} \end{bmatrix} \quad (8.30)$$

where K_{ij} are the elements of $\mathbf{K}_{bar,}$ $c_2^2 = \cos^2 q_2$, $s_2^2 = \sin^2 q_2$, $s_{22} = \sin 2q_2$.

8.4.2 *Stiffness Analysis of Parallelogram-Based Chains*

The most complex element from the point of view of the stiffness analysis in the Orthoglide manipulator is a kinematic parallelogram. It is usually approximated by a linear model (8.30) or even by a bar element with double stiffness. To evaluate the accuracy and validity of both approximations, first let us obtain the stiffness matrices for both of the models and then apply them for the parallel manipulator. For the purposes of comparison, the VJM-based approach developed in this work will be complemented by FEA-based analysis. The latter allows us to determine the accuracy of the proposed technique and to detect critical forces that may provoke buckling phenomena.

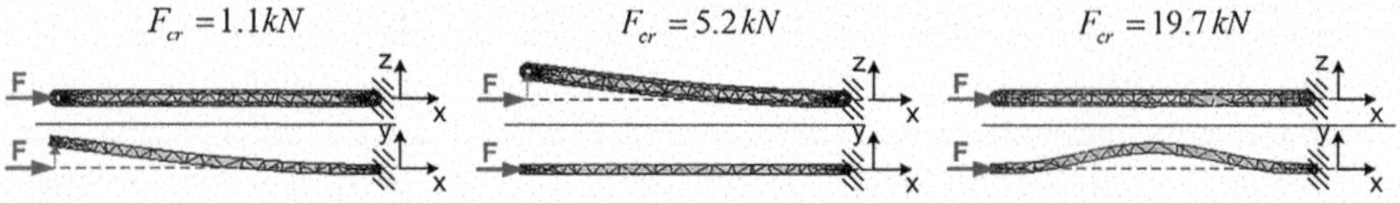

Fig. 8.20 Critical forces and buckling configurations of a bar element employed in the parallelograms of Orthoglide obtained by FEA [14]1

The considered manipulators include three parallelogram-type linkages where the main flexibility source is concentrated in the bar elements of length 310 mm. The stiffness matrix of the bar element was evaluated in Sect. 7.1. This bar-element has also been evaluated with respect to the structural stability under compression, and the FEA modelling produced three possible buckling configurations presented in Fig. 8.20. It is obvious that these configurations (in different combinations, for two bar elements) are potentially dangerous for the compressed parallelogram. However, the parallelogram may also produce some other types of buckling because of the presence of the passive joints.

For the parallelogram-base linkage incorporating the above bar elements, the expression (8.30) yields the following stiffness matrix

$$\mathbf{K}_{p\,\mathrm{lg}} = 2 \cdot \left[\begin{array}{ccc|ccc} 2.20 \cdot 10^4 & 0 & 0 & 0 & 0 & 0 \\ 0 & 1.81 \cdot 10^1 & 0 & 0 & 0 & -2.84 \cdot 10^3 \\ 0 & 0 & 0 & 0 & 0 & 0 \\ \hline 0 & 0 & 0 & 5.64 \cdot 10^4 & 0 & 1.25 \cdot 10^4 \\ 0 & 0 & 0 & 0 & 2.64 \cdot 10^7 & 0 \\ 0 & -2.84 \cdot 10^3 & 0 & 1.25 \cdot 10^4 & 0 & 5.92 \cdot 10^5 \end{array}\right] \tag{8.31}$$

corresponding to the straight configuration of the linkage (i.e. to $q = 0$). Being in good agreement with the physical sense, this matrix is rank deficient and incorporates exactly one zero row/column corresponding to the z-translation, where the parallelogram is completely non-resistant due to the specific arrangement of passive joints. Also, as it follows from comparison with the doubled stiffness matrix of a bar element (see Sect. 7.1) that may be used as a reference, the parallelogram demonstrates essentially higher rotational stiffness that mainly depends on the translational stiffness parameters of the bar (moreover, the rotational stiffness parameter K_{55} of the bar is completely eliminated by the passive joints). The most significant here is the parallelogram width d that was explicitly presented in the rotational sub-block of $\mathbf{K}_{p\,\mathrm{lg}}$.

To investigate the applicability range of the linear model based on the stiffness matrix (8.30), a non-linear "force–deflection" relation corresponding to the parallelogram compression in the x-direction was computed. This computation has been performed using an iterative algorithm presented in Chap. 6. As it follows from the obtained results, the matrix (8.30) ensures a rather accurate description of the parallelogram stiffness in a small deflection area. However, for large deflections, the corresponding VJM-model detects a geometrical buckling that limits the applicability of the matrix (8.30).

A similar analysis was performed using the FEA technique, which yielded almost the same "force–deflection" plot for the small deflections but detected several types of buckling, with critical forces that may be both lower and higher than in the VJM modelling. Corresponding numerical values and parallelogram configurations are presented in Fig. 8.21.

It is worth mentioning that a particular type of buckling that may appear in the manipulator highly depends on the type of joints that precede and succeed the parallelogram. This statement is confirmed by a separate stiffness modelling of the entire manipulators with parallelogram-based linkage. Besides, critical forces essentially depend on the parallelogram joint variable q defining its non-loaded configuration (Table 8.8). Hence, with respect to translational deflections, the linearity zone where the matrix (8.30) may be used without caution is defined by the value $\pm$ 0.1 mm, which is in the frame of reasonable technical specifications for this case study.

A similar analysis has also been performed for the rotational displacement around the x-axis (Fig. 8.22). The obtained results confirmed both validities of the linear model with the derived stiffness matrix (for very large linearity range, deflections up to 15°), and essential advantages of the parallelograms compared to equivalent bar-type links with doubled cross-section area. In particular, for the width parameter

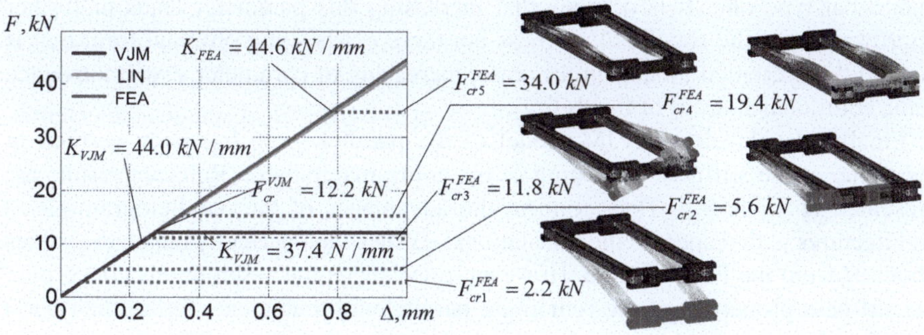

Fig. 8.21 Force–deflection relations and buckling configurations for parallelogram compression (modelling methods: VJM, FEA, and LIN—linear model with proposed stiffness matrix) [14]

Table 8.8 Critical forces corresponding to different types of buckling for two configurations of the parallelogram-based linkage

Critical force	Straight configuration (q = 0) (kN)	Inclined configuration (q = 30°) (kN)
F_{cr1}	2.2	2.3
F_{cr2}	5.6	10.2
F_{cr3}	11.8	12.7
F_{cr4}	19.4	19.7
F_{cr5}	34.0	35.0

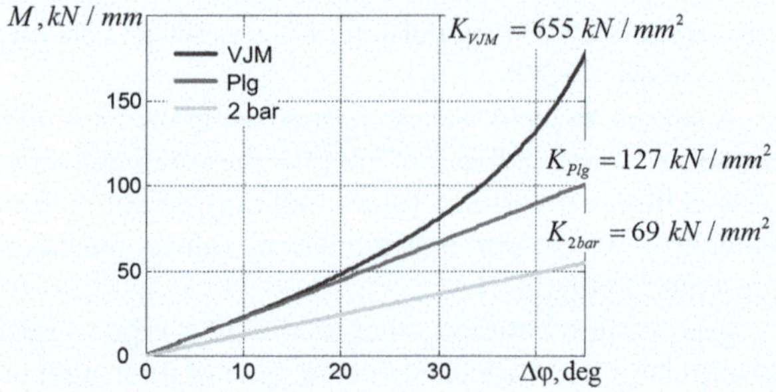

Fig. 8.22 Torque-rotation relations for parallelogram twisting

Table 8.9 Stiffness coefficients of the Orthoglide manipulator for different assumptions concerning parallelogram linkage [14]

Model of linkage	VJM-model		FEA-model	
	K_{tran} [N/mm]	K_{rot} [N/rad]	K_{tran} [N/mm]	K_{rot} [N/rad]
2x-Bar linkage	$3.35{\cdot}10^3$	$0.13{\cdot}10^6$	–	–
Parallelogram linkage (solid axis)	$3.23{\cdot}10^3$	$4.33{\cdot}10^6$	$3.33{\cdot}10^3$	$4.10{\cdot}10^6$
Parallelogram linkage (flexible axis)	$3.08{\cdot}10^3$	$4.06{\cdot}10^6$	$3.31{\cdot}10^3$	$4.07{\cdot}10^6$

d = 80 mm, this type of linkage ensures at least twice better rotational stiffness than a bar-type one. It is obvious that increasing this parameter leads to further improvement of the rotational stiffness but there exists some physical limit that is beyond the presented model (in this case, increasing of parameter *d* will introduce some additional sources of flexibility).

Finally, the developed VJM model of the parallelogram has been verified in the frame of the stiffness modelling of the entire manipulator. Relevant results are presented in Table 8.9. They confirm the advantages of the parallelogram-based architectures with respect to the translational stiffness and perfectly match the values obtained from the FEA method. However, releasing some assumptions concerning the stiffness properties of the remaining parallelogram elements (other than bars) modifies the values of the translational and rotational stiffness. The latter gives a new prospective research direction that targets a more general stiffness model of the parallelogram.

8.4.3 Kinetostatic Singularity in the Neighbourhood of the Flat Configuration

It is evident that the lowest stiffness of the considered translational manipulators corresponds to the near-flat configuration, where the parallel is approaching geometrical singularity (flat configuration corresponds to geometrical singularity when all kinematic chains are laying in the single plane). Thus, to evaluate the specificity of the near-flat configuration, let us consider this case in detail. In order to provide comparability of all case studies which correspond to translational parallel manipulators, let us keep the same assumption concerning the force orientation (parallel to the coordinate frame bisector) but consider both of two possible directions, i.e. towards inside and outside of the cubic workspace.

The simulation results for this manipulator posture are presented in Fig. 8.23 and Table 8.10 where d_0 denotes the initial distance from the flat singularity, K_0 is the translational stiffness for the unloaded mode, $(\Delta_{cr}^{+},\ F_{cr}^{+})$ and $(\Delta_{cr}^{-},\ F_{cr}^{-})$ are respectively the critical deflection and the critical force for the opposite directions of the displacement. As it follows from these results, in the neighbourhood of the flat singularity the stiffness properties are essentially non-symmetrical with respect to the force direction. This conclusion is valid for both of the examined models (3-PUU and 3-PU$_c$U$_c$), which demonstrate almost similar behaviours. In particular, for the inside direction of the loading, the force increases non-linearly but monotonically while the deflection augments. However, for the outside direction, initially, the manipulator reacts to external loading in the same way: increasing the deflection leads to increasing the resisting elastic force. But after achieving the critical value, the reacting force begins to decrease, the configuration becomes unstable and the manipulator abruptly changes its posture to the symmetrical ones (in kinematics, it is treated as an alternative assembling mode). After that, the manipulator demonstrates stable behaviour with respect to the loading.

The simplest model that explains the above-described phenomenon is presented in Fig. 8.24. It is derived via generalization of the "toggle-frame" construction known from the structural mechanics, with relevant modifications motivated by the Orthoglide architecture and relative stiffness properties of its elements. Here, the elasticity is concentrated at the basis of the manipulator legs and it is presented by linear springs with the parameter K_θ. It is assumed that the initial distance between the end-point and the singularity plane is $d_0 = L \sin \varphi_0$, where φ_0 is the corresponding angle between the leg and the plane. The derived expressions

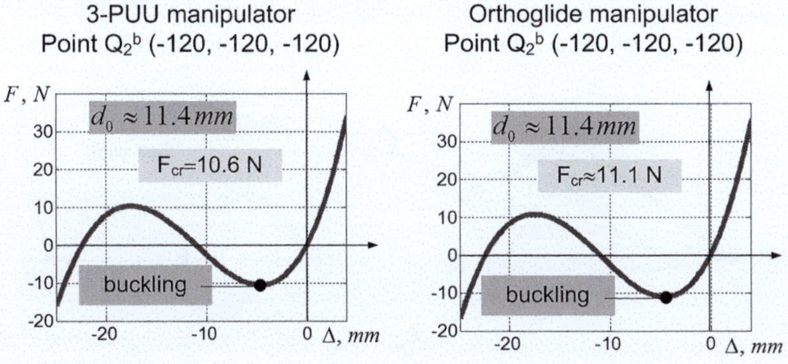

Fig. 8.23 Force–displacement relations for both of the examined manipulators (neighbourhood of the flat singularity; distance to the singularity is 11.4 mm) [12]

Table 8.10 Summary of stiffness analysis in the neighbourhood of the flat singularity [12]

Configuration	d_0, [mm]	Stiffness for unloaded mode, [kN/mm]	F_{cr}^{-}, [kN]	Δ_{cr}^{-}, [mm]	F_{cr}^{+}, [kN]	Δ_{cr}^{+}, [mm]
Model A: 3-PUU manipulator						
Point Q_2	91.7	0.402	−3.61	−12.6	4.80	9.8
Point $Q_2{}^a$	46.0	0.090	−0.67	−16.8	3.20	18.6
Point $Q_2{}^b$	11.4	0.005	−0.01	−4.6	2.50	36.6
Model B: Orthoglide manipulator						
Point Q_2	91.7	0.418	−2.06	−5.7	2.20	4.9
Point $Q_2{}^a$	46.0	0.095	−0.70	−17.1	1.45	11
Point $Q_2{}^b$	11.4	0.006	−0.01	−4.6	0.92	24
$Q_2 = (-76.35,\ -76.35,\ -76.35)$; $Q_2^a = (-100,\ -100,\ -100)$; $Q_2^b = (-120,\ -120,\ -120)$						

$$F = 3K_\theta L_0(1 - \cos\varphi_0/\cos\varphi)\sin\varphi; \quad \Delta d = L_0(\sin\varphi_0 - \sin\varphi) \tag{8.32}$$

perfectly describe the shape of the force–deflection curves obtained from the complete stiffness models.

Besides, a more detailed analysis shows an extremely fast reduction of the stiffness while approaching this singularity. Corresponding expressions derived for small value of φ_0 yield a linear relation for the critical deflection $\Delta_{cr} \approx 0.42d_0$ and cubic relation for the critical force $F_{cr} \approx K_\theta/\sqrt{3}L^2 \cdot d_0^3$. Hence, this simplified model is in perfect agreement with the simulation results and justifies conventional kinematic design objectives (velocity transmission factors, condition number, etc.) that preserve the manipulator from approaching the flat posture. In addition, it proves the common notion of the "distance-to-singularity" that is used in kinematics must be revised in kinetostatics taking into account the fact that loading essentially reduces the margin of the manipulator structural stability.

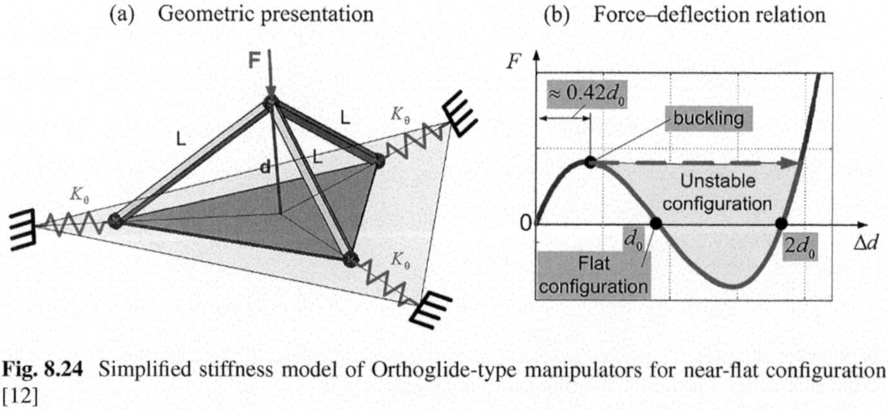

Fig. 8.24 Simplified stiffness model of Orthoglide-type manipulators for near-flat configuration [12]

8.5 Stiffness Model of Serial Chain with Internal Preloading

Let us illustrate the efficiency of stiffness modelling for serial chains with internal preloading in joints on the example of a 2D model of an Orthoglide-type manipulator. The model is obtained by the simplification of the Orthoglide manipulator which has been developed in IRCCyN (Nantes, France) [13, 15]. The under-study Orthoglide model and its typical configurations are present in Fig. 8.25.

Let us carry out stiffness analyses for three types of models which will differ by the kinetostatic models of its chains:

(a) classical model with two passive joints, rigid link and compliance actuator joint;
(b) model with additional linear spring in the first passive joint;
(c) model with additional nonlinear spring in the first passive joint.

The stiffness analyses have been carried out for the square workspace $(-p, -p; p, p)$ with $p = 0.45L$, where L is the length of the link, which satisfies the transmission factor in the range between 0.5 and 2. Stiffness maps for the manipulator with perfect passive joints in the unloaded and loaded modes are presented in Fig. 8.26.

The results show that stiffness in point Q_2(p, p) is much lower than in isotropic point Q_0(0,0) or point Q_1(−p,−p). Besides, the buckling effect in point Q_2 was detected for rather small external loading (see Fig. 8.27).

Therefore liner springs with $\varphi_0 = 0$ have been used to remove the buckling effect and improve the stiffness of the manipulator. Force–deflection relations have been carried out for the additional springs with different stiffness in the three typical points Q_0 (0,0), Q_1(p, p), Q_2 (−p, −p). For the points, Q_1 and Q_2 static control procedures have been used to put the end-effector in the right position. Results are presented in Fig. 8.27. The main stiffness analysis results are summarized in Table 8.11.

The results show that additional springs allow avoid the buckling effect in point Q_2 and increase stiffness for the unloaded mode from $0.240K_{AC}$ to $0.548K_{AC}$ (128.3%) for the additional spring $K_S = 0.1K_{AC}L^2$. Such springs require shifting the actuators on the 0.108 L mm more. This spring increased the stiffness of the manipulator by 9.9% and 4.7% for points Q_0 and Q_1 respectively.

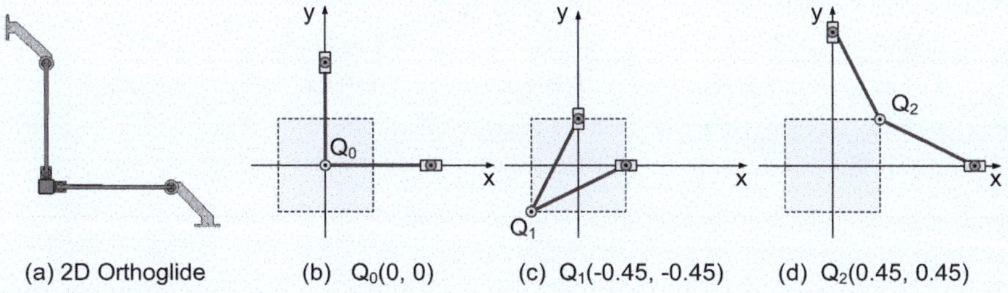

Fig. 8.25 2D Orthoglide and its three typical configurations

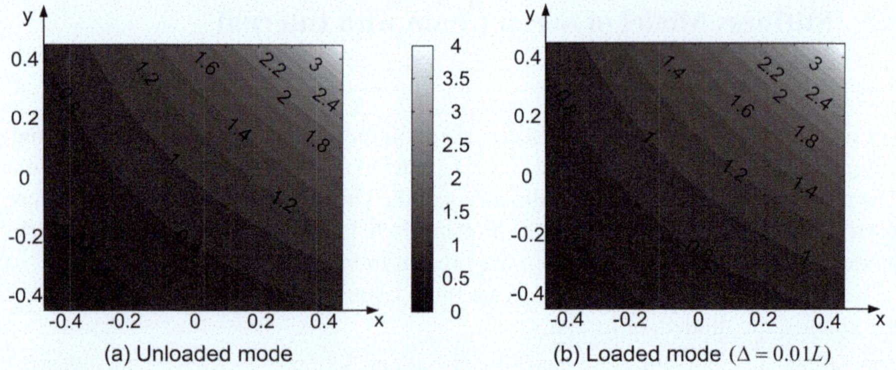

Fig. 8.26 Compliance maps for the 2D Orthoglide without additional springs in passive joints

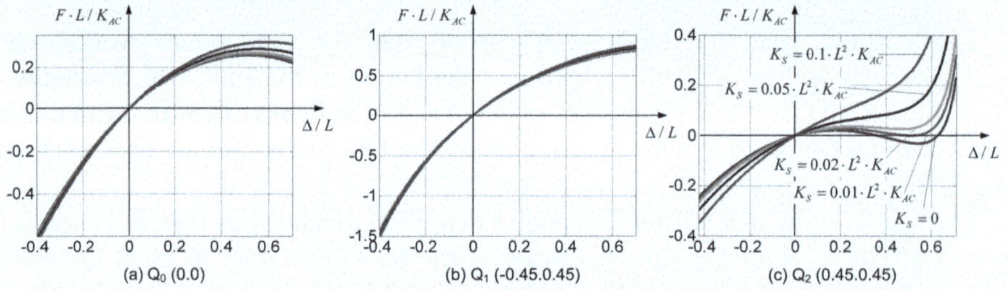

Fig. 8.27 Force–deflection relations for the typical points of 2D Orthoglide manipulator with linear additional springs in the passive joints with $\varphi_0 = 0$ [16]

Table 8.11 Kinetostatic parameters for the 2D Orthoglide manipulator [16]

	$K_S = 0\,K_{AC}$	$K_S = 0.01\,K_{AC}$	$K_S = 0.02\,K_{AC}$	$K_S = 0.05\,K_{AC}$	$K_S = 0.10\,K_{AC}$
Point Q0 (0,0)					
ρ	L	L	L	L	L
K_C	K_{AC}	$1.01\,K_{AC}$	$1.02\,K_{AC}$	$1.05\,K_{AC}$	$1.10\,K_{AC}$
Point Q1 (0.454, 0.454)					
ρ	$0.437\,L$	$0.433\,L$	$0.430\,L$	$0.419\,L$	$0.402\,L$
K_C	$2.276\,K_{AC}$	$2.286\,K_{AC}$	$2.297\,K_{AC}$	$2.329\,K_{AC}$	$2.382\,K_{AC}$
Point Q2 (−0.454, −0.454)					
ρ	$1.345\,L$	$1.356\,L$	$1.367\,L$	$1.399\,L$	$1.453\,L$
K_C	$0.24\,K_{AC}$	$0.27\,K_{AC}$	$0.30\,K_{AC}$	$0.39\,K_{AC}$	$0.55\,K_{AC}$
F_{cr}	$0.020\,K_{AC}\,L$	$0.027\,K_{AC}\,L$	$0.036\,K_{AC}\,L$	–	–

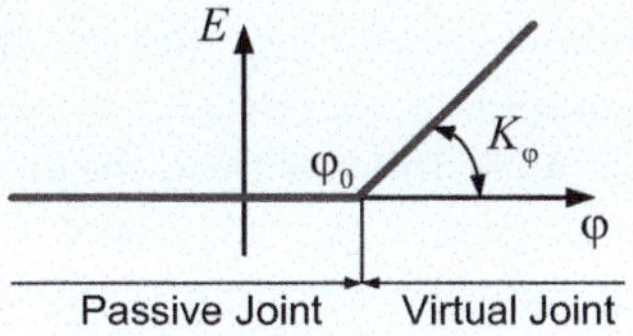

Fig. 8.28 Energy diagram for nonlinear spring with activated angle φ_0

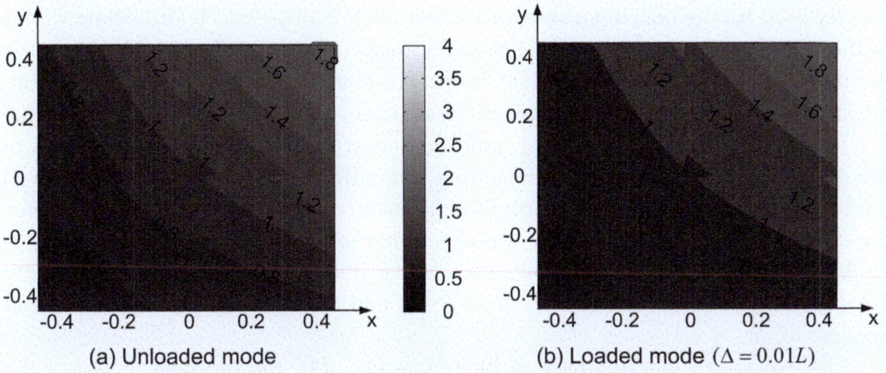

Fig. 8.29 Compliance maps for the 2D Orthoglide with additional springs in passive joints

Compliance maps in Fig. 8.26 show that stiffness is decreased from point Q_1 to point Q_2, therefore it is reasonable to use nonlinear springs (see Fig. 8.28) to improve stiffness in the I quadrant. To make stiffness higher than $0.5K_{AC}$ nonlinear springs with $\varphi_0 = 0$ deg and stiffness $K_S = 0.1K_{AC}\,L^2$ have been used. New compliance maps for unloaded and loaded modes are present on the Fig. 8.29.

Results show that using nonlinear springs stiffness in point Q_2 improved from $4\,K_{AC}$ to $2\,K_{AC}$ both for unloaded and loaded modes, while stiffness in the points Q_0 and Q_1 did not change.

8.6 Stiffness Analysis of Orthoglide and 3-PUU Manipulators

8.6.1 Stiffness Analysis of 3-PUU Parallel Manipulator

For a 3-PUU model, the stiffness analysis was performed for vertices $Q_1 \ldots Q_4$ of the dexterous cubical workspace $200 \times 200 \times 200$ mm^3 corresponding to the Orthoglide design specification with the range of the velocity transmission factors [0.5, 2.0]. Besides, a similar study has been done for the isotropic point Q_0 where the transmission factors are equal to 1.0, similar to a conventional Cartesian machine [17]. For all of these points, there were computed non-linear curves of the force–deflection relations assuming that the displacement is directed along the bisecting line of the coordinate system.

As it follows from the obtained results (Fig. 8.30, Table 8.12), for all considered points $Q_0 \ldots Q_4$, the force–deflection relations are essentially non-linear and include the buckling points where the stiffness reduces radically. Corresponding critical force varies from 4.8 kN for point Q_2 to 15.0 kN for point Q_1, while the critical deflections are within the range of 2.1…9.8 mm. The isotropic point Q_0 that is usually used for the benchmarks of such manipulators possesses intermediate values of these indices (critical force 9.4 kN and critical deflection 2.7 mm). For all cases, the translational stiffness coefficient in the pre-buckling mode is almost constant, but it essentially varies throughout the workspace. In particular, the highest value 7.1 kN/mm is achieved at Point Q_1 and the lowest one 0.4 kN/mm corresponds to the point Q_2. However, after the buckling, the stiffness abruptly reduces down to $0.05 \ldots 0.09$ kN/mm. Hence, the traditional linear stiffness analysis (which ignores the influence of external loading) provides rather reasonable indicators for the pre-buckling mode but it does not allow us to evaluate the range of the loading for which the deflections may be treated as 'small' ones and the buckling phenomena do not appear yet.

It is also worth mentioning that for points $Q_0 \ldots Q_3$, the physical nature of the buckling is perfectly explained by the results from the previous subsection. In these cases, the buckling is caused by the sudden change in the configuration of one of the serial chains (from S- or Z- to Π-configuration, for instance). For the point Q_2 is close to the parallel singularity of the 'flat' type, an additional analysis has been presented in Sect. 8.4.3.

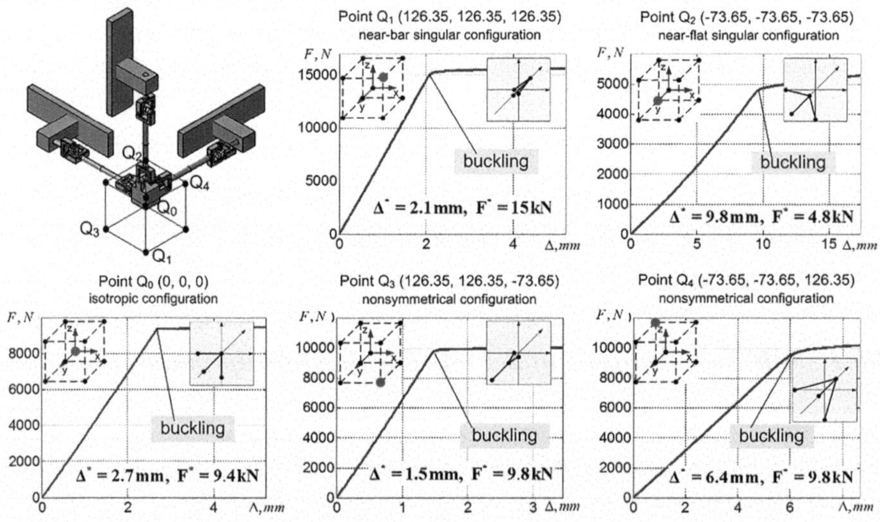

Fig. 8.30 Stiffness of the 3-PUU manipulator in critical points of the cubic workspace

Table. 8.12 Summary of stiffness analysis for 3-PUU manipulator in different work points along the bisecting line of the coordinate system

Configuration	Stiffness for unloaded mode, [kN/mm]	Critical force, [kN]	Critical deflection, [mm]	Stiffness around the buckling. [kN/mm]		Stiffness for large deformations, [kN/mm]
				$F < F_{cr}$	$F > F_{cr}$	
Point Q_0	3.35	9.35	2.68	3.63	0.05	0.05
Point Q_1	7.10	15.05	2.09	7.17	0.18	0.07
Point Q_2	0.40	4.83	9.80	0.58	0.12	0.05
Point Q_3	6.26	9.77	1.48	7.00	0.10	0.05
Point Q_4	1.55	9.78	6.40	1.62	0.40	0.09

8.6.2 *Stiffness Analysis of Orthoglide Parallel Manipulator*

For the 3-PU_cU_c model, the stiffness analysis has been performed for the same set of workspace points $Q_0 \ldots Q_4$ where both of the considered manipulators are equivalent with respect to the kinematic performances. Similarly, for all of these points, the force–deflection relations have been computed and the critical values defining the buckling conditions were evaluated.

The obtained results (Fig. 8.31, Table 8.13) are qualitatively analogous to the previous ones; in particular, the translational stiffness of the unloaded manipulator varies from 16.6 kN/mm for point Q_2 to 0.4 kN/mm for point Q_1. However, in this case, the numerical values of the critical forces are apparently lower, which slightly reduces the advantages of the parallelogram-based architecture demonstrated in [14], where a comparison was done based on the unloaded stiffness analysis. Thus, in point Q_0, the translational stiffness coefficient is almost the same as for the 3-PUU case (3.2 kN/mm) but the critical force and deflection are almost twice lower. Similar conclusions are also valid for the remaining points $Q_1 \ldots Q_4$. Therefore, the parallelogram-based architecture is less robust for the buckling associated with the translational deflections but, as it follows from the supplementary study, its rigidity is essentially higher with respect to the rotational deformations.

The physical nature of the buckling detected for most of the cases (points $Q_0 \ldots Q_3$) can also be explained using results for the serial kinematic chains presented in Sect. 8.2. However, the parallelograms introduce some particularities that lead to additional bifurcations and salient points on the force–deflection curves. For the area of flat singularity (point Q_2), a complimentary analysis has been performed in Sect. 8.4.3.

It should be noted that the stiffness of 3-PUU and Orthoglide translational parallel manipulators are dependent upon the end-effector location and external loading, force–deflection relations are rather non-linear and have bifurcation points, which correspond to buckling. Besides, it was first discovered that in robotic manipulators, along with geometrical singularities, kinetostatic ones may appear.

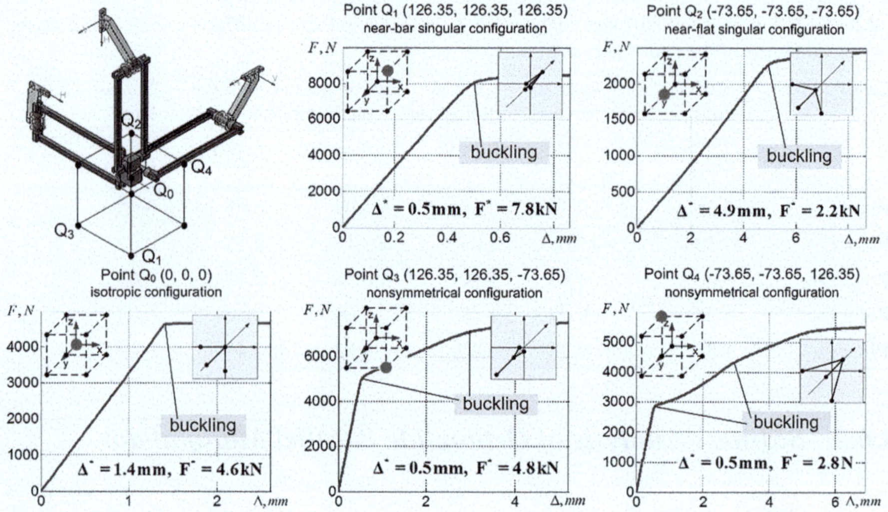

Fig. 8.31 Stiffness of the 3-PU_cU_c manipulator in critical points of the cubic workspace [9]

Table 8.13 Summary of stiffness analysis for Orthoglide manipulator in different work points along the bisecting line of the coordinate system [9]

Configuration	Stiffness for unloaded mode, [kN/mm]	Critical force, [kN]	Critical deflection, [mm]	Stiffness around the buckling, [kN/mm]		Stiffness for large deformations, [kN/mm]
				$F < F_{cr}$	$F > F_{cr}$	
Point Q_0	3.23	4.65	1.41	3.36	0.03	0.03
Point Q_1	16.60	7.85	0.48	16.40	0.42	0.31
Point Q_2	0.42	2.24	4.85	0.48	0.06	0.03
Point Q_3	9.95	4.75	0.48	9.60	0.84	0.11
Point Q_4	5.41	2.78	0.52	5.09	0.65	0.06

References

1. Gosselin C (1990) Stiffness mapping for parallel manipulators. Robot Autom IEEE Trans 6:377–382
2. Chen S-F, Kao I. Simulation of conservative congruence transformation. Conservative properties in the joint and Cartesian spaces. In: Robotics and Automation, 2000. Proceedings. ICRA'00. IEEE International Conference on, pp 1283–1288. IEEE
3. Merlet J-P, Gosselin C (2008) Parallel Mechanisms and Robots. In: Siciliano B, Khatib O (eds) Springer handbook of robotics. Springer, Berlin Heidelberg, pp 269–285
4. Klimchik A, Pashkevich A (2017) Serial versus quasi-serial manipulators: comparison analysis of elasto-static behaviors. Mech Mach Theory 107, 46–70
5. Timoshenko S, Goodier JN (1970) Theory of elasticity. McGraw-Hill, Singapore
6. Jones RM (2006) Buckling of bars, plates, and shells. Bull Ridge Corporation
7. Carricato M, Duffy J, Parenti-Castelli V (2002) Catastrophe analysis of a planar system with flexural pivots. Mech Mach Theory 37:693–716

8. Hines R, Marsh D, Duffy J (1998) Catastrophe analysis of the planar two-spring mechanism. Int J Robot Res 17:89–101
9. Pashkevich A, Klimchik A, Chablat D (2011) Enhanced stiffness modeling of manipulators with passive joints. Mech Mach Theory 46:662–679
10. Connor JJ (1976) Analysis of structural member systems. Ronald Press Company
11. Anatol P, Alexandr K, Damien C (2010) Enhanced stiffness modeling of serial manipulators with passive joints. In: Ernest H (ed.) Advances in robot manipulators, pp Ch. 17. IntechOpen, Rijeka
12. Klimchik A, Chablat D, Pashkevich A (2015) Static stability of manipulator configuration: influence of the external loading. Eur J Mech A Solids 51:193–203
13. Chablat D, Wenger P (2003) Architecture optimization of a 3-DOF translational parallel mechanism for machining applications, the orthoglide. Robot Autom IEEE Trans 19:403–410
14. Pashkevich A, Klimchik A, Caro S, Chablat D (2010) Stiffness modelling of parallelogram-based parallel manipulators. In: Pisla D, Ceccarelli M, Husty M, Corves B (eds) New trends in mechanism science, vol 5. Springer, Netherlands, pp 675–682
15. Majou F, Wenger P, Chablat D (2002) Design of 2-Dof Parallel mechanisms for machining applications. In: Lenarčič J, Thomas F (eds) Advances in robot kinematics: theory and applications. Springer, Netherlands, Dordrecht, pp 319–328
16. Pashkevich A, Klimchik A, Chablat D (2010) Stiffness analysis of parallel manipulators with preloaded passive joints. In: Lenarcic J, Stanisic MM (eds) Advances in robot kinematics: motion in man and machine. Springer, Netherlands, pp 465–474
17. Ceccarelli M, Carbone G (2002) A stiffness analysis for CaPaMan (Cassino parallel manipulator). Mech Mach Theory 37:427–439

Conclusions

This book presents **three main stiffness modelling approaches** currently widely used in robotics and concentrates on their applications to parallel over-constrained manipulators with multiple cross-linkages. The presented techniques are based on the **extensions** of conventional **FEA, MSA and VJM** methods developed by the authors and other researchers. The book explains in detail how these methods can be applied to different parallel kinematic structures, which assumptions are used and how manipulator components can be described in the frame of the adopted stiffness model. It contains both theoretical background and application examples demonstrating the utilisation of the presented methods.

These methods have proved to be **very efficient** for the stiffness analysis of considered parallel robots and **allow for** computing of desired the Cartesian stiffness matrix, estimating the manipulator deflections for a given external load, finding singular manipulator configurations with respect to the stiffness properties, detecting non-linear behaviour under the loading as well as design related compliance error compensation technique. However, despite numerous essential advantages, the presented stiffness modelling techniques have several **limitations** that are caused by the adopted stiffness modelling assumptions. The most significant of them are the following:

(i) For the **VJM-based technique**, the manipulator structure is assumed to be serial or strictly parallel, i.e. composition of serial chains assembled in the end-point or at the end-platform. Because of the specificity of links stiffness models, presented as pseudo-rigid bodies with multidimensional lumped virtual springs at the ends, some types of non-linear stiffness behaviours cannot be detected, such as contact and component bucklings.

(ii) Both the **VJM- and MSA-based techniques** explicitly assume that both the elasticity and loadings are lumped while the forces are applied at model nodes corresponding to the virtual spring locations. This contradiction between the distributed nature of the links' elasticity and their lamped models essentially influences the modelling accuracy. For this reason, the accuracy of the VJM- and MSA-based models is lower compared to the FEA-based ones.

A. Klimchik et al., *Stiffness Modeling of Parallel Robots*, Parallel Robots: Theory and Applications, https://doi.org/10.1007/978-3-031-76320-5

(iii) Current versions of **MSA-based techniques** can be used for **linear** stiffness modelling only with relatively small loadings. However, in practice, the loading may be high enough to cause substantial influence on the manipulator configuration leading to force-dependent components in the Cartesian stiffness matrix.

(iv) All considered techniques implicitly assume that all manipulator components obey Hooke's law, which treats similar tension and compression. However, in some types of manipulators incorporating tensegrity structures and cable-driven mechanisms, the tension and compression response differ fundamentally. For instance, here the component elasticity is zero if the cable is not in tension, besides even in the tension mode the cable sagging also violates Hooke's law.

However, in practice, these limitations are not critical for most industrial manipulators. The presented modelling techniques are rather general and are quite suitable for stiffness analysis of both overconstrained and underconstrained parallel manipulators; they provide reasonable accuracy, which is sufficient for engineering practice. These techniques were successfully used for the design of a large variety of parallel manipulators with the desired stiffness properties and the development of dedicated compliance error compensation algorithms.

The manufacturer's authorised representative in the EU is Springer Nature Customer Service Centre GmbH, Europaplatz 3, 69115 Heidelberg, Germany. If you have any concerns regarding our products, please contact ProductSafety@springernature.com

Printed and bound by CPI Group (UK) Ltd, Croydon, CR0 4YY
01/10/2026
02225965-0003